T0305787

# Application of Numerical Methods in Engineering Problems Using MATLAB®

*Application of Numerical Methods in Engineering Problems Using MATLAB®* presents an analysis of structures using numerical methods and mathematical modeling. This structural analysis also includes beam, plate, and pipe elements, and examines deflection and frequency or buckling loads. The various engineering theories of beams/plates/shells are comprehensively presented, and the relationships between stress and strain, and the governing equations of the structure are extracted. To solve governing equations with numerical methods, there are two general types, including methods based on derivatives or integrals. Derivative-based methods have the advantage of flexibility in modeling boundary conditions, low analysis time, and a very high degree of accuracy. Therefore, the book explains numerical methods based on derivatives, especially the differential quadrature method.

Features:

- Examines the application of numerical methods to obtain the deflection, frequency, and buckling loads.
- Discusses the application of numerical methods for solving motion equations.
- Includes numerous practical and applicable examples throughout.

# Emerging Materials and Technologies

## Series Editor: Boris I. Kharissov

The *Emerging Materials and Technologies* series is devoted to highlighting publications centered on emerging advanced materials and novel technologies. Attention is paid to those newly discovered or applied materials with potential to solve pressing societal problems and improve quality of life, corresponding to environmental protection, medicine, communications, energy, transportation, advanced manufacturing, and related areas.

The series takes into account that, under present strong demands for energy, material, and cost savings, as well as heavy contamination problems and worldwide pandemic conditions, the area of emerging materials and related scalable technologies is a highly interdisciplinary field, with the need for researchers, professionals, and academics across the spectrum of engineering and technological disciplines. The main objective of this book series is to attract more attention to these materials and technologies and invite conversation among the international R&D community.

**Carbon-Based Conductive Polymer Composites**
Processing, Properties, and Applications in Flexible Strain Sensors
*Dong Xiang*

**Nanocarbons**
Preparation, Assessments, and Applications
*Ashwini P. Alegaonkar and Prashant S. Alegaonkar*

**Emerging Applications of Carbon Nanotubes and Graphene**
*Edited by Bhanu Pratap Singh and Kiran M. Subhedar*

**Micro to Quantum Supercapacitor Devices**
Fundamentals and Applications
*Abha Misra*

**Application of Numerical Methods in Engineering Problems Using MATLAB®**
*M.S.H. Al-Furjan, M. Rabani Bidgoli, R. Kolahchi, A. Farrokhian, and M.R. Bayati*

**Advanced Functional Metal-Organic Frameworks**
Fundamentals and Applications
*Edited by Jay Singh, Nidhi Goel, Ranjana Verma and Ravindra Pratap Singh*

**Nanoparticles in Diagnosis, Drug Delivery and Nanotherapeutics**
*Edited by Divya Bajpai Tripathy, Anjali Gupta, Arvind Kumar Jain, Anuradha Mishra and Kuldeep Singh*

For more information about this series, please visit: www.routledge.com/Emerging-Materials-and-Technologies/book-series/CRCEMT

# Application of Numerical Methods in Engineering Problems Using MATLAB®

M.S.H. Al-Furjan, M. Rabani Bidgoli, R. Kolahchi,
A. Farrokhian, and M.R. Bayati

CRC Press is an imprint of the
Taylor & Francis Group, an informa business

Designed cover image: Shutterstock | Bilanol

First edition published 2023
by CRC Press
6000 Broken Sound Parkway NW, Suite 300, Boca Raton, FL 33487-2742

and by CRC Press
4 Park Square, Milton Park, Abingdon, Oxon, OX14 4RN

*CRC Press is an imprint of Taylor & Francis Group, LLC*



ISBN: 978-1-032-39391-9 (hbk)
ISBN: 978-1-032-39392-6 (pbk)
ISBN: 978-1-003-34952-5 (ebk)

DOI: 10.1201/9781003349525

Typeset in Times
by Apex CoVantage, LLC

# Contents

# Preface to the First Edition

The dramatic increase in the use of analytical and numerical methods for solving the engineering problems and the number of research papers published in the last three decades attest to the fact that there has been a major effort to obtain the governing equations of engineering problems and solving those with analytical and numerical methods.

The motivation for the present book came from many years of the authors' research and teaching in numerical methods in engineering and theory of elasticity as well as from the fact there does not exist a book that contains detailed coverage of various beam/plate/shell theories, using the energy method based on Hamilton's principle for deriving the governing equations, and the numerical methods for solving the motion equations. The book is largely based on the authors' work on the differential quadrature numerical method that they have developed over the last two decades.

# Foreword

In this book, mathematical modeling, and the various solution methods for structural elements such as beams/plates/shells are presented. The main purpose of this book is to discuss various theories of beams/plates/shells for mechanical analysis such as buckling, vibration, and dynamic response. This is an essential and basic book for MSc and PhD students for developing their project in the field of mathematical modeling and analytical and numerical analyses of structures. In Chapter 1, various theories for beams, plates, and shells, such as classical, first order, third order, Reddy, and sinusoidal, are presented. In Chapter 2, various solutions methods for example analytical methods (Galerkin and Navier) and derivative based numerical methods for space domain such as differential quadrature, harmonic differential quadrature, discrete singular convolution, and differential cubature as well as numerical methods for time domain, such as Newmark, Poincaré–Lindstedt, multiple scale, and the second-order three-time scale expansion, are presented. The governing equations for various examples are derived based on energy method and Hamilton's principle in Chapters 3 through 14. The analytical and numerical solutions are presented for static and dynamic solutions of the beams, plates, and shells in Chapters 3 through 5, 6 through 10 and 11 through 14, respectively. The key findings of this book are the basic relations and deriving the governing equations for beams, plates, and shells, as well as the mathematical modeling and mechanical analysis of them.

# About the Authors

**M.S.H. Al-Furjan** is an Associate Professor of Mechanical Engineering at Nanjing University of Aeronautics and Astronautics. He received his PhD degree from Zhejiang University, China. His research includes structural mechanics, smart materials and structures, and numerical methods. He has been listed in the World Top 2% Scientists ranked by Stanford University.

**M. Rabani Bidgoli** received his PhD degree from the University of Semnan, Iran. He is the technical reviewer of many top-level journals. His research includes the numerical modeling of concrete structures, numerical methods, and the theory of elasticity and new materials. He is currently an Assistant Professor and the Head of Islamic Azad University, Jasb Branch, Iran.

**R. Kolahchi** is an Associate Professor of Engineering. He received his PhD degree from University of Kashan, Iran. His research includes nanomechanics, continuum mechanics, mechanical analyses of structures, smart materials, and numerical methods. He is among the World Top 2% Scientists based on the database developed by Stanford University.

**A. Farrokhian** is an Assistant Professor of Engineering at Isfahan University of Technology. He received his Ph.D. degree from Isfahan University of science and technology, Iran. His research includes composites, plate and shell theories, and numerical methods.

**M.R. Bayati** is a researcher, he has high quality papers published in well-known journals. His research includes buckling, optimization, new materials, and composite structures.

# Acknowledgments


"Be soulful. Be Kind. Be in Love"—Rumi

The authors would like to express their wishes to be peace in the hearts, nations, and the world.

"Talented People"

The authors would appreciate and thank to the editors and reviewers who provided specific comments for this book.

The authors would like to express their appreciation to many students, researchers, and faculties whom comments have helped to improve the book, especially Amin Shagholani Loor for their research collaboration on the subject. Besides, we owe a major debt to our former teachers, especially the following:

Professor WJ Weng, Zhejiang University
Dr. Ahmad Fakhar, Islamic Azad University, Kashan Branch, Iran
Dr. Mohsen Nasihatgozar, Islamic Azad University, Kashan Branch, Iran
Dr. Seyed Abbas Hoseini, Islamic Azad University, Kashan Branch, Iran

Also, this book wouldn't have been possible without the help and support of Nanjing University of Aeronautics and Astronautics, Islamic Azad University-Jasb Branch, that allowed us to develop our experiences and works.

"Knowledgeable Team"

We would like to thank CRC Press (Taylor & Francis Group) for giving opportunity to publish our book. Thanks to knowledgeable and talented team of CRC Press (Taylor & Francis Group), all editing and production aspects of the book were handled with skill and expertise.

"Parents"

They are like a beacon that lights our way in sailing through our life and the support to improve our acknowledgment. We highly appreciate and thank them for their encouragement, support, and staying up nights. Also, we dedicate this book to fathers' soul of Al-Furjan and Rabani Bidgoli.

"Families are the Compass That Guides Us. They Are the Inspiration to Reach Great Heights, and Our Comfort When We Occasionally Falter"—B. Henry

It has taken thousands of hours for us to write this book over the last years. These hours were mostly taken up by evenings and holidays that could have been spent with family. There are no words of gratitude that can replace the time lost far from the family, but it is important to acknowledge the authors' gratitude to their wives, sons, and daughters. All the support from all our families, and all our friends over the years, we appreciate it.

# 1 Basic Theories

## 1.1 INTRODUCTION

Beams, plates, and shells are three types of structural elements that may be used in many mechanical, civil, and aerospace applications. However, the modeling and mechanical analysis of these structures are essential. In this chapter, most beam, plate, and shell theories applied in many structural problems are presented.

## 1.2 STRAIN–DISPLACEMENT EQUATIONS

When a continuous body is subjected to a physical action, it is assumed that it changes continuously. It should be mentioned that changes are continuous, which means no fracture is considered. Consider an elastic body at time $t = 0$, in which an arbitrary point of the deformable body such as A occupies a position $\mathbf{X}$ in the reference configuration. After the body has deformed, point A has changed its position and moved to point B, it occupies a new position $\mathbf{x}$. In continuum mechanics, two regular descriptions of motion and deformation (Lagrangian and Eulerian descriptions) can be applied. In the referential or Lagrangian description, the motion of the body is referred to the original undeformed configuration, that is, the one that the body occupies at time $t = 0$. The Lagrangian description also is called material description, in which the current coordinates $x_i$ are written in terms of the reference coordinates $X_i$ and time $t$ as follows:

$$\mathbf{x} = \mathbf{x}(\mathbf{X},t), \mathbf{x}(\mathbf{X},0) = \mathbf{X}. \tag{1–1}$$

In the spatial or Eulerian description, the motion of the body is referred to as the current of the deformed configuration. Therefore, a typical variable $\chi$ is depicted in terms of the current position as

$$\chi = \chi(\mathbf{x},t), \mathbf{X} = \mathbf{X}(\mathbf{x},t). \tag{1–2}$$

When the configuration of a body changes due to external forces, the displacement vector $\mathbf{u}$ is written as

$$\mathbf{u} = \mathbf{x} - \mathbf{X}. \tag{1–3}$$

In the present book, the Lagrangian description is applied due to its simplicity and convenience. Hence, to obtain the Lagrangian (or Green's) strain tensor, the components of the displacement vector $u_i$ are functions of $x_i$.

DOI: 10.1201/9781003349525-1

Consider two neighboring points, $A(X_1, X_2, X_3)$ and $A'(X_1 + dX_1, X_2 + dX_2, X_3 + dX_3)$, that relate to an infinitesimal line. The square of the line length can be calculated in the undeformed configuration as follows:

$$\overline{AA'}^2 = dS^2 = dX_1^2 + dX_2^2 + dX_3^2. \tag{1–4}$$

After deformation, points $A$ and $A'$ change their position and become $B(x_1, x_2, x_3)$ and $B'(x_1 + dx_1, x_2 + dx_2, x_3 + dx_3)$ in the deformed configuration. The square of the line length connecting $B$ to $B'$ is written as

$$\overline{BB'}^2 = ds^2 = dx_1^2 + dx_2^2 + dx_3^2. \tag{1–5}$$

The differentials $dx_i$ can be written with respect to the original coordinate system $X_i$ as

$$dx_i = \frac{\partial x_i}{\partial X_1} dX_1 + \frac{\partial x_i}{\partial x_2} dX_2 + \frac{\partial x_i}{\partial X_3} dX_3. \tag{1–6}$$

Using Equation 1–6, Equation 1–5 can be rewritten as follows:

$$ds^2 = \sum_{k=1}^{3} \sum_{i=1}^{3} \sum_{j=1}^{3} \frac{\partial x_k}{\partial X_i} \frac{\partial x_k}{\partial X_j} dX_i dX_j. \tag{1–7}$$

Using Equations 1–4 and 1–7, the difference between the squares of the length of the $BB'$ and $AA'$ can as

$$ds^2 - dS^2 = \sum_{k=1}^{3} \sum_{i=1}^{3} \sum_{j=1}^{3} \left( \frac{\partial x_k}{\partial X_i} \frac{\partial x_k}{\partial X_j} - \delta_{ij} \right) dX_i dX_j, \tag{1–8}$$

in which $\delta_{ij}$ is the Kronecker delta. Considering Equation 1–8, Green's strain tensor $\varepsilon_{ij}$ is obtained as

$$ds^2 - dS^2 = 2 \sum_{i=1}^{3} \sum_{j=1}^{3} E_{ij} dx_i dx_j. \tag{1–9}$$

The following equation may be written using Equation 1–7:

$$E_{ij} = \frac{1}{2} \left( \sum_{k=1}^{3} \frac{\partial x_k}{\partial X_i} \frac{\partial x_k}{\partial X_j} - \delta_{ij} \right). \tag{1–10}$$

Equation 1–1 also can be rewritten as follows:

$$u_k = x_k - X_k \Rightarrow \frac{\partial x_k}{\partial X_i} = \frac{\partial u_k}{\partial X_i} + \delta_{ki}. \tag{1–11}$$

Therefore, the Green (Green–Lagrange) strain tensor can be obtained by substituting Equation 1–11 into Equation 1–10:

$$E_{ij}=\frac{1}{2}\left[\left(\sum_{k=1}^{3}\frac{\partial u_k}{\partial x_i}+\delta_{ki}\right)\left(\sum_{k=1}^{3}\frac{\partial u_k}{\partial x_i}+\delta_{ki}\right)-\delta_{ij}\right]=\frac{1}{2}\left(\frac{\partial u_i}{\partial x_j}+\frac{\partial u_j}{\partial x_i}+\sum_{k=1}^{3}\frac{\partial u_k}{\partial x_i}\frac{\partial u_k}{\partial x_j}\right). \tag{1–12}$$

Expanding Equation 1–12 the following components of strain are given by

$$E_{11}=\frac{\partial u_1}{\partial x_1}+\frac{1}{2}\left[\left(\frac{\partial u_1}{\partial x_1}\right)^2+\left(\frac{\partial u_2}{\partial x_1}\right)^2+\left(\frac{\partial u_3}{\partial x_1}\right)^2\right], \tag{1–13a}$$

$$E_{22}=\frac{\partial u_2}{\partial x_2}+\frac{1}{2}\left[\left(\frac{\partial u_1}{\partial x_2}\right)^2+\left(\frac{\partial u_2}{\partial x_2}\right)^2+\left(\frac{\partial u_3}{\partial x_2}\right)^2\right], \tag{1–13b}$$

$$E_{33}=\frac{\partial u_3}{\partial x_3}+\frac{1}{2}\left[\left(\frac{\partial u_1}{\partial x_3}\right)^2+\left(\frac{\partial u_2}{\partial x_3}\right)^2+\left(\frac{\partial u_3}{\partial x_3}\right)^2\right], \tag{1–13c}$$

$$E_{12}=\frac{1}{2}\left(\frac{\partial u_1}{\partial x_2}+\frac{\partial u_2}{\partial x_1}+\frac{\partial u_1}{\partial x_1}\frac{\partial u_1}{\partial x_2}+\frac{\partial u_2}{\partial x_1}\frac{\partial u_2}{\partial x_2}+\frac{\partial u_3}{\partial x_1}\frac{\partial u_3}{\partial x_2}\right), \tag{1–13d}$$

$$E_{13}=\frac{1}{2}\left(\frac{\partial u_1}{\partial x_3}+\frac{\partial u_3}{\partial x_1}+\frac{\partial u_1}{\partial x_1}\frac{\partial u_1}{\partial x_3}+\frac{\partial u_2}{\partial x_1}\frac{\partial u_2}{\partial x_3}+\frac{\partial u_3}{\partial x_1}\frac{\partial u_3}{\partial x_3}\right), \tag{1–13e}$$

$$E_{23}=\frac{1}{2}\left(\frac{\partial u_2}{\partial x_3}+\frac{\partial u_3}{\partial x_2}+\frac{\partial u_1}{\partial x_2}\frac{\partial u_1}{\partial x_3}+\frac{\partial u_2}{\partial x_2}\frac{\partial u_2}{\partial x_3}+\frac{\partial u_3}{\partial x_2}\frac{\partial u_3}{\partial x_3}\right). \tag{1–13f}$$

The infinitesimal strain tensor can be obtained from the Green–Lagrange strain tensor due to the fact that the displacement gradients and their squares are so small. Therefore, one can write the following equation for the infinitesimal strain tensor:

$$\varepsilon_{ij}=\frac{1}{2}\left(\frac{\partial u_i}{\partial x_j}+\frac{\partial u_j}{\partial x_i}\right). \tag{1–14}$$

Expanding Equation 1–14 yields the linear infinitesimal strain components:

$$\varepsilon_{11}=\frac{\partial u_1}{\partial x_1}, \tag{1–15a}$$

$$\varepsilon_{22}=\frac{\partial u_2}{\partial x_2}, \tag{1–15b}$$

$$\varepsilon_{33}=\frac{\partial u_3}{\partial x_3}, \tag{1–15c}$$

$$2\varepsilon_{12} = \gamma_{12} = \frac{\partial u_1}{\partial x_2} + \frac{\partial u_2}{\partial x_1}, \tag{1–15d}$$

$$2\varepsilon_{13} = \gamma_{13} = \frac{\partial u_1}{\partial x_3} + \frac{\partial u_3}{\partial x_1}, \tag{1–15e}$$

$$2\varepsilon_{23} = \gamma_{23} = \frac{\partial u_2}{\partial x_3} + \frac{\partial u_3}{\partial x_2}. \tag{1–15f}$$

## 1.3 BEAM THEORIES

### 1.3.1 Introduction

Beams are defined as plane structural elements with a small thickness compared to planar dimensions. Typically, the thickness-to-width ratio is less than 0.1. Using beam theories has the advantage of reducing a two-dimensional problem to a one-dimensional problem. Also, the deformation and stresses in a plate subjected to external loads can be calculated using beam theories.

In this section, different governing beam theories are introduced. For this purpose, in Section 1.3.2, the governing assumptions are expressed, and then using these assumptions, beam theories are concluded.

### 1.3.2 Preliminaries

The components of a displacement vector of an arbitrary point of the beam's middle surface are depicted by $u$, $v$, and $w$ in $x$, $y$, and $z$ directions, respectively. Also, $u_1$, $u_2$, and $u_3$ are the displacement components of an arbitrary point of the beam at distance $z$ form the middle surface of the plate in $x$, $y$, and $z$ directions for rectangular plates, respectively.

In order to study different beam theories, it is necessary to express the governing assumptions and theories. The Love–Kirchhoff's assumptions are as follows [1–5]:

1. Straight lines perpendicular to the mid-surface remain straight after deformation.
2. Straight lines perpendicular to the mid-surface remain perpendicular to the mid-surface after deformation.
3. The straight lines do not experience elongation; in other words, the beam's thickness does not change during deformation.

The first and third assumptions imply that the transverse displacement $w$ is independent of the transverse coordinate $z$; therefore, the transverse normal strain $\varepsilon_{33}$ is equal to zero.

However, it is worth mentioning that the results obtained by linear theories are not accurate when the deflection of beam $w$ is of the same order of beam thickness magnitude $h$. Therefore, the following assumptions are made for large deflections of beam:

1. The beam's thickness is small compared to beam's length and width ($h << a$ and $h << b$). In other words, the beam is assumed to be thin.

2. The magnitude of transverse deflection $w$ is small compared to the beam's length and width, and it is the same order as the beam's thickness $h$.
3. The slope is small at each point: $|\partial w/\partial x| << 1$, $|\partial w/\partial y| << 1$.
4. Linear elasticity can be used due to the fact that all strain components are small.
5. The Kirchhoff's assumptions, mentioned earlier, hold.
6. According to von Karman assumptions, in the strain–displacement relations, those nonlinear terms dependent on in-plane displacements $u$ and $v$ are negligible since they are infinitesimal. Indeed, only those nonlinear terms dependent on transverse displacement $w$ should be considered.

If the von Karman assumption is used, the following nonlinear strain-displacement relations can be concluded from Equation 1–13:

$$\varepsilon_{11} = \frac{\partial u_1}{\partial x_1} + \frac{1}{2}\left(\frac{\partial u_3}{\partial x_1}\right)^2, \tag{1–16a}$$

$$\varepsilon_{22} = \frac{\partial u_2}{\partial x_2} + \frac{1}{2}\left(\frac{\partial u_3}{\partial x_2}\right)^2, \tag{1–16b}$$

$$\varepsilon_{12} = \frac{1}{2}\left(\frac{\partial u_1}{\partial x_2} + \frac{\partial u_2}{\partial x_1} + \frac{\partial u_3}{\partial x_1}\frac{\partial u_3}{\partial x_2}\right), \tag{1–16c}$$

$$\varepsilon_{13} = \frac{1}{2}\left(\frac{\partial u_1}{\partial x_3} + \frac{\partial u_3}{\partial x_1}\right), \tag{1–16d}$$

$$\varepsilon_{23} = \frac{1}{2}\left(\frac{\partial u_2}{\partial x_3} + \frac{\partial u_3}{\partial x_2}\right), \tag{1–16e}$$

$$\varepsilon_{33} = \frac{\partial u_3}{\partial x_3}. \tag{1–16f}$$

Therefore, the nonlinear strain–displacement relations for a structure can be written as follows:

$$\varepsilon_{xx} = \frac{\partial u}{\partial x} + \frac{1}{2}\left(\frac{\partial u_3}{\partial x}\right)^2, \tag{1–17a}$$

$$\varepsilon_{yy} = \frac{\partial u_2}{\partial y} + \frac{1}{2}\left(\frac{\partial u_3}{\partial y}\right)^2, \tag{1–17b}$$

$$\varepsilon_{zz} = \frac{\partial u_3}{\partial z}, \tag{1–17c}$$

$$\varepsilon_{xy} = \frac{1}{2}\left(\frac{\partial u_1}{\partial y} + \frac{\partial u_2}{\partial x} + \frac{\partial u_3}{\partial x}\frac{\partial u_3}{\partial y}\right), \tag{1–17d}$$

$$\varepsilon_{xz} = \frac{1}{2}\left(\frac{\partial u_1}{\partial z} + \frac{\partial u_3}{\partial x}\right), \tag{1–17e}$$

$$\varepsilon_{yz} = \frac{1}{2}\left(\frac{\partial u_2}{\partial z} + \frac{\partial u_3}{\partial y}\right). \tag{1–17f}$$

### 1.3.3 Euler–Bernoulli Theory

In the Euler–Bernoulli theory, Kirchhoff's suppositions are assumed. It is mentioned that the first and third assumptions conclude that $\varepsilon_{zz}$ is equal to zero. In addition, the second hypothesis implies zero-transverse shear strains of $\varepsilon_{xz} = 0$, $\varepsilon_{yz} = 0$. Therefore, the displacement for Euler–Bernoulli theory can be written as follows:

$$u_1(x,z) = U(x) - z\frac{\partial W(x)}{\partial x}, \tag{1–18a}$$

$$u_2(x,z) = 0, \tag{1–18b}$$

$$u_3(x,z) = W(x), \tag{1–18c}$$

Substituting Equations 1–18 into Equation 1–17 yields the following nonlinear strain–displacement relations:

$$\varepsilon_x = \left(\frac{\partial U}{\partial x}\right) + \frac{1}{2}\left(\frac{\partial W}{\partial x}\right)^2 - z\left(\frac{\partial^2 W}{\partial x^2}\right). \tag{1–19}$$

It is obvious from Equation 1–19 that $\varepsilon_{xz} = \varepsilon_{yz} = \varepsilon_{zz} = 0$.

### 1.3.4 Timoshenko Beam Theory

In the first-order shear deformation theory of Timoshenko, it is assumed that straight lines perpendicular to the mid-surface do not remain perpendicular to the mid-surface after deformation. Indeed, the second hypothesis of Kirchhoff assumptions is removed. However, the most prominent difference between the Euler–Bernoulli and Timoshenko theories is the effect of considering transverse shear deformations. The displacement field of the Timoshenko theory can be written as follows:

$$u_1(x,z) = U(x) + z\psi(x), \tag{1–20a}$$

$$u_2(x,z) = 0, \tag{1–20b}$$

$$u_3(x,z) = W(x), \tag{1–20c}$$

in which $\psi$ is the rotation. Substituting Equation 1–20 into Equation 1–17 yields the nonlinear strain–displacement relations:

$$\varepsilon_{xx} = \frac{\partial U}{\partial x} + z\frac{\partial \psi}{\partial x} + \frac{1}{2}\left(\frac{\partial W}{\partial x}\right)^2, \tag{1–21a}$$

$$\gamma_{xz} = \frac{\partial W}{\partial x} + \psi. \tag{1–21b}$$

### 1.3.5 Sinusoidal Shear Deformation Theory

In both the Timoshenko and Euler theories, the inextensibility and straightness of transverse normal can be removed. Such extensions lead to the second- and higher order theories of plates. In sinusoidal shear deformation theory, the straight, normal changes to the middle surface lead to cubic curves after deformation. Therefore, the displacement field for this theory can be written as follows:

$$u_1(x,z,t) = u(x,t) - z\frac{\partial w(x,t)}{\partial x} + f\Psi(x,t), \tag{1–22a}$$

$$u_2(x,z,t) = 0, \tag{1–22b}$$

$$u_3(x,z,t) = w(x,t), \tag{1–22c}$$

in which $\psi$ represents the rotation of the cross section about the $y$-axis; $f = h/\pi \sin(\pi z/h)$. Substituting Equation 1–22 into Equation 1–17 yields the strain–displacement relations for sinusoidal shear deformation theory as follows:

$$\varepsilon_{xx} = \frac{\partial u}{\partial x} - z\frac{\partial^2 w}{\partial x^2} + \frac{1}{2}\left(\frac{\partial w}{\partial x}\right)^2 + f\frac{\partial \psi}{\partial x}, \tag{1–23a}$$

$$\varepsilon_{xz} = \cos\left(\frac{\pi z}{h}\right)\psi. \tag{1–23b}$$

### 1.3.6 Hyperbolic Shear Deformation Beam Theory

The displacement field of the hyperbolic shear deformation beam theory is written as follows:

$$u_1(x,z) = U(x) - z\frac{\partial W(x)}{\partial x} + \Phi(z)\left(\frac{\partial W(x)}{\partial x} - \varphi(x)\right), \tag{1–24a}$$

$$u_2(x,z) = 0, \tag{1–24b}$$

$$u_3(x,z) = W(x), \tag{1–24c}$$

where $U$, $V$, and $W$ are the respective translation displacements of a point at the mid-plane of the beam in the longitudinal $x$, transverse $y$, and thickness $z$ directions. Also, $\phi$ denotes the rotation of the cross-sectional area, and

$\Phi(z) = h\sinh(z/h) - z\cosh(1/2)$, is the shape function of the beam. Substituting Equation 1–24 into Equation 1–17 yields the following:

$$\varepsilon_{xx} = \left(\frac{\partial U}{\partial x}\right) + \left(\frac{1}{2}\left(\frac{\partial W}{\partial x}\right)^2\right) - z\left(\frac{\partial^2 W}{\partial x^2}\right) + \Phi(z)\left(\frac{\partial^2 W}{\partial x^2} - \frac{\partial \varphi}{\partial x}\right) \tag{1–25a}$$

$$\gamma_{xz} = \frac{\partial \Phi(z)}{\partial z}\left(\frac{\partial W}{\partial x} - \varphi\right). \tag{1–25b}$$

### 1.3.7 Exponential Shear Deformation Beam Theory

The displacement field of exponential shear deformation beam theory is described as

$$u_1(x,z) = U(x) - z\frac{\partial W(x)}{\partial x} + \Phi(z)\left(\frac{\partial W(x)}{\partial x} - \varphi(x)\right), \tag{1–26a}$$

$$u_2(x,z) = 0, \tag{1–26b}$$

$$u_3(x,z) = W(x), \tag{1–26c}$$

where $U$, $V$, and $W$ are the respective translation displacements of a point at the mid-plane of the beam in the longitudinal $x$, transverse $y$, and thickness $z$ directions. Also, $\phi$ denotes the rotation of the cross-sectional area and $\Phi(z) = ze^{-2(z/h)^2}$ is the shape function of the beam. Substituting Equation 1–26 into Equation 1–17, the following strain–displacement relations for this theory can be obtained:

$$\varepsilon_{xx} = \left(\frac{\partial U}{\partial x}\right) + \left(\frac{1}{2}\left(\frac{\partial W}{\partial x}\right)^2\right) - z\left(\frac{\partial^2 W}{\partial x^2}\right) + \Phi(z)\left(\frac{\partial^2 W}{\partial x^2} - \frac{\partial \varphi}{\partial x}\right), \tag{1–27a}$$

$$\gamma_{xz} = \frac{\partial \Phi(z)}{\partial z}\left(\frac{\partial W}{\partial x} - \varphi\right). \tag{1–27b}$$

## 1.4 PLATE THEORIES

### 1.4.1 Classical Theory

Based on classical plate theory, using Kirchhoff–Law assumptions, the displacement components can be written as

$$u_1(x,y,z,t) = u(x,y,t) - z\frac{\partial w(x,y,t)}{\partial x}, \tag{1–28a}$$

$$u_2(x,y,z,t) = v(x,y,t) - z\frac{\partial w(x,y,t)}{\partial y}, \tag{1–28b}$$

$$u_3(x,y,z,t) = w(x,y,t), \tag{1–28c}$$

where $(u_1, u_2, u_3)$ denotes the displacement components at an arbitrary point $(x, y, z)$ in the plate and $(u, v, w)$ are the displacement components of the middle surface of the plate in the $x$-, $y$-, and $z$-directions, respectively. Also, $z$ is the distance from an arbitrary point to the middle surface. The strain–displacement relationships may be written as

$$\varepsilon_{xx} = \frac{\partial u}{\partial x} - z\frac{\partial^2 w}{\partial x^2}, \tag{1–29a}$$

$$\varepsilon_{yy} = \frac{\partial v}{\partial y} - z\frac{\partial^2 w}{\partial y^2}, \tag{1–29b}$$

$$\varepsilon_{xy} = \frac{1}{2}\left(\frac{\partial u}{\partial y} + \frac{\partial v}{\partial x}\right) - z\frac{\partial^2 w}{\partial x \partial y}. \tag{1–29c}$$

### 1.4.2 First-Order Shear Deformation Theory

Based on first-order shear deformation theory (FSDT), the displacement field can be expressed as

$$u(x,y,z,t) = u(x,y,t) + z\phi_x(x,y,t), \tag{1–30a}$$

$$v(x,y,z,t) = v(x,y,t) + z\phi_y(x,y,t), \tag{1–30b}$$

$$w(x,y,z,t) = w(x,y,t), \tag{1–30c}$$

where $(u(x,y,z,t), v(x,y,z,t), w(x,y,z,t))$ denote the displacement components at an arbitrary point $(x, y, z)$ in the plate and $(u(x,y,t), v(x,y,t), w(x,y,t))$ are the displacement of a material point at $(x, y)$ on the mid-plane (i.e., $z = 0$) of the shell along the $x$-, $y$-, and $z$-directions, respectively; $\varphi_x$ and $\phi_y$ are the rotations of the normal to the mid-plane in the $x$- and $y$-directions, respectively. Based on the preceding relations, the strain–displacement equations may be written as

$$\varepsilon_{xx} = \frac{\partial u}{\partial x} + z\frac{\partial \phi_x}{\partial x}, \tag{1–31a}$$

$$\varepsilon_{yy} = \frac{\partial v}{\partial y} + z\frac{\partial \phi_y}{\partial y}, \tag{1–31b}$$

$$\gamma_{xy} = \frac{\partial v}{\partial x} + \frac{\partial u}{\partial y} + z\left(\frac{\partial \phi_y}{\partial x} + \frac{\partial \phi_x}{\partial y}\right), \tag{1–31c}$$

$$\gamma_{xz} = \phi_x + \frac{\partial w}{\partial x}, \tag{1–31d}$$

$$\gamma_{zy} = \frac{\partial w}{\partial y} + \phi_y, \tag{1–31e}$$

where $(\varepsilon_{xx}, \varepsilon_{yy})$ are the normal strain components and $(\gamma_{yz}, \gamma_{xz}, \gamma_{xy})$ are the shear strain components.

### 1.4.3 Reddy Theory

Based on the Reddy theory, the displacement field can be expressed as

$$u_1(x,y,z,t) = u(x,y,t) + z\phi_x(x,y,t) + c_1 z^3\left(\phi_x + \frac{\partial w}{\partial x}\right), \tag{1–32a}$$

$$u_2(x,y,z,t) = v(x,y,t) + z\phi_y(x,y,t) + c_1 z^3\left(\phi_y + \frac{\partial w}{\partial y}\right), \tag{1–32b}$$

$$u_3(x,y,z,t) = w(x,y,t), \tag{1–32c}$$

where $(u_1(x,y,z,t), u_2(x,y,z,t), u_3(x,y,z,t))$ denote the displacement components at an arbitrary point $(x,y,z)$ in the plate and $(u(x,y,t), v(x,y,t), w(x,y,t))$ are the displacement of a material point at $(x,y)$ on the mid-plane (i.e., $z = 0$) of the plate along the $x$ and $y$ directions, respectively; $\varphi_x$ and $\varphi_y$ are the rotations of the normal to the mid-plane in the $x$- and $y$- directions, respectively; Also, $c_1 = 4/3h^2$. Based on the preceding relations, the strain–displacement equations may be written as

$$\begin{pmatrix} \varepsilon_{xx} \\ \varepsilon_{yy} \\ \gamma_{xy} \end{pmatrix} = \begin{pmatrix} \varepsilon_{xx}^0 \\ \varepsilon_{yy}^0 \\ \gamma_{xy}^0 \end{pmatrix} + z\begin{pmatrix} \varepsilon_{xx}^1 \\ \varepsilon_{yy}^1 \\ \gamma_{xy}^1 \end{pmatrix} + z^3\begin{pmatrix} \varepsilon_{xx}^3 \\ \varepsilon_{yy}^3 \\ \gamma_{xy}^3 \end{pmatrix}, \tag{1–33a}$$

$$\begin{pmatrix} \gamma_{yz} \\ \gamma_{xz} \end{pmatrix} = \begin{pmatrix} \gamma_{yz}^0 \\ \gamma_{xz}^0 \end{pmatrix} + z^2\begin{pmatrix} \gamma_{yz}^2 \\ \gamma_{xz}^2 \end{pmatrix}, \tag{1–33b}$$

where

$$\begin{pmatrix} \varepsilon_{xx}^0 \\ \varepsilon_{yy}^0 \\ \gamma_{xy}^0 \end{pmatrix} = \begin{pmatrix} \frac{\partial u}{\partial x} \\ \frac{\partial v}{\partial y} \\ \frac{\partial u}{\partial y} + \frac{\partial v}{\partial x} \end{pmatrix}, \quad \begin{pmatrix} \varepsilon_{xx}^1 \\ \varepsilon_{yy}^1 \\ \gamma_{xy}^1 \end{pmatrix} = \begin{pmatrix} \frac{\partial \varphi_x}{\partial x} \\ \frac{\partial \varphi_y}{\partial y} \\ \frac{\partial \varphi_x}{\partial y} + \frac{\partial \varphi_y}{\partial x} \end{pmatrix}, \quad \begin{pmatrix} \varepsilon_{xx}^3 \\ \varepsilon_{yy}^3 \\ \gamma_{xy}^3 \end{pmatrix}$$

$$= c_1 \begin{pmatrix} \frac{\partial \varphi_x}{\partial x} + \frac{\partial^2 w}{\partial x^2} \\ \frac{\partial \varphi_y}{\partial y} + \frac{\partial^2 w}{\partial y^2} \\ \frac{\partial \varphi_x}{\partial y} + \frac{\partial \varphi_y}{\partial x} + 2\frac{\partial^2 w}{\partial x \partial y} \end{pmatrix}, \tag{1–34a}$$

$$\begin{pmatrix} \gamma_{yz}^0 \\ \gamma_{xz}^0 \end{pmatrix} = \begin{pmatrix} \varphi_y + \frac{\partial w}{\partial y} \\ \varphi_x + \frac{\partial w}{\partial x} \end{pmatrix}, \quad \begin{pmatrix} \gamma_{yz}^2 \\ \gamma_{xz}^2 \end{pmatrix} = c_2 \begin{pmatrix} \varphi_y + \frac{\partial w}{\partial y} \\ \varphi_x + \frac{\partial w}{\partial x} \end{pmatrix}, \tag{1–34b}$$

where $c_2 = 3c_1$.

### 1.4.4 Sinusoidal Shear Deformation Theory

Based on sinusoidal shear deformation, the displacement field can be expressed as

$$U_1(x,y,z,t) = U(x,y,t) - z\frac{\partial W_b}{\partial x} - \left( z - (\frac{h}{\pi}\sin\frac{\pi z}{h}) \right)\frac{\partial W_s}{\partial x}, \tag{1–35a}$$

$$U_2(x,y,z,t) = V(x,y,t) - z\frac{\partial W_b}{\partial y} - \left( z - (\frac{h}{\pi}\sin\frac{\pi z}{h}) \right)\frac{\partial W_s}{\partial y}, \tag{1–35b}$$

$$U_3(x,y,z,t) = W_b(x,y,t) + W_s(x,y,t), \tag{1–35c}$$

where $(U, V, W_b, W_s)$ denote the displacement components. Based on the preceding relations, the strain–displacement equations may be written as

$$\begin{pmatrix} \varepsilon_{xx} \\ \varepsilon_{yy} \\ \gamma_{xy} \end{pmatrix} = \begin{pmatrix} \varepsilon_{xx}^0 \\ \varepsilon_{yy}^0 \\ \gamma_{xy}^0 \end{pmatrix} + z\begin{pmatrix} k_{xx}^b \\ k_{yy}^b \\ k_{xy}^b \end{pmatrix} + \left( z - \frac{h}{\pi}\sin\frac{\pi z}{h} \right)\begin{pmatrix} k_{xx}^s \\ k_{yy}^s \\ k_{xy}^s \end{pmatrix}, \tag{1–36a}$$

$$\begin{pmatrix} \gamma_{yz} \\ \gamma_{xz} \end{pmatrix} = \cos\left(\frac{\pi z}{h}\right)\begin{pmatrix} \gamma_{yz}^{s} \\ \gamma_{xz}^{s} \end{pmatrix}, \tag{1–36b}$$

where

$$\begin{pmatrix} \varepsilon_{xx}^{0} \\ \varepsilon_{yy}^{0} \\ \gamma_{xy}^{0} \end{pmatrix} = \begin{pmatrix} \dfrac{\partial U}{\partial x} \\ \dfrac{\partial V}{\partial y} \\ \dfrac{\partial U}{\partial y} + \dfrac{\partial V}{\partial x} \end{pmatrix}, \quad \begin{pmatrix} k_{xx}^{b} \\ k_{yy}^{b} \\ k_{xy}^{b} \end{pmatrix} = \begin{pmatrix} -\dfrac{\partial^2 W_b}{\partial x^2} \\ -\dfrac{\partial^2 W_b}{\partial y^2} \\ -2\dfrac{\partial^2 W_b}{\partial x \partial y} \end{pmatrix}, \quad \begin{pmatrix} k_{xx}^{s} \\ k_{yy}^{s} \\ k_{xy}^{s} \end{pmatrix} = \begin{pmatrix} -\dfrac{\partial^2 W_s}{\partial x^2} \\ -\dfrac{\partial^2 W_s}{\partial y^2} \\ -2\dfrac{\partial^2 W_s}{\partial x \partial y} \end{pmatrix}, \tag{1–37a}$$

$$\begin{pmatrix} \gamma_{yz}^{s} \\ \gamma_{xz}^{s} \end{pmatrix} = \frac{-4}{h^3}\begin{pmatrix} \dfrac{\partial W_s}{\partial y} \\ \dfrac{\partial W_s}{\partial x} \end{pmatrix}. \tag{1–37b}$$

## 1.5 SHELL THEORIES

### 1.5.1 Classical Shell Theory

In order to calculate the middle-surface strain and curvatures, using the Kirchhoff–Law assumptions, the displacement components of a cylindrical shell in the axial $x$, circumferential $\theta$, and radial $z$ directions can be written as

$$u_1(x,\theta,z,t) = u(x,\theta,t) - z\frac{\partial w(x,\theta,t)}{\partial x}, \tag{1–38a}$$

$$u_2(x,\theta,z,t) = v(x,\theta,t) - \frac{z}{R}\frac{\partial w(x,\theta,t)}{\partial \theta}, \tag{1–38b}$$

$$u_3(x,\theta,z,t) = w(x,\theta,t), \tag{1–38c}$$

where $(u_1, u_2, u_3)$ denote the displacement components at an arbitrary point $(x,\theta,z)$ in the shell and $(u,v,w)$ are the displacement components of the middle surface of the shell in the axial, circumferential, and radial directions, respectively. Also, $z$ is the distance from an arbitrary point to the middle surface. Using Donnell's linear theory and applying Equation 1–3, the strain–displacement relationships may be written as

$$\varepsilon_{xx} = \frac{\partial u}{\partial x} - z\frac{\partial^2 w}{\partial x^2}, \tag{1–39a}$$

$$\varepsilon_{\theta\theta} = \frac{\partial v}{R\partial \theta} + \frac{w}{R} - \frac{z}{R^2}\frac{\partial^2 w}{\partial \theta^2}, \tag{1–39b}$$

$$\varepsilon_{xy}=\frac{1}{2}\left(\frac{\partial u}{R\partial\theta}+\frac{\partial v}{\partial x}\right)-z\frac{\partial^2 w}{R\partial x\partial\theta}, \tag{1–39c}$$

where $(\varepsilon_{xx},\varepsilon_{\theta\theta})$ are the normal strain components and $(\varepsilon_{x\theta})$ is the shear strain component.

### 1.5.2 FSDT or the Mindlin Theory

According to the Mindlin theory, the displacement field is given as follows:

$$u(x,\theta,z,t)=u(x,\theta,t)+z\psi_x(x,\theta,t), \tag{1–40a}$$

$$v(x,\theta,z,t)=v(x,\theta,t)+z\psi_\theta(x,\theta,t), \tag{1–40b}$$

$$w(x,\theta,z,t)=w(x,\theta,t), \tag{1–40c}$$

in which $\psi_x$ and $\psi_\theta$ are the angle of rotation around the $x$ and $\theta$ axes, respectively. By substituting Equations 1–40a and 1–40b into the strain–displacement relations, we have

$$\varepsilon_{xx}=\frac{\partial u}{\partial x}+\frac{1}{2}\left(\frac{\partial w}{\partial x}\right)^2+z\frac{\partial\psi_x}{\partial x}, \tag{1–41a}$$

$$\varepsilon_{\theta\theta}=\frac{\partial v}{R\partial\theta}+\frac{w}{R}+\frac{1}{2}\left(\frac{\partial w}{R\partial\theta}\right)^2+z\frac{\partial\psi_\theta}{R\partial\theta}, \tag{1–41b}$$

$$\gamma_{x\theta}=\frac{\partial v}{\partial x}+\frac{\partial u}{R\partial\theta}+\frac{\partial w}{\partial x}\frac{\partial w}{R\partial\theta}+z\left(\frac{\partial\psi_x}{R\partial\theta}+\frac{\partial\psi_\theta}{\partial x}\right). \tag{1–41c}$$

$$\gamma_{xz}=\frac{\partial w}{\partial x}+\psi_x, \tag{1–41d}$$

$$\gamma_{\theta z}=\frac{\partial w}{R\partial\theta}-\frac{v}{R}+\psi_\theta. \tag{1–41e}$$

### 1.5.3 Reddy Theory

Based on the Reddy shell theory, the displacement field can be expressed as

$$u_x(x,\theta,z,t)=u\ (x,\theta,t)+z\psi_x(x,\theta,t)-\frac{4z^3}{3h^2}\left(\psi_x(x,\theta,t)+\frac{\partial}{\partial x}w\ (x,\theta,t)\right), \tag{1–42a}$$

$$u_\theta(x,\theta,z,t)=v(x,\theta,t)+z\psi_\theta(x,\theta,t)-\frac{4z^3}{3h^2}\left(\psi_\theta(x,\theta,t)+\frac{\partial}{R\partial\theta}w\ (x,\theta,t)\right), \tag{1–42b}$$

$$u_z(x,\theta,z,t)=w(x,\theta,t), \tag{1–42c}$$

where $(u_x,u_\theta,u_z)$ denote the displacement components at an arbitrary point $(x,\theta,z)$ in the pipe, and $(u,v,w)$ are the displacement of a material point at $(x,\theta)$ on the mid-plane (i.e., $z = 0$) of the pipe along the $x$-, $\theta$-, and $z$-directions, respectively; $\psi_x$ and $\psi_\theta$ are the rotations of the normal to the mid-plane about $\theta$- and $x$-directions, respectively. The von Kármán strains associated with the previous displacement field can be expressed in the following form:

$$\begin{Bmatrix} \varepsilon_{xx} \\ \varepsilon_{\theta\theta} \\ \varepsilon_{x\theta} \\ \varepsilon_{xz} \\ \varepsilon_{\theta z} \end{Bmatrix} = \begin{Bmatrix} \varepsilon_{xx}^0 \\ \varepsilon_{\theta\theta}^0 \\ \varepsilon_{x\theta}^0 \\ \varepsilon_{xz}^0 \\ \varepsilon_{\theta z}^0 \end{Bmatrix} + z \begin{Bmatrix} \varepsilon_{xx}^1 \\ \varepsilon_{\theta\theta}^1 \\ \varepsilon_{x\theta}^1 \\ \varepsilon_{xz}^1 \\ \varepsilon_{\theta z}^1 \end{Bmatrix} + z^2 \begin{Bmatrix} \varepsilon_{xx}^2 \\ \varepsilon_{\theta\theta}^2 \\ \varepsilon_{x\theta}^2 \\ \varepsilon_{xz}^2 \\ \varepsilon_{\theta z}^2 \end{Bmatrix} + z^3 \begin{Bmatrix} \varepsilon_{xx}^3 \\ \varepsilon_{\theta\theta}^3 \\ \varepsilon_{x\theta}^3 \\ \varepsilon_{xz}^3 \\ \varepsilon_{\theta z}^3 \end{Bmatrix}, \tag{1–43}$$

where

$$\begin{Bmatrix} \varepsilon_{xx}^0 \\ \varepsilon_{\theta\theta}^0 \\ \varepsilon_{x\theta}^0 \\ \varepsilon_{xz}^0 \\ \varepsilon_{\theta z}^0 \end{Bmatrix} = \begin{Bmatrix} \dfrac{\partial u}{\partial x} + \dfrac{1}{2}\left(\dfrac{\partial w}{\partial x}\right)^2 \\ \dfrac{\partial v}{R\partial\theta} + \dfrac{w}{R} + \dfrac{1}{2}\left(\dfrac{\partial w}{R\partial\theta}\right)^2 \\ \dfrac{\partial v}{\partial x} + \dfrac{\partial u}{R\partial\theta} + \dfrac{\partial w}{\partial x}\dfrac{\partial w}{R\partial\theta} \\ \psi_x + \dfrac{\partial w}{\partial x} \\ \psi_\theta + \dfrac{\partial w}{R\partial\theta} \end{Bmatrix}, \tag{1–44a}$$

$$\begin{Bmatrix} \varepsilon_{xx}^1 \\ \varepsilon_{\theta\theta}^1 \\ \varepsilon_{x\theta}^1 \\ \varepsilon_{xz}^1 \\ \varepsilon_{\theta z}^1 \end{Bmatrix} = \begin{Bmatrix} \dfrac{\partial \psi_x}{\partial x} \\ \dfrac{\partial \psi_\theta}{R\partial\theta} \\ \dfrac{\partial \psi_x}{R\partial\theta} + \dfrac{\partial \psi_\theta}{\partial x} \\ 0 \\ 0 \end{Bmatrix}, \tag{1–44b}$$

$$\begin{Bmatrix} \varepsilon_{xx}^2 \\ \varepsilon_{\theta\theta}^2 \\ \varepsilon_{x\theta}^2 \\ \varepsilon_{xz}^2 \\ \varepsilon_{\theta z}^2 \end{Bmatrix} = \begin{Bmatrix} 0 \\ 0 \\ 0 \\ \dfrac{-4}{h^2}\left(\psi_x + \dfrac{\partial w}{\partial x}\right) \\ \dfrac{-4}{h^2}\left(\psi_\theta + \dfrac{\partial w}{R\partial\theta}\right) \end{Bmatrix}, \tag{1–44c}$$

$$\begin{Bmatrix} \varepsilon_{xx}^{3} \\ \varepsilon_{\theta\theta}^{3} \\ \varepsilon_{x\theta}^{3} \\ \varepsilon_{xz}^{3} \\ \varepsilon_{\theta z}^{3} \end{Bmatrix} = \begin{Bmatrix} \dfrac{-4}{3h^2}\left(\dfrac{\partial \psi_x}{\partial x} + \dfrac{\partial^2 w}{\partial x^2}\right) \\ \dfrac{-4}{3h^2}\left(\dfrac{\partial \psi_\theta}{R\partial \theta} + \dfrac{\partial^2 w}{R^2\partial \theta^2}\right) \\ \dfrac{-4}{3h^2}\left(\dfrac{\partial \psi_\theta}{\partial x} + \dfrac{\partial \psi_x}{R\partial \theta} + 2\dfrac{\partial^2 w}{R\partial x\partial \theta}\right) \\ 0 \\ 0 \end{Bmatrix}, \tag{1–44d}$$

where $(\varepsilon_{xx}, \varepsilon_{\theta\theta})$ are the normal strain components and $(\gamma_{\theta z}, \gamma_{xz}, \gamma_{x\theta})$ are the shear strain components.

## REFERENCES


[1] Vodenitcharova T, Zhang LC. Bending and local buckling of a nanocomposite beam reinforced by a single-walled carbon nanotube. *Int J Solids Struct* 2006;43:3006–3024.

[2] Reddy JN. *Mechanics of laminated composite plates and shells: Theory and analysis.* Second edition. CRC Press, Boca Raton, FL, 2002.

[3] Bowles JE. *Foundation analysis and design.* McGraw-Hill Education, New York, 1988.

[4] Brush O, Almorth B. *Buckling of bars, plates and shells.* McGraw-Hill, New York, 1975.

[5] Reddy JN. A simple higher order theory for laminated composite plates. *J Appl Mech* 1984;51:745–752.

# 2 Solution Methods

## 2.1 ANALYTICAL METHODS

### 2.1.1 NAVIER METHOD

This method is an analytical method that can be used for beam/plate/shell with only simply supported boundary conditions. Steady-state solutions to the governing equations of the system with five variables of deflections in *x*-, *y*- and *z*- directions (i.e., *u*, *v*, *w*) and rotations (i.e., $\phi_x$ and $\phi_y$) for the simply supported boundary conditions can be assumed as follows [1]:

$$u(x, y, t) = u_0 \cos(\frac{n\pi x}{L})\sin(\frac{m\pi y}{b})e^{i\omega t}, \tag{2–1}$$

$$v(x, y, t) = v_0 \sin(\frac{n\pi x}{L})\cos(\frac{m\pi y}{b})e^{i\omega t}, \tag{2–2}$$

$$w(x, y, t) = w_0 \sin(\frac{n\pi x}{L})\sin(\frac{m\pi y}{b})e^{i\omega t}, \tag{2–3}$$

$$\phi_x(x, y, t) = \psi_{x0} \cos(\frac{n\pi x}{L})\sin(\frac{m\pi y}{b})e^{i\omega t}, \tag{2–4}$$

$$\phi_y(x, y, t) = \psi_{y0} \sin(\frac{n\pi x}{L})\cos(\frac{m\pi y}{b})e^{i\omega t}. \tag{2–5}$$

Substituting Equations 2–1 through 2–5 into motion equations yields

$$[K]\begin{bmatrix} u_0 \\ v_0 \\ w_0 \\ \psi_{x0} \\ \psi_{y0} \end{bmatrix} = 0, \tag{2–6}$$

where $K_{ij}$ are stiffness constants. Finally, for the solution, the determinant of matrix in Equation 2–6 should be equal to zero.

### 2.1.2 GALERKIN METHOD

Let us present the Galerkin method through the following general form for the Poisson problem [2]:

$$-\nabla^2 \tilde{u} = f, \tag{2–7}$$

DOI: 10.1201/9781003349525-2

where $\tilde{u}$ is a function and $f$ is the external force. By multiplying both sides of the previous relation to test function ($X$), which satisfies the boundary conditions, we have

$$-X\nabla^2\tilde{u} = Xf. \quad (2\text{–}8)$$

Now, we should integrate from Equation 2–8 over the domain as follows:

$$-\int_\Omega X\nabla^2\tilde{u}\,dV = \int_\Omega Xf\,dV, \quad (2\text{–}9)$$

where $\Omega$ is the domain of the problem. Assuming $X = \sum_i x_i\varphi_i$, Equation 2–9 can be written as

$$\int_\Omega \nabla\left(\sum_{i=1}^{n} x_i\varphi_i(x)\right).\nabla\left(\sum_{j=1}^{n} x_j\varphi_j(x)\right)dV = \sum_{i=1}^{n}\sum_{j=1}^{n} X\left(\int_\Omega \nabla\varphi_i(x).\nabla\varphi_j(x)\,dV\right)u_j. \quad (2\text{–}10)$$

with the following definitions:

$$A_{ij} := \int_\Omega \nabla\varphi_i(x).\nabla\varphi_j(x)\,dV, \quad (2\text{–}11)$$

$$b_i := \int_\Omega \nabla\varphi_i(x).f\,dV, \quad (2\text{–}12)$$

$$\underline{X} := (X_1, X_2, \ldots, X_n)^T, \quad (2\text{–}13)$$

$$\underline{u} := (u_1, u_2, \ldots, u_n)^T. \quad (2\text{–}14)$$

Finally we have

$$A\underline{u} = \underline{b}. \quad (2\text{–}15)$$

However, with the solution of the preceding relation, the main output of a special problem may be obtained.

## 2.2 NUMERICAL METHODS FOR SPACE DOMAIN

### 2.2.1 Differential Quadrature Method

There are a lot of numerical methods to solve the initial and/or boundary value problems that occur in an engineering domain. One of the best numerical methods is the differential quadrature method (DQM). This method has several advantages compared to other numerical methods:

1. DQM is a precise method for solving nonlinear differential equations in approximating the derivatives.
2. DQM can satisfy a variety of boundary conditions and needs much less formulation and programming effort.
3. The accuracy and convergence of the DQM are high.

Due to the mentioned outstanding merits of the DQM, in recent years, the method has become increasingly popular in the numerical solution of problems in analyzing structural and dynamical problems. In this method, the derivative of the function may be defined as follows [3]:

$$\frac{d^n f_x(x_i,\theta_j)}{dx^n} = \sum_{k=1}^{N_x} A_{ik}^{(n)} f(x_k,\theta_j) \qquad n = 1,\ldots,N_x - 1. \tag{2–16}$$

$$\frac{d^m f_y(x_i,\theta_j)}{d\theta^m} = \sum_{l=1}^{N_\theta} B_{jl}^{(m)} f(x_i,\theta_l) \qquad m = 1,\ldots,N_\theta - 1. \tag{2–17}$$

$$\frac{d^{n+m} f_{xy}(x_i,\theta_j)}{dx^n d\theta^m} = \sum_{k=1}^{N_x}\sum_{l=1}^{N_\theta} A_{ik}^{(n)} B_{jl}^{(m)} f(x_k,\theta_l), \tag{2–18}$$

where $N_x$ and $N_\theta$ denote the number of points in $x$ and $\theta$ directions, $f(x,\theta)$ is the function, and $A_{ik}$, $B_{jl}$ are the weighting coefficients defined as

$$A_{ij}^{(1)} = \begin{cases} \dfrac{M(x_i)}{(x_i - x_j)M(x_j)} & for \quad i \neq j, \quad i,j = 1,2,\ldots,N_x, \\ -\sum\limits_{\substack{j=1 \\ i\neq j}}^{N_x} A_{ij}^{(1)} & for \quad i = j, \quad i,j = 1,2,\ldots,N_x, \end{cases} \tag{2–19}$$

$$B_{ij}^{(1)} = \begin{cases} \dfrac{P(\theta_i)}{(\theta_i - \theta_j)P(\theta_j)} & for \quad i \neq j, \quad i,j = 1,2,\ldots,N_\theta, \\ -\sum\limits_{\substack{j=1 \\ i\neq j}}^{N_\theta} B_{ij}^{(1)} & for \quad i = j, \quad i,j = 1,2,\ldots,N_\theta, \end{cases} \tag{2–20}$$

where $M$ and $P$ are Lagrangian operators defined as

$$M(x_i) = \prod_{\substack{j=1 \\ j\neq i}}^{N_x} (x_i - x_j), \tag{2–21}$$

$$P(\theta_i) = \prod_{\substack{j=1 \\ j\neq i}}^{N_\theta} (\theta_i - \theta_j), \tag{2–22}$$

and for higher order derivatives, we have

$$A_{ij}^{(n)} = n\left( A_{ii}^{(n-1)} A_{ij}^{(1)} - \frac{A_{ij}^{(n-1)}}{(x_i - x_j)} \right). \tag{2–23}$$

$$B_{ij}^{(m)} = m\left( B_{ii}^{(m-1)} B_{ij}^{(1)} - \frac{B_{ij}^{(m-1)}}{(\theta_i - \theta_j)} \right). \tag{2–24}$$

The Chebyshev polynomials are used as follows for selecting sampling grid points:

$$X_i = \frac{L}{2}\left[1 - \cos\left(\frac{i-1}{N_x - 1}\right)\pi\right] \qquad i = 1,\ldots,N_x, \tag{2–25}$$

$$\theta_i = \frac{2\pi}{2}\left[1 - \cos\left(\frac{i-1}{N_\theta - 1}\right)\pi\right] \qquad i = 1,\ldots,N_\theta. \tag{2–26}$$

Applying preceding equations to the motion equations, the matrix form of the governing equations can be written as

$$\left(\left[\underbrace{K_L + K_{NL}}_{K}\right] + \Omega[C] + \Omega^2[M]\right)\begin{Bmatrix}\{d_b\}\\ \{d_d\}\end{Bmatrix} = \begin{Bmatrix}\{0\}\\ \{0\}\end{Bmatrix}, \tag{2–27}$$

where $K_L$, $K_{NL}$, $C$, $M$, $d_b$, and $d_d$ represent the linear stiffness matrix, the nonlinear stiffness matrix, the damping matrix, the mass matrix, the boundary points, and domain points, respectively. Finally, based on the eigenvalue problem, the previous relation can be solved.

### 2.2.2 Harmonic Differential Quadrature Method

The harmonic differential quadrature method (HDQM) is one of the numerical methods in which the governing differential equations turn into a set of first-order algebraic equations by applying the weighting coefficients so that at a given discrete point, a derivative of a function with respect to a spatial variable will be expressed as a weighted linear sum of the function values at all discrete points chosen in the solution domain of that variable and in the direction of the axes of the coordinate system [3]. In these methods, the one-dimensional and two-dimensional derivatives of the function may be defined as follows:

$$\frac{d^n f_x(x_i, y_j)}{dx^n} = \sum_{k=1}^{N_x} A_{ik}^{(n)} f(x_k, y_j) \qquad n = 1,\ldots,N_x - 1. \tag{2–28}$$

$$\frac{d^m f_y(x_i, y_j)}{dy^m} = \sum_{k=1}^{N_x} B_{jl}^{(m)} f(x_i, y_l) \qquad m = 1,\ldots,N_y - 1. \tag{2–29}$$

$$\frac{d^{n+m} f_{xy}(x_i, y_j)}{dx^n dy^m} = \sum_{k=1}^{N_y}\sum_{l=1}^{N_y} A_{ik}^{(n)} B_{jl}^{(m)} f(x_k, y_l). \tag{2–30}$$

So it is apparent that the two most important factors in determining the accuracy of HDQM, are the selection of sampling grid points and weighting coefficients. For choosing sampling grid points, the Chebyshev polynomials are used as follows:

$$x_i = \frac{L}{2}\left[1-\cos\left(\frac{i-1}{N_x-1}\right)\pi\right] \qquad i=1,\dots,N_x, \tag{2–31}$$

$$y_i = \frac{b}{2}\left[1-\cos\left(\frac{i-1}{N_y-1}\right)\pi\right] \qquad i=1,\dots,N_y. \tag{2–32}$$

The weighting coefficients can be obtained by the following simple algebraic relations:

$$A_{ij}^{(1)} = \begin{cases} \dfrac{(\pi/2)M(y_i)}{M(x_j)\sin[(x_i-x_j)/2]\pi} & for \quad i\neq j, \quad i,j=1,2,\dots,N_x, \\ \sum\limits_{\substack{j=1 \\ i\neq j}}^{N_x} A_{ij}^{(1)} & for \quad i=j, \quad i,j=1,2,\dots,N_x, \end{cases} \tag{2–33}$$

$$B_{ij}^{(1)} = \begin{cases} \dfrac{(\pi/2)P(y_i)}{P(y_j)\sin[(y_i-y_j)]\pi} & for \quad i\neq j, \quad i,j=1,2,\dots,N_y, \\ -\sum\limits_{\substack{j=1 \\ i\neq j}}^{N_\theta} B_{ij}^{(1)} & for \quad i=j, \quad i,j=1,2,\dots,N_y, \end{cases} \tag{2–34}$$

in which

$$M(x_i) = \prod_{\substack{j=1 \\ j\neq i}}^{N_x} \sin\left(\frac{(x_i-x_j)\pi}{2}\right) \tag{2–35}$$

$$P(y_i) = \prod_{\substack{j=1 \\ j\neq i}}^{N_y} \sin\left(\frac{(y_i-y_j)\pi}{2}\right) \tag{2–36}$$

and for higher order derivatives, we have

$$A_{ij}^{(n)}=n\left(A_{ii}^{(n\text{-}1)}A_{ij}^{(1)}-\pi ctg\left(\frac{x_i-x_j}{2}\right)\pi\right), \tag{2–37}$$

$$B_{ij}^{(m)}=m\left(B_{ii}^{(m\text{-}1)}B_{ij}^{(1)}-\pi ctg\left(\frac{y_i-y_j}{2}\right)\pi\right), \tag{2–38}$$

The final form and solution of the problem are the same as Equation 2–27 presented in the DQM section.

### 2.2.3 Discrete Singular Convolution Method

Based on this method and considering the distribution and the space element of the test functions as $T$ and $\eta(t)$, we have the following singular convolution relation [4]:

$$F(t)=\int_{-\infty}^{\infty}T(t-x)\eta(x)\,dx, \tag{2–39}$$

where $T(t-x)$ is the singular kernel that, for linear and tomography problems, can be assumed as $T(x)=1/x$ and $T(x)=1/x^2$, respectively. In addition, for curves and surfaces, it can be considered as $T(x)=\delta^{(n)}(x), (n=0,1,2,...)$. The most important kernels are presented as

Dirichlet kernel: $$\frac{\sin\left[\left(l+\frac{1}{2}\right)(x-x')\right]}{2\pi\sin\left[\frac{1}{2}(x-x')\right]}, \tag{2–40}$$

modified Dirichlet kernel: $$\frac{\sin\left[\left(l+\frac{1}{2}\right)(x-x')\right]}{2\pi\tan\left[\frac{1}{2}(x-x')\right]}, \tag{2–41}$$

Vallée Poussin kernel: $$\frac{\sin\left[\left(l+\frac{1}{2}\right)(x-x')\right]}{2\pi\tan\left[\frac{1}{2}(x-x')\right]}. \tag{2–42}$$

Shannon's kernel: $$\frac{\sin\left[\alpha(x-x')\right]}{\pi(x-x')}\rightarrow\frac{-\sin\frac{x}{\Delta}(x-x_k)}{\frac{x}{\Delta}(x-x_k)}. \tag{2–43}$$

For example, for a surface, we have the following truncated singular kernel:

$$f^{(n)}(x) \approx \sum_{k=-W}^{W} \delta_{\alpha,\sigma}^{(n)}(x - x_k) f(x_k), \quad (n = 0,1,2,...), \tag{2–44}$$

where for $x \neq x_k$ we have

$$\delta_{\pi/\Delta,\sigma}^{(1)}(x_m - x_k) = \frac{\cos\frac{\pi}{\Delta}(x - x_k)}{(x - x_k)} \exp\left(\frac{(x - x_k)^2}{2\sigma^2}\right) - \frac{\sin\frac{\pi}{\Delta}(x - x_k)}{\frac{\pi}{\Delta}(x - x_k)^2}$$
$$\exp\left(-\frac{(x - x_k)^2}{2\sigma^2}\right) - \frac{\sin\frac{\pi}{\Delta}(x - x_k)}{\frac{\pi}{\Delta}\sigma^2} \exp\left(-\frac{(x - x_k)^2}{2\sigma^2}\right), \tag{2–45}$$

$$\delta_{\pi/\Delta,\sigma}^{(2)}(x_m - x_k) = -\frac{\frac{\pi}{\Delta}\sin\frac{\pi}{\Delta}(x - x_k)}{(x - x_k)} \exp\left(\frac{(x - x_k)^2}{2\sigma^2}\right) - 2\frac{\cos\frac{\pi}{\Delta}(x - x_k)}{(x - x_k)^2}$$
$$\exp\left(-\frac{(x - x_k)^2}{2\sigma^2}\right) - 2\frac{\cos\frac{\pi}{\Delta}(x - x_k)}{\sigma^2} \exp\left(-\frac{(x - x_k)^2}{2\sigma^2}\right) + 2\frac{\sin\frac{\pi}{\Delta}(x - x_k)}{\frac{\pi}{\Delta}(x - x_k)^3}$$
$$\exp\left(-\frac{(x - x_k)^2}{2\sigma^2}\right) + \frac{\sin\frac{\pi}{\Delta}(x - x_k)}{\frac{\pi}{\Delta}(x - x_k)\sigma^2} \exp\left(-\frac{(x - x_k)^2}{2\sigma^2}\right) \tag{2–46}$$
$$+ \frac{\sin\frac{\pi}{\Delta}(x - x_k)}{\frac{\pi}{\Delta}\sigma^4}(x - x_k)\exp\left(-\frac{(x - x_k)^2}{2\sigma^2}\right),$$

### 2.2.4 Differential Cubature Method

The differential cubature method (DCM) is a numerical procedure expressing a calculus operator $\Re$ value of the function $f(x, y)$ at a discrete point in the solution domain as a weighted linear sum of discrete function values chosen within the overall domain of a problem. For a two-dimensional problem, supposing that there are $N$ arbitrarily located grid points, the cubature approximation at the $i$th discrete point can be expressed as [3]

$$\Re f(x, y)_i = \sum_{j=1}^{N} C_{ij} f(x_j, y_j), \tag{2–47}$$

where $C_{ij}$ and $N$ are the cubature weighting coefficients and the total number of grid points in the solution domain, respectively. The computation of the weighting coefficients can be done using the following expression:

$$\Re\left\{x^{\nu-\mu}y^{\mu}\right\}_i=\sum_{j=1}^{N}C_{ij}f(x_j^{\nu-\mu},y_j^{\mu}),\quad \mu=0,1,2,\ldots,\nu,$$
$$\nu=0,1,2,\ldots,N-1,\quad i=1,2,\ldots,N. \tag{2–48}$$

Equation 2–48 may be written in matrix form as

$$\left[x_j^{\nu-\mu}y_j^{\mu}\right]\begin{bmatrix}C_{i1}\\C_{i2}\\ \cdot\\ \cdot\\ \cdot\\ C_{in}\end{bmatrix}=\left[\Re\left\{x^{\nu-\mu}y^{\mu}\right\}_i\right]. \tag{2–49}$$

The coefficient matrix $\left[x_j^{\nu-\mu}y_j^{\mu}\right]$ can be expanded with $j$ beam-wise and one row of each pair of $(\nu,\mu)$. Also, each pair of $(\nu,\mu)$ is required to fill the beam on the right of the equal sign. The cubature weighting coefficients may be obtained by solving Equation 2–30 repeatedly $i=1,2,\ldots,N$, respectively.

## 2.3 NUMERICAL METHODS FOR TIME DOMAIN

### 2.3.1 Newmark Method

The Newmark-$\beta$ method can be employed to obtain the time response of the structure. Based on this method, we have [5]

$$K^*(d_{i+1})=Q_{i+1}, \tag{2–50}$$

where $K^*(d_{i+1})$ and $Q_{i+1}$ are the effective stiffness matrix and the effective load vector in time of $i+1$, respectively, which can be presented as

$$K^*(d_{i+1})=K_L+K_{NL}(d_{i+1})+\alpha_0 M+\alpha_1 C, \tag{2–51}$$

$$Q_{i+1}^*=Q_{i+1}+M\left(\alpha_0 d_i+\alpha_2\dot{d}_i+\alpha_3\ddot{d}_i\right)+C\left(\alpha_1 d_i+\alpha_4\dot{d}_i+\alpha_5\ddot{d}_i\right), \tag{2–52}$$

where $K_L$, $K_{NL}$, $M$, and $C$ are the linear stiffness, nonlinear stiffness, mass, and damp matrices, respectively, and

$$\alpha_0=\frac{1}{\chi\Delta t^2},\quad \alpha_1=\frac{\gamma}{\chi\Delta t},\quad \alpha_2=\frac{1}{\chi\Delta t},\quad \alpha_3=\frac{1}{2\chi}-1,\quad \alpha_4=\frac{\gamma}{\chi}-1,$$
$$\alpha_5=\frac{\Delta t}{2}\left(\frac{\gamma}{\chi}-2\right),\ \alpha_6=\Delta t\left(1-\gamma\right),\ \alpha_7=\Delta t\gamma, \tag{2–53}$$

where $\gamma = 0.5$ and $\chi = 0.25$ and the acceleration and velocity are

$$\ddot{d}_{i+1} = \alpha_0(d_{i+1} - d_i) - \alpha_2\dot{d}_i - \alpha_3\ddot{d}_i, \tag{2–54}$$

$$\dot{d}_{i+1} = \dot{d}_i + \alpha_6\ddot{d}_i + \alpha_7\ddot{d}_{i+1}. \tag{2–55}$$

The modified velocity and acceleration in Equations 2–54 and 2–55 are considered in the next time step, and all the procedures mentioned earlier are repeated.

### 2.3.2 Poincaré–Lindstedt Method

In this method, which is applicable to initial value problems, we have the following assumption [5]:

$$\rho = \omega t, \tag{2–56}$$

where $\rho$ and $\omega$ are strained coordinate and frequency, respectively. Using the preceding relation, the first and second derivative transforms can be written as

$$\frac{d}{dt} = \frac{d}{d\rho}\frac{d\rho}{dt} = \omega\frac{d}{d\rho}, \tag{2–57}$$

$$\frac{d^2}{dt^2} = \omega\frac{d^2}{dtd\rho} = \omega\frac{d\rho}{dt}\frac{d^2}{d\rho^2} = \omega^2\frac{d^2}{d\rho^2}. \tag{2–58}$$

Then, we can write the deflection ($x$) and the frequency in terms of $\varepsilon$ as

$$\omega = 1 + \varepsilon\omega_1 + \varepsilon^2\omega_2 + ..., \tag{2–59}$$

$$x = x_0(\rho) + \varepsilon x_1(\rho) + \varepsilon^2 x_2(\rho) + ... \tag{2–60}$$

For example, for the equation $\omega^2 x'' + 2\omega\varepsilon x' + x = 0$ with boundary conditions of $x(0) = 1, x'(0) = 0$, we have

$$(1 + \varepsilon\omega_1 + \varepsilon^2\omega_2 + ...)^2(x_0'' + \varepsilon x_1'' + \varepsilon^2 x_2'' + ...) + 2\varepsilon(1 + \varepsilon\omega_1 + \varepsilon^2\omega_2 + ...)$$
$$(x_0' + \varepsilon x_1' + \varepsilon^2 x_2' + ...) + (x_0 + \varepsilon x_1 + \varepsilon^2 x_2 ...) = 0, \tag{2–61}$$

$$x_0(0) + \varepsilon x_1(0) + \varepsilon^2 x_2(0) + ... = 1, \qquad x_0'(0) + \varepsilon x_1(0) + \varepsilon^2 x_2(0) + ... = 0. \tag{2–62}$$

By simplifying the previous equation, we have

$$x_0'' + x_0 + \varepsilon(x_1'' + 2\omega_1 x_0'' + 2x_0' + x_1)$$
$$+\varepsilon^2(x_2'' + x_2 + 2\omega_2 x_0'' + \omega_1^2 x_0'' + 2\omega_1 x_1'' + 2\omega_1 x_0' + 2x_1' + ...) = 0$$

$$x_0(0) = 1, \qquad x_0'(0) = 0, \tag{2–63a}$$
$$x_0'' + x_0 = 0,$$

$$x_1(0)=0, \qquad x_1'(0)=0, \tag{2–63b}$$

$$x_2''+x_2=-2\omega_2 x_0''-\omega_1^2 x_1''-2\omega_1 x_0'-2x_1',$$

$$x_2(0)=0, \qquad x_2'(0)=0. \tag{2–63c}$$

However, with the solution of the preceding relation, we have

$$x(\rho)=x_0(\rho)+\varepsilon x_1(\rho)=\cos\rho+\varepsilon(\sin\rho-\rho\cos\rho). \tag{2–64}$$

### 2.3.3 Multiple Scale Method

In this method, the time scales ($T$) may be assumed as [6]

$$T_0=t, \qquad T_1=\varepsilon t, \quad \ldots, \; T_n=\varepsilon^n t. \tag{2–65}$$

Using this relation, the first and second derivative transforms can be written as

$$\frac{d}{dt}=\frac{\partial}{\partial T_0}\frac{\partial T_0}{\partial t}+\frac{\partial}{\partial T_1}\frac{\partial T_1}{\partial t}+\frac{\partial}{\partial T_2}\frac{\partial T_2}{\partial_t}+\ldots$$

$$=\frac{\partial}{\partial T_0}+\varepsilon\frac{\partial}{\partial T_1}+\varepsilon^2\frac{\partial}{\partial T_2}+\ldots, \tag{2–66}$$

$$\frac{d^2}{dt^2}=\frac{\partial^2}{\partial T_0^2}+2\varepsilon\frac{\partial^2}{\partial T_0 \partial T_1}+\varepsilon^2\left(\frac{\partial^2}{\partial T_0\partial T_2}+\frac{\partial^2}{\partial T_1^2}\right)+\ldots \tag{2–67}$$

For example, in the equation $x''+2\varepsilon x'+x=0$ with boundary conditions of $x(0)=1, x'(0)=0$, we have

$$\frac{\partial^2 x}{\partial T_0^2}+2\varepsilon\frac{\partial^2 x}{\partial T_0\partial T_1}+\varepsilon^2\left(\frac{\partial^2 x}{\partial T_0\partial T_2}+\frac{\partial^2 x}{\partial T_1^2}\right)+2\varepsilon\left(\frac{\partial x}{\partial T_0}+\varepsilon\frac{\partial x}{\partial T_1}+\varepsilon^2\frac{\partial x}{\partial T_2}\right)$$

$$+x+\ldots=0. \tag{2–68}$$

Note that in $t=0$, all variables of $T_0, T_1, ..$ are zero. Finally, the deflection can be in the following form:

$$x(t)=x(T_0,T_1,\ldots,T_n;\varepsilon)\sim(T_0,T_1,\ldots,T_n)+\varepsilon x_1(T_0,T_1,\ldots,T_n)$$

$$+\varepsilon^2 x_2(T_0,T_1,\ldots,T_n)+\ldots \tag{2–69}$$

### 2.3.4 First-Order Two-Scale Expansion Method

In this method, it is assumed that there are only two-time scales of $T_0$ and $T_1$ in the following form [6]:

$$T_0=t, \qquad T_1=\varepsilon t. \tag{2–70}$$

In addition, this method focuses on determining $x$ as a function of $T_0, T_1$ instead of $x$ as a function of $t$ the same as a multiple scale method. Based on this method, for the equation $x'' + 2\varepsilon x' + x = 0$ with boundary conditions of $x(0) = 1, x'(0) = 0$, we have

$$\frac{\partial^2 x}{\partial T_0^2} + 2\varepsilon \frac{\partial^2 x}{\partial T_0 \partial T_1} + \varepsilon^2 \frac{\partial^2 x}{\partial T_1^2} + 2\varepsilon \left( \frac{\partial x}{\partial T_0} + \varepsilon \frac{\partial x}{\partial T_1} \right) + x + \ldots = 0. \tag{2–71}$$

We assume an approximation for $x$ in the following form:

$$x(t) \equiv x(T_0, T_1; \varepsilon) \sim x_0\ (T_0, T_1) + \varepsilon x_1 (T_0, T_1). \tag{2–72}$$

Substituting Equation 2–72 into Equation 2–71, we have

$$\frac{\partial^2 x_0}{\partial T_0^2} + x_0 + \varepsilon \left( \frac{\partial^2 x_1}{\partial T_0^2} + 2 \frac{\partial^2 x_0}{\partial T_0 \partial T_1} + 2 \frac{\partial x_0}{\partial T_0} + x_1 \right) = 0. \tag{2–73}$$

By separating Equation 2–73, we have

$$O(1) \qquad \frac{\partial^2 x_0}{\partial T_0^2} + x_0 = 0,$$

$$O(\varepsilon): \qquad \frac{\partial^2 x_1}{\partial T_0^2} + x_1 = -2 \frac{\partial^2 x_0}{\partial T_0 \partial T_1} - 2 \frac{\partial x_0}{\partial T_0}. \tag{2–74}$$

With the solution of the preceding equations, $x$ as a function of $T_0, T_1$ is

$$x_0 = A_0(T_1) \cos T_0 + B_0(T_1) \sin T_0, \tag{2–75}$$

where $A_0$ and $B_0$ are constant, which can be determined by the boundary conditions. Finally, we have the following deflection:

$$x = e^{-\varepsilon t} \cos t + O(\varepsilon). \tag{2–76}$$

### 2.3.5 Second-Order Three-Time Scale Expansion Method

In this method, the deflection can be assumed as follows [6]:

$$x(t) = x(T_0, T_1, T_n; \varepsilon) \sim (T_0, T_1, T_2) \sim x_0 + \varepsilon x_1 (T_0, T_1, T_2) + \varepsilon^2 x_2 (T_0, T_1, T_2), \tag{2–77}$$

where the time scales are

$$T_0 = t, \qquad T_1 = \varepsilon t, \qquad T_2 = \varepsilon^n t. \tag{2–78}$$

Substituting Equations 2–77 and 2–78 into $x'' + 2\varepsilon x' + x = 0$ yields

$$\frac{\partial^2 x_0}{\partial T_0^2}+x_0+\varepsilon\left(\frac{\partial^2 x_1}{\partial T_0^2}+2\frac{\partial^2 x_0}{\partial T_0 \partial T_1}+2\frac{\partial x_0}{\partial T_0}+x_1\right)$$

$$+\varepsilon^2\left(\frac{\partial^2 x_2}{\partial T_0^2}+2\frac{\partial^2 x_1}{\partial T_0 \partial T_1}+2\frac{\partial^2 x_0}{\partial T_0 \partial T_2}+\frac{\partial^2 x_0}{\partial T_1^2}+2\frac{\partial^2 x_1}{\partial T_0}+x_2\right)=0. \tag{2–79}$$

Equating constants of like powers of 1, $\varepsilon$ and $\varepsilon^2$, we have

$$O(1): \qquad \frac{\partial^2 x_0}{\partial T_0^2}+x_0=0, \tag{2–80}$$

$$O(\varepsilon): \qquad \frac{\partial^2 x_1}{\partial T_0^2}+x_1=-2\frac{\partial^2 x_0}{\partial T_0 \partial T_1}-\frac{\partial x_1}{\partial T_0}, \tag{2–81}$$

$$O(\varepsilon^2): \qquad \frac{\partial^2 x_0}{\partial T_0^2}+x_2=-2\frac{\partial^2 x_0}{\partial T_0 \partial T_2}-\frac{\partial^2 x_0}{\partial T_1^2}$$

$$-2\frac{\partial^2 x_0}{\partial T_0 \partial T_2}-\frac{\partial^2 x_0}{\partial T_1^2}-2\frac{\partial x_0}{\partial T_1}-2\frac{\partial x_1}{\partial T_0}. \tag{2–82}$$

The initial conditions for solving the preceding equations are

$$x_0=1, \qquad \frac{\partial x_0}{\partial T_0}=0, \tag{2–83}$$

$$x_2=0, \qquad \frac{\partial x_2}{\partial T_0}=-\frac{\partial x_1}{\partial T_1}-\frac{\partial x_0}{\partial T_2}. \tag{2–84}$$

In final, the solution of the preceding equation can be written as

$$x=e^{\varepsilon t}\left[\cos\left(t-\frac{1}{2}\varepsilon^2 t\right)+\varepsilon\sin\left(t-\frac{1}{2}\varepsilon^2 t\right)\right]. \tag{2–85}$$

## REFERENCES


[1] Kolahchi R, RabaniBidgoli M, Beygipoor GH, Fakhar MH. A nonlocal nonlinear analysis for buckling in embedded FG-SWCNT-reinforced microplates subjected to magnetic field. *J Mech Sci Tech* 2015;29:3669–3677.

[2] Simsek M. Non-linear vibration analysis of a functionally graded Timoshenko beam under action of a moving harmonic load. *Compos Struct* 2010;92:2532–2546.

[3] Wei GW. *Discrete singular convolutional methods in applied mechanics.* Wiley, New York, 1984.

[4] Qian LW, Wei GW. A note on regularized Shannons sampling formulac. *J Approx Theory* 1999;22:134–144.

[5] Wei GW, Zhang DS, Kouri DJ, Hoffman DK. Lagrange distributed approximating functionals. *Phys Rev Lett* 1997;79:775–779.

[6] Wei GW. Quasi-wavelets and quasi interpolating wavelets. *Chem Phys Let* 1998; 296:215–222.

# 3 Buckling of Nanoparticle-Reinforced Beams Exposed to Fire

## 3.1 INTRODUCTION

In the past few decades, many researchers have used a wide range of supplementary materials like nanoparticles. Carbon nanotubes, silicon dioxide ($SiO_2$), Nanoclay, aluminum oxide ($AL_2O_3$), and titanium dioxide ($TiO_2$) are some naturally occurring nanoparticles. The use of additional cementitious materials due to economic, technical, and environmental considerations has become very common in modern construction.

The buckling of beams exposed to fire has been investigated by many researchers. Tan and Yao [1] developed a simple and rational method to predict the fire resistance of reinforced beams subjected to four-face heating. Fire tests and calculation methods for circular beams were presented by Franssen and Dotreppe [2]. A numerical model, in the form of a computer program, for tracing the behavior of high-performance beams exposed to fire was presented by Kodur et al. [3]. Bratina et al. [4] used a two-step finite element formulation for the thermo-mechanical nonlinear analysis of the behavior of the reinforced beams in fire. The importance of capillary pressure and adsorbed water in the behavior of heat and moisture transport when exposed to high temperatures was explored by Davie et al. [5], who incorporated their behavior explicitly into a computational model. A nonlinear structural analysis of cross sections of three-dimensional reinforced frames exposed to fire was studied by Capua and Mari [6]. A two-step formulation, consisting of separate thermal and mechanical analyses, was presented by Bratina and Saje [7] for the thermo-mechanical analysis of reinforced planar frames subject to fire conditions. The buckling of restrained steel beams due to fire conditions was investigated by Hozjan et al. [8]. Rodrigues et al. [9] presented the results of a research program on the behavior of fiber-reinforced beams in fire. Several fire resistance tests on fiber-reinforced beams with restrained thermal elongation were carried out. The buckling of axially restrained steel beams in fire was presented by Shepherd and Burgess [10]. Fire analysis of steel-composite beams with an interlayer slip was developed by Hozjan et al. [11]. Sixteen fire tests conducted on slender circular hollow section beams filled with normal and high strength, subjected to concentric axial loads were done by Romero et al. [12]. Yu and Kodur [13] presented results from a set of numerical studies on the influence of critical factors on the fire response of beams reinforced with fiber reinforced polymer (FRP) rebars. The macroscopic finite element model used in the analysis accounts for high-temperature properties of the constitutive materials, realistic load, and restraint conditions, as well as the temperature-induced slip between FRP rebars. Bidgoli and Saeidifar [14] presented the time-dependent

DOI: 10.1201/9781003349525-3

buckling behavior of embedded straight concrete columns reinforced with Silicon dioxide nanoparticles under fire exposure. The mathematical simulation employs the Timoshenko beam model to represent the column's structural characteristics.

It can be observed from the literature that theoretical research on the time-dependent buckling of beams armed with $SiO_2$ nanoparticles exposed to fire is rare. The main goal of this chapter is to present a mathematical model for beams exposed to fire and discuss effects of the nanotechnology. For this end, the beam is modeled with the Timoshenko beam model. The heat and mass transfer are described, considering the transfer of free water, water vapor, and dry air caused by pressure and concentration gradients and the conversion of energy. The foundation is simulated with spring and shear constants. Applying the energy method and Hamilton's principle, the governing equations are derived. The differential quadrature method (DQM) is used for obtaining the critical buckling load and time of the structure. The effects of different parameters, such as the volume percent of $SiO_2$ nanoparticles, geometrical parameters, an elastic foundation, and porosity on the time-dependent buckling of beams, are discussed.

## 3.2 MATHEMATICAL MODELING

Figure 3.1 shows an $SiO_2$ nanoparticle-reinforced beam with length $L$ and thickness $h$ embedded in a foundation exposed to fire. The surrounding foundation is described by the spring constant $k_w$ and shear layer $k_g$.

### 3.2.1 Energy Method

The beam is modeled with a Timoshenko beam. The displacements of an arbitrary point in the beam are based on the Timoshenko beam theory presented in Section 1.3.4. The potential energy of the structure can be expressed as

$$U = \frac{1}{2}\int_0^L \int_A (\sigma_{xx}\varepsilon_{xx} + 2\sigma_{xz}\varepsilon_{xz})\, dA dx. \tag{3–1}$$

Submitting Equations 1–21a and 1–21b into Equation 3–1 gives

$$U = \frac{1}{2}\int_0^L \int_A \left\{\sigma_x \left[\frac{\partial U}{\partial x} + z\frac{\partial \psi}{\partial x} + \frac{1}{2}\left(\frac{\partial W}{\partial x}\right)^2 - \alpha_x T\right] + \sigma_{xz}\left[\frac{\partial W}{\partial x} + \psi\right]\right\} dA\, dx, \tag{3–2}$$

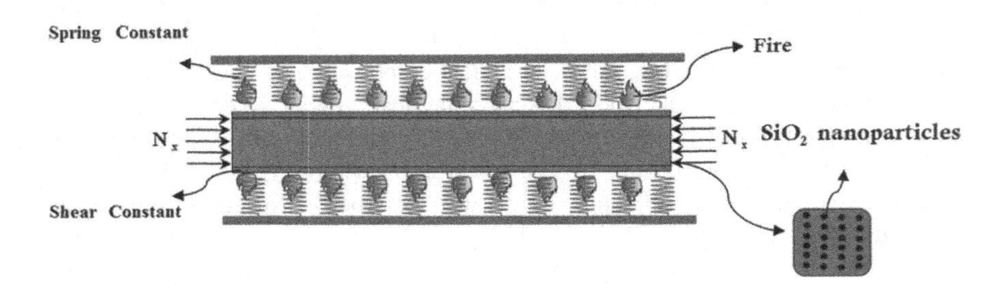


**FIGURE 3.1** Geometry of the SiO2 nanoparticle–reinforced beam exposed to fire.

where the resultant force ($N_x$, $Q_x$) and bending moment $M_x$, are defined as

$$N_x = \int_A \sigma_{xx} \, dA, \tag{3–3}$$

$$M_x = \int_A \sigma_{xx} z \, dA, \tag{3–4}$$

$$Q_x = K_s \int_A \sigma_{xz} \, dA, \tag{3–5}$$

where $K_s$ is a shear correction factor. The kinetic energy of the structure can be expressed as

$$K_{tube} = \frac{1}{2}\rho \int_0^l \int_A \left[ \left( \frac{\partial U}{\partial t} + z \frac{\partial \psi}{\partial t} \right)^2 + \left( \frac{\partial W}{\partial t} \right)^2 \right] dA \, dx, \tag{3–6}$$

where $\rho$ is the density of the structure. The force induced by the foundation is denoted by [16, 17]

$$F_{Elastic\ Medium} = k_w W - k_g \nabla^2 W. \tag{3–7}$$

The governing mass conservation equations to describe heat and moisture transport in containing free water, water vapor, and dry air can be defined as follows [5]:

$$\frac{\partial(\varepsilon_G \hat{\rho}_V)}{\partial t} = -\nabla.\mathrm{J}_V + \dot{E}_{EW}, \tag{3–8}$$

$$\frac{\partial(\varepsilon_G \hat{\rho}_A)}{\partial t} = -\nabla.\mathrm{J}_A, \tag{3–9}$$

$$\frac{\partial(\varepsilon_{FW} \rho_L)}{\partial t} = -\nabla.\mathrm{J}_{FW} - \dot{E}_{FW} + \frac{\partial(\varepsilon_D \rho_L)}{\partial t}, \tag{3–10}$$

where $\varepsilon_i$ is the volume fraction of phase $i$, $\rho_i$ is the density of phase $i$, $\hat{\rho}_i$ is the mass of phase $i$ per unit volume of gaseous material, $\mathrm{J}_i$ is the mass flux of phase $i$, $\dot{E}_{FW}$ is the rate of evaporation of free water (including desorption), $t$ is time, and $i = FE, V, A, D$ are, respectively, free water, water vapor, dry air, and dehydrated water phases. The energy conservation for the system can be defined as

$$\rho C \frac{\partial T}{\partial t} = -\nabla.(-k\nabla T) - (\rho C\mathrm{v})\nabla T - \lambda_E \dot{E}_{FW} - \lambda_D \frac{\partial(\varepsilon_D \rho_L)}{\partial t}, \tag{3–11}$$

where $\rho C$ is the heat capacity, $k$ is the thermal conductivity, $\rho C\mathrm{v}$ relates to the energy transferred by fluid flow, $\lambda_E$ is the specific heat of evaporation (or of desorption when appropriate), $\lambda_D$ is the specific heat of dehydration, and $T$ is the absolute temperature. Substituting $\dot{E}_{FW}$ from Equation 3–11 into Equations 3–8 through 3–10 yields

$$\frac{\partial\left(\varepsilon_G\hat{\rho}_A\right)}{\partial t}=-\nabla.\mathrm{J}_A, \tag{3–12}$$

$$\frac{\partial\left(\varepsilon_G\hat{\rho}_V\right)}{\partial t}+\frac{\partial\left(\varepsilon_{FW}\rho_L\right)}{\partial t}-\frac{\partial\left(\varepsilon_D\rho_L\right)}{\partial t}=-\nabla.\left(\mathrm{J}_{FW}+\mathrm{J}_V\right), \tag{3–13}$$

$$\rho C\frac{\partial T}{\partial t}-\lambda_E\frac{\partial\left(\varepsilon_{FW}\rho_L\right)}{\partial t}+\left(\lambda_E+\lambda_D\right)\frac{\partial\left(\varepsilon_D\rho_L\right)}{\partial t}=\nabla.\left(k\nabla T\right)$$
$$+\lambda_E\nabla.\mathrm{J}_{FW}-\left(\rho C\mathrm{v}\right).\nabla T. \tag{3–14}$$

Based on Fick's laws, the mass fluxes of dry air, water vapor, and free water can be expressed in terms of pressure and concentration gradients as

$$\mathrm{J}_A=\varepsilon_G\hat{\rho}_A\mathrm{v}_G-\varepsilon_G\hat{\rho}_G D_{AV}\nabla\left(\frac{\hat{\rho}_A}{\hat{\rho}_G}\right), \tag{3–15}$$

$$\mathrm{J}_V=\varepsilon_G\hat{\rho}_V\mathrm{v}_G-\varepsilon_G\hat{\rho}_G D_{VA}\nabla\left(\frac{\hat{\rho}_V}{\hat{\rho}_G}\right), \tag{3–16}$$

$$\mathrm{J}_{FW}=\varepsilon_{FW}\rho_L\mathrm{v}_L, \tag{3–17}$$

where $D_{AV}$ and $D_{VA}$ are, respectively, the diffusion coefficients of dry air in water vapor and water vapor in dry air within the porous beam (which are subsequently assumed to be equal [5]), and $\mathrm{v}_G$ and $\mathrm{v}_L$ are the velocities of the gas and liquid water phases resulting from pressure-driven flow as given by Darcy's law:

$$\mathrm{v}_G=-\frac{\bar{K}K_G}{\mu_G}\nabla P_G, \tag{3–18}$$

$$\mathrm{v}_L=-\frac{\bar{K}K_L}{\mu_L}\nabla P_L, \tag{3–19}$$

where $\bar{K}$ is the intrinsic permeability of the dry air, $K_G$ and $K_L$ are the relative permeabilities of the gas and liquid phases, $\mu_G$ and $\mu_L$ are their dynamic viscosities, and $P_G$ and $P_L$ are the corresponding pressures which can be assumed equal to each other's [6]. After extensive algebraic manipulation, the system of governing differential equations can be written in the form [18]

$$C_{TT}\frac{\partial T}{\partial t}+C_{TP}\frac{\partial P_G}{\partial t}+C_{TV}\frac{\partial\hat{\rho}_V}{\partial t}=\nabla.\left(K_{TT}\nabla T+K_{TP}\nabla P_G+K_{TV}\nabla\hat{\rho}_V\right), \tag{3–20}$$

$$C_{AT}\frac{\partial T}{\partial t}+C_{AP}\frac{\partial P_G}{\partial t}+C_{AV}\frac{\partial\hat{\rho}_V}{\partial t}=\nabla.\left(K_{AT}\nabla T+K_{AP}\nabla P_G+K_{AV}\nabla\hat{\rho}_V\right), \tag{3–21}$$

$$C_{MT}\frac{\partial T}{\partial t}+C_{MP}\frac{\partial P_G}{\partial t}+C_{MV}\frac{\partial\hat{\rho}_V}{\partial t}=\nabla.\left(K_{MT}\nabla T+K_{MP}\nabla P_G+K_{MV}\nabla\hat{\rho}_V\right), \tag{3–22}$$

where $C_{ij}$ and $K_{ij}$ are constants. However, the work done by the elastic medium and fire can be written as

$$W = \int_0^L \left( \underbrace{F_{Elastic\ medium} + LP_G}_{q} \right) W dx. \tag{3–23}$$

### 3.2.2 Hamilton's Principle

The governing equations of the structure can be derived from Hamilton's principle as

$$\int_{t_0}^{t_1} \left[ \delta U - \left( \delta K + \delta W \right) \right] = 0. \tag{3–24}$$

Using Equation 3–24, the governing equations may be derived as

$$C_{11} \frac{\partial^2 U}{\partial x^2} + C_{11} \frac{\partial^2 W}{\partial x^2} \frac{\partial W}{\partial x} = \rho h \frac{\partial^2 U}{\partial t^2}, \tag{3–25}$$

$$\begin{aligned} & C_{55} A \left[ \frac{\partial^2 W}{\partial x^2} + \frac{\partial \psi}{\partial x} \right] + N_x^M \frac{\partial^2 W}{\partial x^2} - C_{11} A \alpha_x \frac{\partial T}{\partial x} \frac{\partial W}{\partial x} \\ & - C_{11} A \alpha_x T \frac{\partial^2 W}{\partial x^2} - LP_G - K_w W + G_p \nabla^2 W = \rho h \frac{\partial^2 W}{\partial t^2}, \end{aligned} \tag{3–26}$$

$$C_{11} I \frac{\partial^2 \psi}{\partial x^2} + K_s C_{55} A \left[ \frac{\partial W}{\partial x} + \psi \right] = \frac{\rho h^3}{12} \frac{\partial^2 \psi}{\partial t^2}, \tag{3–27}$$

where $N_x^M$ is the axial mechanical load applied to the beam. The mechanical boundary conditions at both ends of beam are clamped–clamped (CC). Hence,

$$x = 0, L \Rightarrow U = W = \psi = \frac{\partial W}{\partial x} = 0. \tag{3–28}$$

Based on the energy conservation equation, the temperature gradient across the boundary can be expressed as

$$x = 0, L \Rightarrow \frac{\partial T}{\partial x} = \frac{h_{qr}}{k} (T_\infty - T). \tag{3–29}$$

For the gas pressure boundary condition, it may be noted that the gas pressure on the boundary will always be equal to the atmospheric pressure, so the gas pressure gradient across the boundary will always be zero:

$$x = 0, L \Rightarrow \frac{\partial P_G}{\partial x} = 0. \tag{3–30}$$

Based on mass conservation, the mass conservation of water vapor on the boundary can be written as

$$x = 0, L \Rightarrow \frac{\partial \hat{\rho}_V}{\partial x} = -\frac{K_{VT}}{K_{VV}}\frac{h_{qr}}{k}(T_\infty - T) + \frac{\beta}{K_{VV}}(\hat{\rho}_{V\infty} - \hat{\rho}_V). \tag{3–31}$$

In addition, the initial boundary conditions can be written as

$$t = 0 \Rightarrow U = W = \psi = 0, T = T_\infty, P_G = 0, \hat{\rho}_V = \hat{\rho}_{V\infty}. \tag{3–32}$$

Since the boundary conditions are nonclassical, assuming $T - T_\infty = \theta$ and $\hat{\rho}_V - \hat{\rho}_{V\infty} = \bar{\rho}_V$ and with respect to the applied beam theory, $T = T(x,t)$ yields

$$C_{11}\frac{\partial^2 U}{\partial x^2} + C_{11}\frac{\partial^2 W}{\partial x^2}\frac{\partial W}{\partial x} = \rho h \frac{\partial^2 U}{\partial t^2}, \tag{3–33}$$

$$\begin{aligned} &C_{55}A\left[\frac{\partial^2 W}{\partial x^2} + \frac{\partial \psi}{\partial x}\right] + N_x^M \frac{\partial^2 W}{\partial x^2} - C_{11}A\alpha_x \frac{\partial \theta}{\partial x}\frac{\partial W}{\partial x} \\ &-C_{11}A\alpha_x T\frac{\partial^2 W}{\partial x^2} - LP_G - K_w W + G_p \nabla^2 W = \rho h \frac{\partial^2 W}{\partial t^2}, \end{aligned} \tag{3–34}$$

$$C_{11}I\frac{\partial^2 \psi}{\partial x^2} + K_s C_{55}A\left[\frac{\partial W}{\partial x} + \psi\right] = \frac{\rho h^3}{12}\frac{\partial^2 \psi}{\partial t^2}, \tag{3–35}$$

$$C_{TT}\frac{\partial T}{\partial t}\theta + C_{TP}\frac{\partial P_G}{\partial t} + C_{TV}\frac{\partial \bar{\rho}_V}{\partial t} - K_{TT}\frac{\partial^2 \theta}{\partial x^2} + K_{TP}\frac{\partial^2 P_G}{\partial x^2} + K_{TV}\frac{\partial^2 \bar{\rho}_V}{\partial x^2} = 0, \tag{3–36}$$

$$C_{AT}\theta + C_{AP}\frac{\partial P_G}{\partial t} + C_{AV}\frac{\partial \bar{\rho}_V}{\partial t} - K_{AT}\frac{\partial^2 T}{\partial x^2}\theta + K_{AP}\frac{\partial^2 P_G}{\partial x^2} + K_{AV}\frac{\partial^2 \bar{\rho}_V}{\partial x^2} = 0, \tag{3–37}$$

$$C_{MT}\frac{\partial T}{\partial t}\theta + C_{MP}\frac{\partial P_G}{\partial t} + C_{MV}\frac{\partial \bar{\rho}_V}{\partial t} - K_{MT}\frac{\partial^2 T}{\partial x^2}\theta + K_{MP}\frac{\partial^2 P_G}{\partial x^2} + K_{MV}\frac{\partial^2 \bar{\rho}_V}{\partial x^2} = 0, \tag{3–38}$$

and the associated boundary conditions are

$$x = 0, L \Rightarrow U = W = \psi = 0, \tag{3–39}$$

$$x = 0, L \Rightarrow \frac{\partial \theta}{\partial x} = \frac{-h_{qr}\theta}{k}, \tag{3–40}$$

$$x = 0, L \Rightarrow \frac{\partial P_G}{\partial x} = 0, \tag{3–41}$$

$$x = 0, L \Rightarrow \frac{\partial \bar{\rho}_V}{\partial x} = \frac{K_{VT}}{K_{VV}}\frac{h_{qr}\theta}{k} - \frac{\beta \bar{\rho}_V}{K_{VV}}, \tag{3–42}$$

$$t=0 \Rightarrow U=W=\psi=0,\ \theta=0,\ P_G=0,\ \bar{\rho}_V=0. \tag{3–43}$$

## 3.3 MORI–TANAKA RULE

In this section, the effective modulus of the beam reinforced by $SiO_2$ nanoparticles is developed. The $SiO_2$ nanoparticles are assumed with the dispersion of uniform in the polymer. The matrix is assumed to be elastic and isotropic, with a Young's modulus $E_m$ and a Poisson's ratio $\upsilon_m$. The constitutive relations for the composite are [19]

$$\begin{Bmatrix} \sigma_{xx} \\ \sigma_{yy} \\ \sigma_{zz} \\ \sigma_{yz} \\ \sigma_{xz} \\ \sigma_{xy} \end{Bmatrix} = \begin{bmatrix} \underbrace{k+m}_{C_{11}} & \underbrace{l}_{C_{12}} & \underbrace{k-m}_{C_{13}} & 0 & 0 & 0 \\ \underbrace{l}_{C_{21}} & \underbrace{n}_{C_{22}} & \underbrace{l}_{C_{23}} & 0 & 0 & 0 \\ \underbrace{k-m}_{C_{31}} & \underbrace{l}_{C_{32}} & \underbrace{k+m}_{C_{33}} & 0 & 0 & 0 \\ 0 & 0 & 0 & \underbrace{p}_{C_{44}} & 0 & 0 \\ 0 & 0 & 0 & 0 & \underbrace{m}_{C_{55}} & 0 \\ 0 & 0 & 0 & 0 & 0 & \underbrace{p}_{C_{66}} \end{bmatrix} \begin{Bmatrix} \varepsilon_{xx} \\ \varepsilon_{yy} \\ \varepsilon_{zz} \\ \gamma_{yz} \\ \gamma_{xz} \\ \gamma_{xy} \end{Bmatrix}, \tag{3–44}$$

where $\sigma_{ij}, \varepsilon_{ij}, \gamma_{ij}, k, m, n, l, p$ are the stress components, the strain components, and the stiffness coefficients, respectively. According to the Mori–Tanaka method, the stiffness coefficients are given by [18]

$$\begin{aligned} k &= \frac{E_m\{E_m c_m + 2k_r(1+v_m)[1+c_r(1-2v_m)]\}}{2(1+v_m)[E_m(1+c_r-2v_m)+2c_m k_r(1-v_m-2v_m^2)]} \\ l &= \frac{E_m\{c_m v_m[E_m+2k_r(1+v_m)]+2c_r l_r(1-v_m^2)]\}}{(1+v_m)[E_m(1+c_r-2v_m)+2c_m k_r(1-v_m-2v_m^2)]} \\ n &= \frac{E_m^2 c_m(1+c_r-c_m v_m)+2c_m c_r(k_r n_r-l_r^2)(1+v_m)^2(1-2v_m)}{(1+v_m)[E_m(1+c_r-2v_m)+2c_m k_r(1-v_m-2v_m^2)]} \\ &\quad + \frac{E_m[2c_m^2 k_r(1-v_m)+c_r n_r(1+c_r-2v_m)-4c_m l_r v_m]}{E_m(1+c_r-2v_m)+2c_m k_r(1-v_m-2v_m^2)} \\ p &= \frac{E_m[E_m c_m+2p_r(1+v_m)(1+c_r)]}{2(1+v_m)[E_m(1+c_r)+2c_m p_r(1+v_m)]} \\ m &= \frac{E_m[E_m c_m+2m_r(1+v_m)(3+c_r-4v_m)]}{2(1+v_m)\{E_m[c_m+4c_r(1-v_m)]+2c_m m_r(3-v_m-4v_m^2)\}} \end{aligned} \tag{3–45}$$

where $C_m$ and $C_r$ are the volume fractions of the $SiO_2$ nanoparticles, respectively, and $k_r$, $l_r$, $n_r$, $p_r$, and $m_r$ are Hill's elastic modulus for the $SiO_2$ nanoparticles [15].

## 3.4 NUMERICAL RESULTS

In this section, a beam with elastic modules of $E_m$ = 20 GPa is considered, which is reinforced with $SiO_2$ nanoparticles with elastic modules of $E_r$ = 75 GPa. All the properties related to fire are chosen from Bajc et al. [14]. Based on DQM (presented in Section 2.2.1), the critical buckling load and the critical buckling time of the structure are calculated.

### 3.4.1 Accuracy of DQM

The effect of the grid point number in DQM on the critical buckling load and the critical buckling time of the beam is demonstrated in Figures 3.2 and 3.3, respectively.

As can be seen, the fast rate of convergence of the method is quite evident, and it is found that 15 DQM grid points can yield accurate results. In addition, by increasing the slenderness ratio ($\lambda$) of the beam, the critical buckling load and the critical buckling time decrease due to a reduction in the stiffness of the system.

### 3.4.2 Validation

In the absence of similar publications in the literature covering the same scope of the problem, one cannot directly validate the results found here. However, the present work could be partially validated based on a simplified analysis suggested by Bajc et al. [14] on the buckling of a beam where the nanoparticles and elastic foundation ($C_r = 0, k_w = k_g = 0$) were ignored. The results are shown in Figures 3.4 and 3.5, in which the buckling load and the critical buckling time versus the slenderness ratio are plotted, respectively. As can be seen, the two analyses agree well.

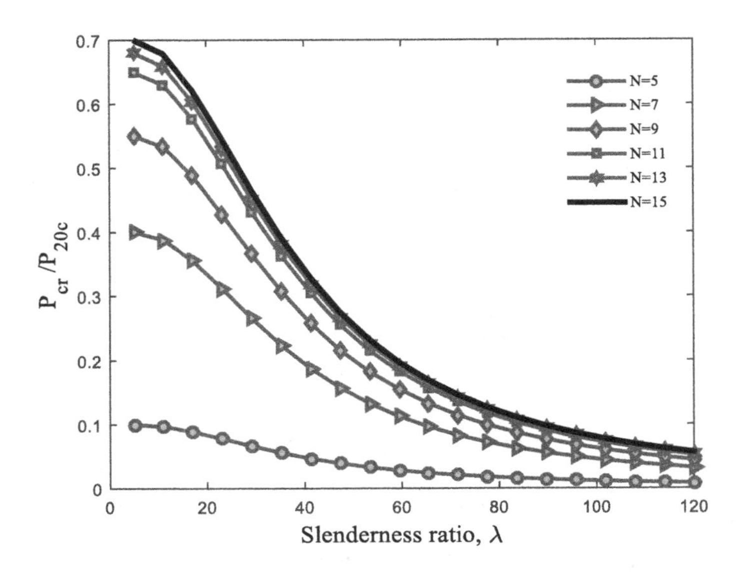


**FIGURE 3.2** Accuracy of DQM for the critical buckling load.

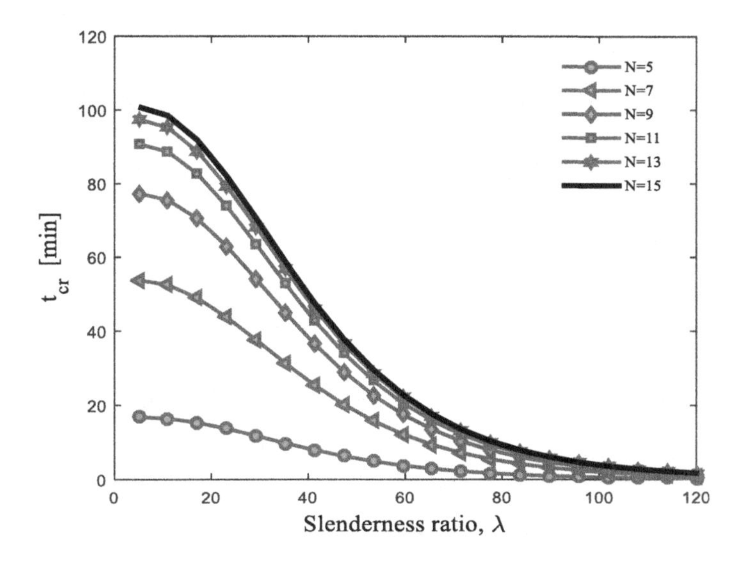


**FIGURE 3.3** Accuracy of DQM for the critical buckling time.

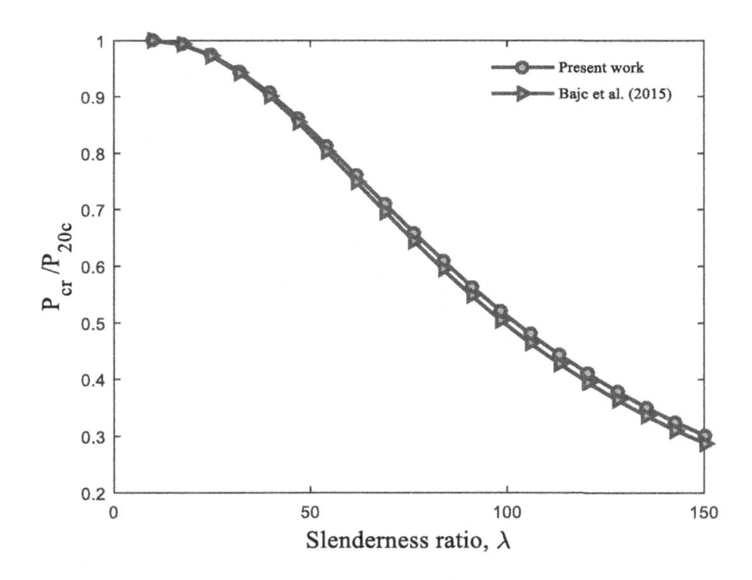


**FIGURE 3.4** Validation of the critical buckling load.

### 3.4.3 Effect of Different Parameters

The effect of the volume percent of $SiO_2$ nanoparticles on the time-dependent critical buckling load and the critical buckling time of beam is illustrated in Figures 3.6 and 3.7, respectively. It can be found that by increasing the volume percent of $SiO_2$

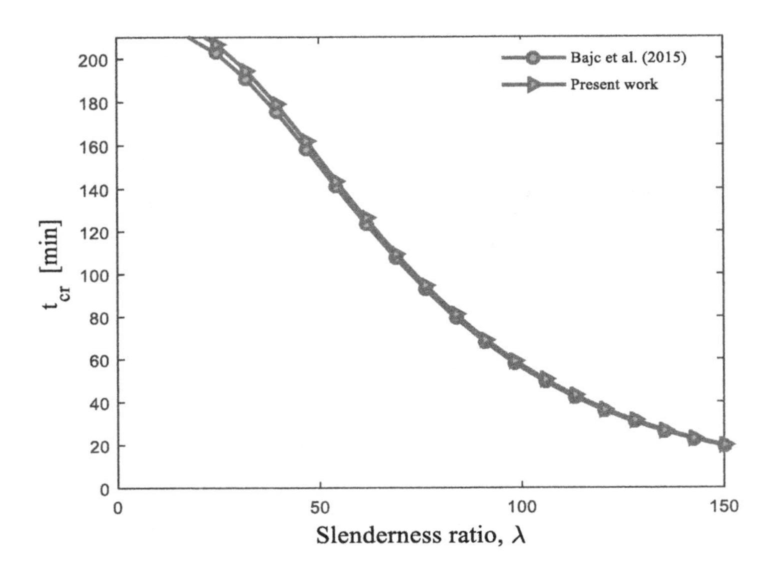


**FIGURE 3.5** Validation of the critical buckling time.

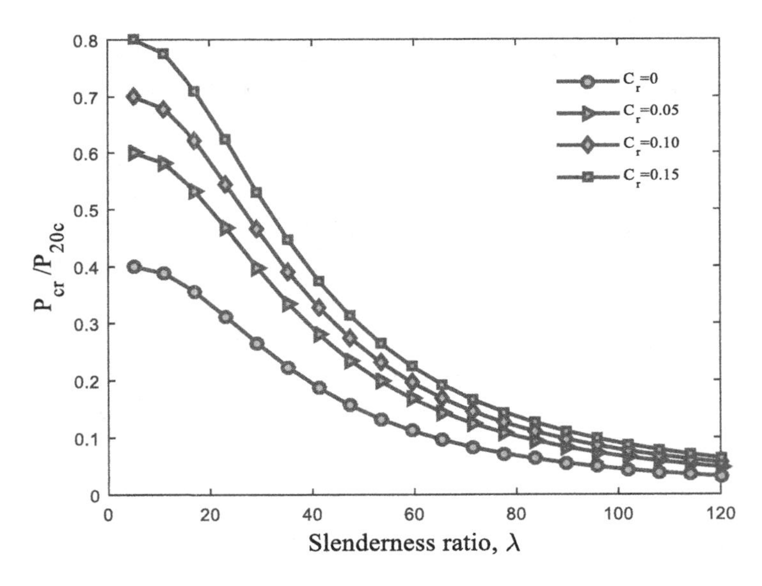


**FIGURE 3.6** Effect of the $SiO_2$ nanoparticle volume percent on the critical buckling load.

nanoparticles, the critical buckling load and the critical buckling time increase. This is due to the fact that by increasing the volume percent of $SiO_2$ nanoparticles, the stiffness of the structure increases. Hence, the $SiO_2$ nanoparticles volume fraction is effective in controlling parameters for the critical buckling load and the critical buckling time of the beam. In addition, the effect of the volume percent of

$SiO_2$ nanoparticles on the critical buckling load and the critical buckling time load becomes prominent for short beams ($\lambda < 50$).

Figures 3.8 and 3.9 illustrate the influence of the elastic medium on the critical buckling load and the critical buckling time along the slenderness ratio, respectively. Obviously, the foundation has a significant effect on the critical buckling load and

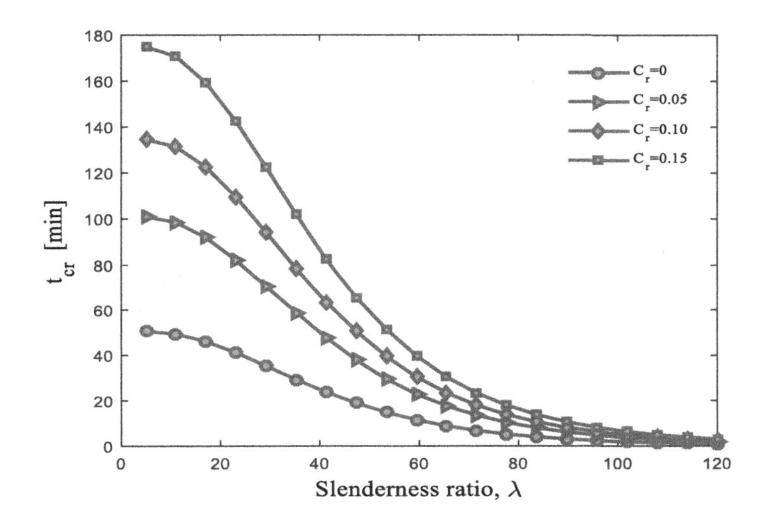


**FIGURE 3.7** Effect of the $SiO_2$ nanoparticle volume percent on the critical buckling load.

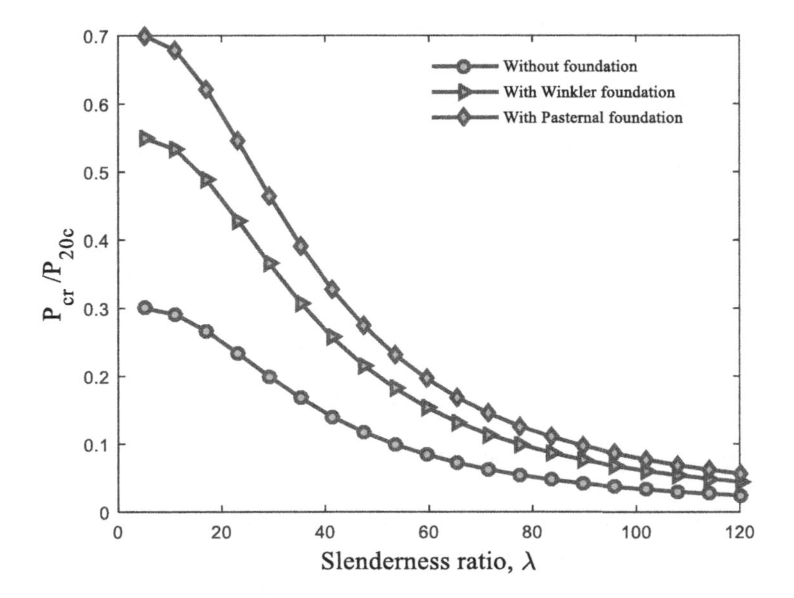


**FIGURE 3.8** Foundation effect on the critical buckling load.

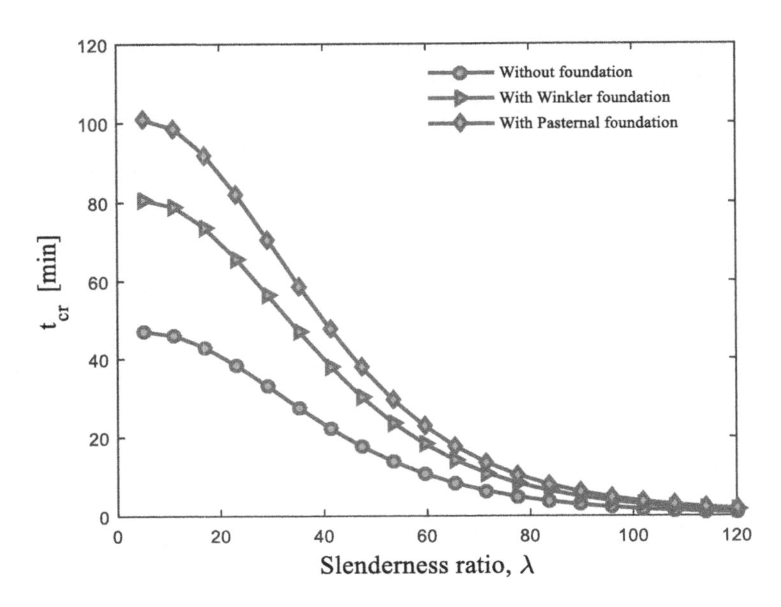


**FIGURE 3.9** Foundation effect on the critical buckling time.

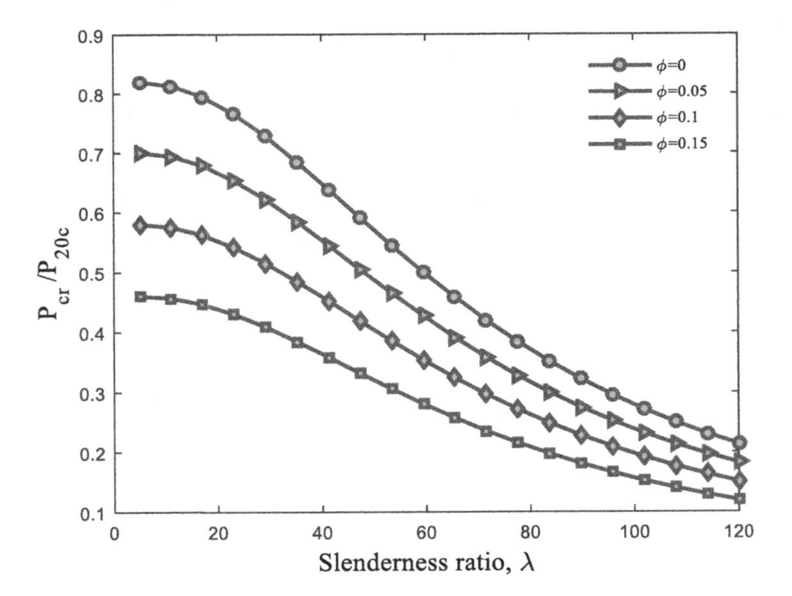


**FIGURE 3.10** Porosity effect on the critical buckling load.

the critical buckling time of the beam, since the critical buckling load and critical buckling time of the system in the case of without foundation are lower than in other cases. It can be concluded that the critical buckling load and the critical buckling time for a Pasternak model (spring and shear constants) are higher than for a Winkler

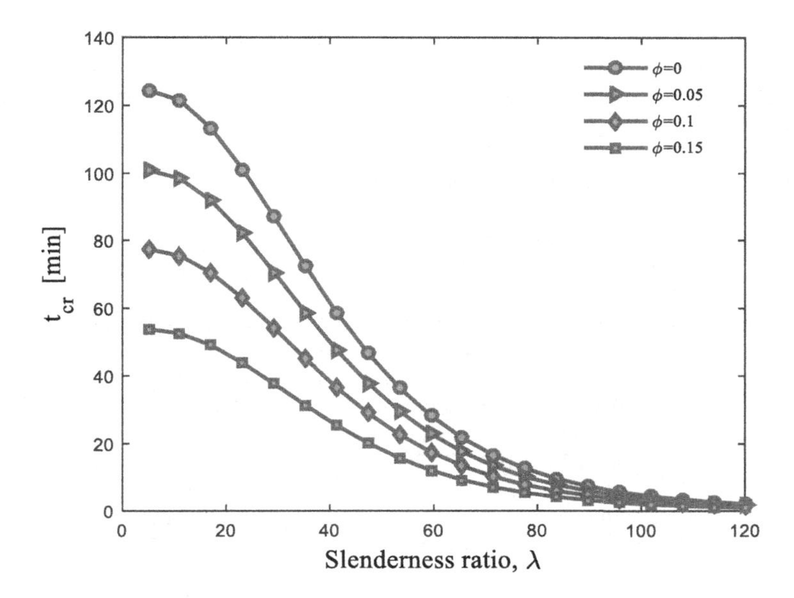


**FIGURE 3.11** Porosity effect on the critical buckling time.

(spring constant) one. The preceding results are reasonable, since the Pasternak medium considers not only the normal stresses (i.e., the Winkler foundation) but also the transverse shear deformation and continuity among the spring elements.

The effect of porosity on the critical buckling load and the critical buckling time of the structure is shown in Figures 3.10 and 3.11, respectively. It can be found that increasing the porosity leads to a higher critical buckling load and critical buckling time. This is reasonable since the stiffness of the structure becomes lower with an increase in porosity.

**Acknowledgments:** This chapter is a slightly modified version of Ref. [14] and has been reproduced here with the permission of the copyright holder.

## REFERENCES

[1] Tan KH, Yao Y. Fire resistance of four-face heated reinforced beams. *J Struct Eng* 2003;129:1220–1229.

[2] Franssen JM, Dotreppe JC. Fire tests and calculation methods for circular beams. *Fire Technol* 2003;39:89–97.

[3] Kodur V, Wang TC, Cheng FP. Predicting the fire resistance behavior of high strength beams. *Cement Concr Compos* 2004;26:141–153.

[4] Bratina S, Čas B, Saje M, Planinc I. Numerical modelling of behaviour of reinforced beams in fire and comparison with Eurocode 2. *Int J Solid Struct* 2005;42:5715–5733.

[5] Davie CT, Pearce CJ, Bićanić N. Coupled heat and moisture transport in at elevated temperatures—effects of capillary pressure and adsorbed water. *Numeric Heat Transfer Part A* 2006;49:733–763.

[6] Capua DD, Mari AR. Nonlinear analysis of reinforced cross-sections exposed to fire. *Fire Saf J* 2007;27:139–149.
[7] Bratina S, Saje M, Planinc I. The effects of different strain contributions on the response of RC beams infire. *Eng Struct* 2007;27:418–430.
[8] Hozjan T, Planinc I, Saje M, Srpčič S. Buckling of restrained steel beams due to fire conditions. *Steel Compos Struct* 2008;8:159–178.
[9] Rodrigues JPC, Laím L, Correia AM. Behaviour of fiber reinforced beams in fire. *Compos Struct* 2010;92:1263–1268.
[10] Shepherd P, Burgess I. On the buckling of axially restrained steel beams in fire. *Eng Struct* 2011;33:2832–2838.
[11] Hozjan T, Saje M, Srpčič S, Planinc I. Fire analysis of steel–composite beam with interlayer slip. *Steel Compos Struct* 2011;89:189–200.
[12] Romero ML, Moliner V, Espinos A, Ibañez C, Hospitaler A. Fire behavior of axially loaded slender high strength-filled tubular beams. *J Constr Steel Res* 2011;67:1953–1965.
[13] Yu B, Kodur VKR. Factors governing the fire response of beams reinforced with FRP rebars. *Compos Struct* 2013;100:257–269.
[14] Bidgoli MR, Saeidifar M. Time-dependent buckling analysis of $SiO_2$ nanoparticles reinforced concrete columns exposed to fire. *Comput Concrete* 2017;(20)2:119–127.
[15] Brush DO, Almroth BO. *Buckling of bars, plates and shells*. McGraw-Hill, New York, 1975.
[16] Ghorbanpour Arani A, Kolahchi R, Khoddami Maraghi Z. Nonlinear vibration and instability of embedded double-walled boron nitride nanotubes based on nonlocal cylindrical shell theory. *Appl Math Model* 2013;37:7685–7707.
[17] Ghorbanpour Arani A, Kolahchi R, Zarei MSH. Visco-surface-nonlocal piezoelasticity effects on nonlinear dynamic stability of graphene sheets integrated with ZnO sensors and actuators using refined zigzag theory. *Compos Struct* 2015;132:506–526.
[18] Mori T, Tanaka K. Average stress in matrix and average elastic energy of materials with misfitting inclusions. *Acta Metallurgica et Materialia* 1973;21:571–574.
[19] Kolahchi R, Rabani Bidgoli M, Beygipoor GH, Fakhar MH. A nonlocal nonlinear analysis for buckling in embedded FG-SWCNT-reinforced microplates subjected to magnetic field. *J Mech Sci Tech* 2015;29:36695–3677.

# 4 Dynamic Response of Nanofiber-Reinforced Beams Subjected to Seismic Ground Excitation

## 4.1 INTRODUCTION

Seismic analysis is a subset of structural analysis in which the dynamic response of a building structure (or nonbuilding structures such as bridges, etc.) against the earthquake is examined. This analysis is a part of the structural engineering, earthquake engineering, and seismic retrofitting of the structures that should be constructed in earthquake-prone zones [1].

Liang and Parra-Montesinos [2] studied the seismic behavior of four reinforced-steel beam under various ground motions using experimental tests. Cheng and Chen [2] and Changwang et al. [3] studied the seismic behavior of a steel-reinforced beam–steel truss beam. They developed a design formula for the shear strength of the structure subjected to seismic activities using experimental tests. The effect of cumulative damage on the seismic behavior of steel tube–reinforced (ST-RC) beams through experimental testing was investigated by Ji et al. [4]. Six large-scale ST-RC beam specimens were subjected to high axial forces and cyclic lateral loading. The effect of plastic hinge relocation on the potential damage of a reinforced frame subjected to different seismic levels was studied by Cao and Ronagh [5] based on current seismic designs. The optimal seismic retrofit method that uses fiber-reinforced polymer (FRP) jackets for shear-critical RC frames was presented by Choi et al. [6]. This optimal method uses nondominated sorting genetic algorithm-II to optimize the two conflicting objective functions of the retrofit cost as well as the seismic performance, simultaneously. They examined various parameters, like failure mode, hysteresis curves, ductility, and a reduction of stiffness. Liu et al. [7] focused on studying the seismic behavior of steel-reinforced special-shaped beam–beam joints. Six specimens, which are designed according to the principle of strong-member and weak-joint core, are tested under low cyclic reversed load.

In none of the above articles, the nanocomposite structure is considered. Wuite and Adali [8] performed stress analysis of carbon nanotube (CNT)–reinforced beams. They concluded that using CNTs during the reinforcing phase can increase the stiffness and stability of the system. Also, Matsunaga [9] examined the stability

DOI: 10.1201/9781003349525-4

of a composite cylindrical shell using third-order shear deformation theory (TSDT). Formica et al. [10] analyzed the vibration behavior of CNT-reinforced composites. They employed an equivalent continuum model based on the Eshelby–Mori–Tanaka model to obtain the material properties of the composite. Liew et al. [11] studied the post-buckling of nanocomposite cylindrical panels. They used the extended rule of mixture to estimate the effective material properties of the nanocomposite structure. They also applied a meshless approach to examine the post-buckling response of a nanocomposite cylindrical panel. In another similar work, Lei et al. [12] studied the dynamic stability of a CNT-reinforced functionally graded (FG) cylindrical panel. They used the Eshelby–Mori–Tanaka model to estimate effective material properties of the resulting nanocomposite structure and employed the Ritz method to distinguish the instability regions of the structure. A static–stress analysis of CNT-reinforced cylindrical shells has been presented by Ghorbanpour Arani et al. [13]. In this work, the cylindrical shell was subjected to non-axisymmetric thermal-mechanical loads and uniform electromagnetic fields. Eventually, the stress distribution in the structure is determined analytically by the Fourier series. A buckling analysis of CNT-reinforced microplates was carried out by Kolahchi et al. [14]. They derived the governing equations of the structure based on Mindlin plate theory and using Hamilton's principle. They obtained buckling load of the structure by applying the differential quadrature method (DQM). The dynamic response of FG circular cylindrical shells is examined by Davar et al. [15]. They developed the mathematical formulation of the structure according to first-order shear deformation theory (FSDT) and Love's first approximation theory. Also, Kolahchi et al. [16] investigated dynamic stability of FG-CNT–reinforced plates. The material properties of the plate were assumed to be a function of temperature, and the structure was considered to be resting on an orthotropic elastomeric medium. Jafarian Arani and Kolahchi [17] presented a mathematical model for a buckling analysis of a CNT-reinforced beam. They simulated the problem based on the Euler–Bernoulli and Timoshenko beam theories. The nonlinear vibration of laminated cylindrical shells was analyzed by Shen and Yang [18]. They examined the influences of temperature variation, shell geometric parameters, and applied voltage on the linear and nonlinear vibration of the structure. Alibeigloo [19] employed the theory of piezo-elasticity to study the bending behavior of FG CNT-reinforced composite cylindrical panels. They used an analytical method to study the effect of CNT volume fraction, temperature variation, and applied voltage on the bending behavior of the system. Rad and Bidgoli [20] explored the dynamic response of a horizontally oriented concrete beam reinforced with nanofibers (NFRP) when subjected to seismic ground excitation. The structural model of the concrete beam employs the hyperbolic shear deformation beam theory (HSDBT). The mathematical formulation is systematically applied to derive the governing equations that describe the behavior of the structure under seismic loading.

For the first time, the dynamic response of NFRP-strengthened beams subjected to seismic excitation is studied in the present research. So the results of this research are of great importance in civil engineering. The beam is modeled by applying the hyperbolic shear deformation beam theory (HSDBT), and the effective material properties of the NFRP layer are obtained based on the Mori–Tanaka model. The dynamic displacement of the structure is calculated by HDQM in conjunction with the Newmark method. The effect of nanotechnology on the dynamic response of the structure can be examined by changing the volume fraction of nanofibers in the resulting composite.

## 4.2 MATHEMATICAL MODEL

In this section, the governing equations of the NFRP-strengthened beams are derived by applying HSDBT to analyze the dynamic behavior of the structure (Figure 4.1).

Figure 4.1 illustrates a hollow circular beam subjected to the earthquake loads with an outer radius of $R_o$, an inner radius of $R_i$, and a thickness of $h$, which was strengthened by an NFRP layer with a thickness of $h_f$. By applying the HSDBT [21], the displacement fields and the strain relations based on Section 1.3.6 are assumed. The constitutive equations of the orthotropic beam are considered as follows:

$$\sigma_{xx}^{c} = C_{11}\varepsilon_{xx}, \tag{4–1}$$

$$\tau_{xz}^{c} = C_{44}\gamma_{xz}, \tag{4–2}$$

where $C_{11}$ and $C_{44}$ are the elastic constants of the beam. Also, the constitutive equations of the NFRP layer are defined as follows:

$$\sigma_{xx}^{f} = Q_{11}\varepsilon_{xx}, \tag{4–3}$$

$$\tau_{xz}^{f} = Q_{44}\gamma_{xz}, \tag{4–4}$$

in which $Q_{11}$ and $Q_{44}$ are the elastic constants of the NFRP layer. To obtain the effective material properties of the NFRP layer and to consider the agglomeration effect, Mori–Tanaka model, which is introduced in the next section, is employed.

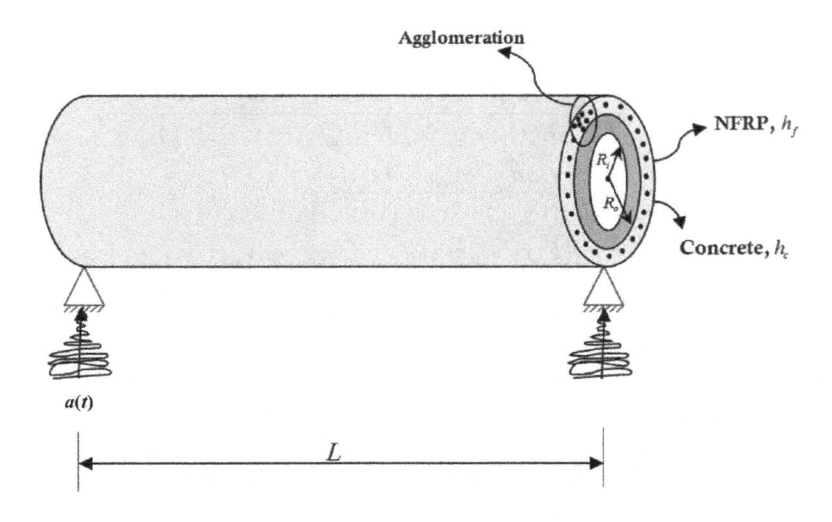


**FIGURE 4.1** Schematic of beams with an NFRP layer under seismic load.

## 4.3 MORI–TANAKA MODEL

In this section, the material properties of a beam reinforced by carbon nanofibers are obtained based on micro-mechanical approach. $E_m$ and $v_m$ are considered as the Young's modulus and the Poisson's ratio of the beam, respectively. The stress–strain relations of the equivalent composite material are given as follows [22]:

$$\begin{Bmatrix} \sigma_{11} \\ \sigma_{22} \\ \sigma_{33} \\ \sigma_{23} \\ \sigma_{13} \\ \sigma_{12} \end{Bmatrix} = \begin{bmatrix} k+m & l & k-m & 0 & 0 & 0 \\ l & n & l & 0 & 0 & 0 \\ k-m & l & k+m & 0 & 0 & 0 \\ 0 & 0 & 0 & p & 0 & 0 \\ 0 & 0 & 0 & 0 & m & 0 \\ 0 & 0 & 0 & 0 & 0 & p \end{bmatrix} \begin{Bmatrix} \varepsilon_{11} \\ \varepsilon_{22} \\ \varepsilon_{33} \\ \gamma_{23} \\ \gamma_{13} \\ \gamma_{12} \end{Bmatrix}, \tag{4–5}$$

where $k$, $l$, $m$, $n$, and $p$ are known as Hill's elastic moduli so that $k$ is plane–strain bulk modulus normal to the fiber direction, $n$ is the uniaxial tension modulus in the longitudinal direction of the fiber, $l$ is the associated cross-modulus, and $m$ and $p$ are the shear moduli in planes normal and parallel to the fiber direction, respectively. It should be noted that the mentioned constants depend on the elastic constants of the material. For example, $Q_{11} = k + m$. By applying the Mori–Tanaka model, Hill's elastic moduli can be obtained as follows [22]:

$$\begin{aligned}
k &= \frac{E_m\{E_m c_m + 2k_r(1+v_m)[1+c_r(1-2v_m)]\}}{2(1+v_m)[E_m(1+c_r-2v_m)+2c_m k_r(1-v_m-2v_m^2)]}, \\
l &= \frac{E_m\{c_m v_m[E_m + 2k_r(1+v_m)] + 2c_r l_r(1-v_m^2)]\}}{(1+v_m)[E_m(1+c_r-2v_m)+2c_m k_r(1-v_m-2v_m^2)]}, \\
n &= \frac{E_m^2 c_m(1+c_r-c_m v_m) + 2c_m c_r(k_r n_r - l_r^2)(1+v_m)^2(1-2v_m)}{(1+v_m)[E_m(1+c_r-2v_m)+2c_m k_r(1-v_m-2v_m^2)]} \\
&\quad + \frac{E_m[2c_m^2 k_r(1-v_m) + c_r n_r(1+c_r-2v_m) - 4c_m l_r v_m]}{E_m(1+c_r-2v_m)+2c_m k_r(1-v_m-2v_m^2)}, \\
p &= \frac{E_m[E_m c_m + 2p_r(1+v_m)(1+c_r)]}{2(1+v_m)[E_m(1+c_r)+2c_m p_r(1+v_m)]}, \\
m &= \frac{E_m[E_m c_m + 2m_r(1+v_m)(3+c_r-4v_m)]}{2(1+v_m)\{E_m[c_m+4c_r(1-v_m)]+2c_m m_r(3-v_m-4v_m^2)\}},
\end{aligned} \tag{4–6}$$

in which $k_r, l_r, n_r, p_r$ and $m_r$ are Hill's elastic moduli of the reinforcing phase of the composite material. Finally, by substituting Equation 4–6 into Equation 4–5, the stiffness of the matrix can be obtained. The experimental results show that a uniform distribution of the nanofibers is rarely achievable [23]. It is observed that most of the nanofibers centralized in regions throughout the matrix. These regions are assumed to be spherical shapes, which are known as "inclusions," with different material properties from the surrounding regions. $V_r$ is the total volume of nanofibers and is defined as

$$V_r = V_r^{inclusion} + V_r^m, \tag{4–7}$$

in which $V_r^{inclusion}$ and $V_r^m$ represent the volume of the CNTs inside the inclusion and the polymer matrix, respectively. The agglomeration effect can be considered based on the micro-mechanical model by introducing the two following parameters:

$$\xi = \frac{V_{inclusion}}{V}, \tag{4–8}$$

$$\zeta = \frac{V_r^{inclusion}}{V_r}. \tag{4–9}$$

The average volume fraction $C_r$ of the nanofibers in the composite material is given as follows:

$$C_r = \frac{V_r}{V}. \tag{4–10}$$

The volume fraction of the nanofibers inside the inclusion and the matrix can be defined as

$$\frac{V_r^{inclusion}}{V_{inclusion}} = \frac{C_r \zeta}{\xi}, \tag{4–11}$$

$$\frac{V_r^m}{V - V_{inclusion}} = \frac{C_r (1-\zeta)}{1-\xi}. \tag{4–12}$$

We assume that the nanofibers are transversely isotropic and that their orientation is random. Hence, the inclusion is considered to be isotropic, and the effective bulk modulus $K$ and shear modulus $G$ may be written as follows:

$$K = K_{out}\left[1 + \frac{\xi\left(\frac{K_{in}}{K_{out}} - 1\right)}{1 + \alpha(1-\xi)\left(\frac{K_{in}}{K_{out}} - 1\right)}\right], \tag{4–13}$$

$$G = G_{out}\left[1 + \frac{\xi\left(\frac{G_{in}}{G_{out}} - 1\right)}{1 + \beta(1-\xi)\left(\frac{G_{in}}{G_{out}} - 1\right)}\right], \tag{4–14}$$

in which $K_{in}$ and $K_{out}$ are the effective bulk modulus of the inclusion and the matrix outside the inclusion, respectively. Also, $G_{in}$ and $G_{out}$ are the effective shear modulus of the inclusion and the matrix outside the inclusion, respectively, and are given as follows:

$$K_{in} = K_m + \frac{(\delta_r - 3K_m\chi_r)C_r\zeta}{3(\xi - C_r\zeta + C_r\zeta\chi_r)}, \tag{4–15}$$

$$K_{out} = K_m + \frac{C_r(\delta_r - 3K_m\chi_r)(1-\zeta)}{3\left[1-\xi - C_r(1-\zeta) + C_r\chi_r(1-\zeta)\right]}, \tag{4–16}$$

$$G_{in} = G_m + \frac{(\eta_r - 3G_m\beta_r)C_r\zeta}{2(\xi - C_r\zeta + C_r\zeta\beta_r)}, \tag{4–17}$$

$$G_{out} = G_m + \frac{C_r(\eta_r - 3G_m\beta_r)(1-\zeta)}{2\left[1-\xi - C_r(1-\zeta) + C_r\beta_r(1-\zeta)\right]}, \tag{4–18}$$

where $\chi_r, \beta_r, \delta_r$ and $\eta_r$ can be obtained as

$$\chi_r = \frac{3(K_m + G_m) + k_r - l_r}{3(k_r + G_m)}, \tag{4–19}$$

$$\beta_r = \frac{1}{5}\left\{\frac{4G_m + 2k_r + l_r}{3(k_r + G_m)} + \frac{4G_m}{(p_r + G_m)} + \frac{2\left[G_m(3K_m + G_m) + G_m(3K_m + 7G_m)\right]}{G_m(3K_m + G_m) + m_r(3K_m + 7G_m)}\right\}, \tag{4–20}$$

$$\delta_r = \frac{1}{3}\left[n_r + 2l_r + \frac{(2k_r - l_r)(3K_m + 2G_m - l_r)}{k_r + G_m}\right], \tag{4–21}$$

$$\eta_r = \frac{1}{5}\left[\frac{2}{3}(n_r - l_r) + \frac{4G_m p_r}{(p_r + G_m)} + \frac{8G_m m_r(3K_m + 4G_m)}{3K_m(m_r + G_m) + G_m(7m_r + G_m)} + \frac{2(k_r - l_r)(2G_m + l_r)}{3(k_r + G_m)}\right], \tag{4–22}$$

where $K_m$ and $G_m$ are the bulk and shear moduli of the matrix phase, which are defined as follows:

$$K_m = \frac{E_m}{3(1 - 2\upsilon_m)}, \tag{4–23}$$

$$G_m = \frac{E_m}{2(1 + \upsilon_m)}. \tag{4–24}$$

Moreover, $\alpha$ and $\beta$ in Equations 4–13 and 4–14 are given as follows:

$$\alpha = \frac{\left(1+\upsilon_{out}\right)}{3\left(1-\upsilon_{out}\right)}, \tag{4–25}$$

$$\alpha = \frac{\left(1+\upsilon_{out}\right)}{3\left(1-\upsilon_{out}\right)}, \tag{4–26}$$

$$\upsilon_{out} = \frac{3K_{out} - 2G_{out}}{6K_{out} + 2G_{out}}. \tag{4–27}$$

Therefore, the effective Young's modulus $E$ and Poisson's ratio $v$ of the composite material are given by

$$E = \frac{9KG}{3K+G}, \tag{4–28}$$

$$\upsilon = \frac{3K-2G}{6K+2G}. \tag{4–29}$$

## 4.4 ENERGY METHOD

To derive the governing equations of the structure by employing the energy method and using Hamilton's principle, the work done by external forces is equated to the strain energy and kinetic energy stored in the structure. The potential strain energy stored in the structure is given as follows:

$$U = \int_{V_c} \left(\sigma_{xx}^{c}\varepsilon_{xx} + \tau_{xz}^{c}\gamma_{xz}\right) dA_c dx + \int_{V_f} \left(\sigma_{xx}^{f}\varepsilon_{xx} + \tau_{xz}^{f}\gamma_{xz}\right) dA_f dx, \tag{4–30}$$

where $A_c$ and $A_f$ are the cross-sectional area of the beam and the NFRP layer, respectively. By substituting strain relations into Equation 4–30, we have

$$\begin{aligned} U = {} & \frac{1}{2}\int_0^L \left[ \int_{A_c} \left[ \sigma_{xx}^{c} \left\{ \left(\frac{\partial U}{\partial x}\right) + \left(\frac{1}{2}\left(\frac{\partial W}{\partial x}\right)^2\right) - z\left(\frac{\partial^2 W}{\partial x^2}\right) + \Phi(z)\left(\frac{\partial^2 W}{\partial x^2} - \frac{\partial \varphi}{\partial x}\right) \right\} \right.\right. \\ & \left.\left. + \tau_{xz}^{c} \left\{ \frac{\partial \Phi(z)}{\partial z}\left(\frac{\partial W}{\partial x} - \varphi\right) \right\} \right] dA_c dx \right. \\ & + \frac{1}{2}\int_0^L \left[ \int_{A_f} \left[ \sigma_{xx}^{f} \left\{ \left(\frac{\partial U}{\partial x}\right) + \left(\frac{1}{2}\left(\frac{\partial W}{\partial x}\right)^2\right) - z\left(\frac{\partial^2 W}{\partial x^2}\right) + \Phi(z)\left(\frac{\partial^2 W}{\partial x^2} - \frac{\partial \varphi}{\partial x}\right) \right\} \right.\right. \\ & \left.\left. + \tau_{xz}^{f} \left\{ \frac{\partial \Phi(z)}{\partial z}\left(\frac{\partial W}{\partial x} - \varphi\right) \right\} \right] dA_f dx. \right. \end{aligned} \tag{4–31}$$

Defining the in-plane stress results in the following:

$$N_x = \int_{A_c} \sigma_{xx}^c dA_c + \int_{A_f} \sigma_{xx}^f dA_f, \tag{4–32}$$

$$M_x = \int_{A_c} \sigma_{xx}^c z dA_c + \int_{A_f} \sigma_{xx}^f z dA_f, \tag{4–33}$$

$$F_x = \int_{A_c} \sigma_{xx}^c \Phi(z) dA_c + \int_{A_f} \sigma_{xx}^f \Phi(z) dA_f, \tag{4–34}$$

$$Q_x = \int_{A_c} \tau_{xz}^c \frac{\partial \Phi(z)}{\partial z} dA_c + \int_{A_f} \tau_{xz}^f \frac{\partial \Phi(z)}{\partial z} dA_f, \tag{4–35}$$

Equation 4–31 can be rewritten as follows:

$$U = \frac{1}{2}\int_0^L \left[ \int \left[ N_x \left( \left( \frac{\partial U}{\partial x} \right) + \left( \frac{1}{2} \left( \frac{\partial W}{\partial x} \right)^2 \right) \right) - M_x \left( \frac{\partial^2 W}{\partial x^2} \right) \right.\right.$$
$$\left.\left. + F_x \left( \frac{\partial^2 W}{\partial x^2} - \frac{\partial \varphi}{\partial x} \right) + Q_x \left( \left( \frac{\partial W}{\partial x} - \varphi \right) \right) \right] dx. \tag{4–36}$$

By substituting Equations 1–25a and 1–25b into Equations 4–32 through 4–35, the stress resultants of the beam take the following form:

$$N_x = A_{11} \left( \frac{\partial U}{\partial x} + \frac{1}{2} \left( \frac{\partial W}{\partial x} \right)^2 \right) - B_{11} \left( \frac{\partial^2 W}{\partial x^2} \right) + E_{11} \left( \frac{\partial^2 W}{\partial x^2} - \frac{\partial \varphi}{\partial x} \right), \tag{4–37}$$

$$M_x = B_{11} \left( \frac{\partial U}{\partial x} + \frac{1}{2} \left( \frac{\partial W}{\partial x} \right)^2 \right) - D_{11} \left( \frac{\partial^2 W}{\partial x^2} \right) + F_{11} \left( \frac{\partial^2 W}{\partial x^2} - \frac{\partial \varphi}{\partial x} \right), \tag{4–38}$$

$$F_x = E_{11} \left( \frac{\partial U}{\partial x} + \frac{1}{2} \left( \frac{\partial W}{\partial x} \right)^2 \right) - F_{11} \left( \frac{\partial^2 W}{\partial x^2} \right) + H_{11} \left( \frac{\partial^2 W}{\partial x^2} - \frac{\partial \varphi}{\partial x} \right), \tag{4–39}$$

$$Q_x = L_{44} \left( \frac{\partial W}{\partial x} - \varphi \right), \tag{4–40}$$

in which

$$A_{11} = \int_{A_c} C_{11}\, dA_c + \int_{A_f} Q_{11}\, dA_f, \tag{4–41}$$

$$B_{11} = \int_{A_c} C_{11}\, z dA_c + \int_{A_f} Q_{11}\, z dA_f, \tag{4–42}$$

$$D_{11} = \int_{A_c} C_{11}\, z^2 dA_c + \int_{A_f} Q_{11}\, z^2 dA_f, \tag{4–43}$$

$$E_{11} = \int_{A_c} C_{11}\,\Phi(z)dA_c + \int_{A_f} Q_{11}\,\Phi(z)dA_f, \tag{4–44}$$

$$F_{11} = \int_{A_c} C_{11}\,z\Phi(z)dA_c + \int_{A_f} Q_{11}\,z\Phi(z)dA_f, \tag{4–45}$$

$$H_{11} = \int_{A_c} C_{11}\,\Phi(z)^2 dA_c + \int_{A_f} Q_{11}\,\Phi(z)^2 dA_f, \tag{4–46}$$

$$L_{44} = \int_{A_c} C_{11}\frac{\partial\Phi(z)}{\partial z}dA_c + \int_{A_f} Q_{11}\frac{\partial\Phi(z)}{\partial z}dA_f. \tag{4–47}$$

The kinetic energy of the structure is defined as follows:

$$K = \frac{\rho}{2}\int\left(\dot{u}_1^{\,2} + \dot{u}_2^{\,2} + \dot{u}_3^{\,2}\right)dV. \tag{4–48}$$

Substituting Equation 1–24 into Equation 4–48, we have

$$K = \frac{\rho}{2}\int\left(\left(\frac{\partial U}{\partial t} - z\frac{\partial^2 W}{\partial x\partial t} + \Phi(z)\left(\frac{\partial^2 W}{\partial x\partial t} - \frac{\partial\varphi}{\partial t}\right)\right)^2 + \left(\frac{\partial W}{\partial t}\right)^2\right)dV. \tag{4–49}$$

Defining the inertia moment terms as

$$\begin{Bmatrix} I_0 \\ I_1 \\ I_2 \\ I_3 \\ I_4 \\ I_5 \end{Bmatrix} = \int_{A_c}\begin{bmatrix} \rho_c \\ \rho_c z \\ \rho_c z^2 \\ \rho_c\Phi(z) \\ \rho_c z\Phi(z) \\ \rho_c\Phi(z)^2 \end{bmatrix} dA_c + \int_{A_f}\begin{bmatrix} \rho_f \\ \rho_f z \\ \rho_f z^2 \\ \rho_f\Phi(z) \\ \rho_f z\Phi(z) \\ \rho_f\Phi(z)^2 \end{bmatrix} dA_f, \tag{4–50}$$

Equation 4–49 can be rewritten as follows:

$$\begin{aligned} K = 0.5\int \Bigg[ & I_0\left(\left(\frac{\partial U}{\partial t}\right)^2 + \left(\frac{\partial W}{\partial t}\right)^2\right) - 2I_1\left(\frac{\partial U}{\partial t}\frac{\partial^2 W}{\partial x\partial t}\right) + I_2\left(\frac{\partial^2 W}{\partial x\partial t}\right)^2 \\ & + I_3\frac{\partial U}{\partial t}\left(\frac{\partial^2 W}{\partial x\partial t} - \frac{\partial\varphi}{\partial t}\right) - I_4\frac{\partial^2 W}{\partial x\partial t}\left(\frac{\partial^2 W}{\partial x\partial t} - \frac{\partial\varphi}{\partial t}\right) + I_5\left(\frac{\partial^2 W}{\partial x\partial t} - \frac{\partial\varphi}{\partial t}\right)^2\Bigg] dx. \end{aligned} \tag{4–51}$$

The external work due the earthquake can be calculated as follows:

$$W = \int \underbrace{(ma(t))}_{F_{Seismic}} W dx, \tag{4–52}$$

where $m$ and $a(t)$ are the mass and acceleration of the earth, respectively. To extract the governing equations of motion, Hamilton's principle is expressed as follows [16,17]:

$$\int_0^t (\delta U - \delta K - \delta W)dt = 0, \tag{4–53}$$

where $\delta$ denotes the variational operator. By considering Equations 4–36, 4–51, and 4–52, the first variations of the potential strain energy, the kinetic energy, and the external work are presented as in the following:

$$\begin{aligned}\delta U = \int_0^L \Bigg[ \int \Bigg[ & N_x \left( \frac{\partial \delta U}{\partial x} + \frac{\partial W}{\partial x}\frac{\partial \delta W}{\partial x} \right) - M_x \left( \frac{\partial^2 \delta W}{\partial x^2} \right) \\ & + F_x \left( \frac{\partial^2 \delta W}{\partial x^2} - \frac{\partial \delta \varphi}{\partial x} \right) + Q_x \left( \left( \frac{\partial \delta W}{\partial x} - \delta \varphi \right) \right) \Bigg] dx, \end{aligned} \tag{4–54}$$

$$\begin{aligned}\delta K = \int_0^L \Bigg[ & I_0 \left( \frac{\partial U}{\partial t}\frac{\partial \delta U}{\partial t} + \frac{\partial W}{\partial t}\frac{\partial \delta W}{\partial t} \right) - I_1 \left( \frac{\partial \delta U}{\partial t}\frac{\partial^2 W}{\partial x \partial t} + \frac{\partial U}{\partial t}\frac{\partial^2 \delta W}{\partial x \partial t} \right) \\ & + I_3 \left( \frac{\partial \delta U}{\partial t}\left( \frac{\partial^2 W}{\partial x \partial t} - \frac{\partial \varphi}{\partial t} \right) + \frac{\partial U}{\partial t}\left( \frac{\partial^2 \delta W}{\partial x \partial t} - \frac{\partial \delta \varphi}{\partial t} \right) \right) + I_2 \left( \frac{\partial^2 W}{\partial x \partial t}\frac{\partial^2 \delta W}{\partial x \partial t} \right) \\ & - I_4 \left( \frac{\partial^2 \delta W}{\partial x \partial t}\left( \frac{\partial^2 W}{\partial x \partial t} - \frac{\partial \varphi}{\partial t} \right) + \frac{\partial^2 W}{\partial x \partial t}\left( \frac{\partial^2 \delta W}{\partial x \partial t} - \frac{\partial \delta \varphi}{\partial t} \right) \right) \\ & + I_5 \left( \frac{\partial^2 W}{\partial x \partial t} - \frac{\partial \varphi}{\partial t} \right)\left( \frac{\partial^2 \delta W}{\partial x \partial t} - \frac{\partial \delta \varphi}{\partial t} \right) \Bigg] dx, \end{aligned} \tag{4–55}$$

$$\delta W = -\int \underbrace{(ma(t))}_{F_{Seismic}} \delta W dx. \tag{4–56}$$

Now, by substituting Equations 4–54 through 4–56 into Equation 4–53, the motion equations of the structure are obtained as follows:

$$\delta U:$$

$$\frac{\partial N_x}{\partial x} = I_0 \frac{\partial^2 U}{\partial t^2} + (I_3 - I_1)\frac{\partial^3 W}{\partial x \partial t^2} - I_3 \frac{\partial^2 \varphi}{\partial t^2}, \tag{4–57}$$

$$\delta W:$$

$$\begin{aligned} & \frac{\partial^2 M_x}{\partial x^2} - \frac{\partial}{\partial x}\left( N_x^M \frac{\partial W}{\partial x} \right) - \frac{\partial^2 F_x}{\partial x^2} + \frac{\partial Q_x}{\partial x} + F_{Seismic} = I_0 \frac{\partial^2 W}{\partial t^2} \\ & + (I_1 - I_3)\frac{\partial^3 U}{\partial x \partial t^2} + (2I_4 - I_2 - I_5)\frac{\partial^4 W}{\partial x^2 \partial t^2} + (I_5 - I_4)\frac{\partial^3 \varphi}{\partial x \partial t^2}, \end{aligned} \tag{4–58}$$

$\delta\varphi$ :

$$Q_x - \frac{\partial F_x}{\partial x} = I_5 \frac{\partial^2 \varphi}{\partial t^2} - I_3 \frac{\partial^2 U}{\partial t^2} + (I_4 - I_5)\frac{\partial^3 W}{\partial x \partial t^2}. \tag{4–59}$$

By substituting Equations 4–37 through 4–40 into Equations 4–57 through 4–59, the governing equations of the system are expressed as follows:

$\delta U$ :

$$A_{11}\left(\frac{\partial^2 U}{\partial x^2} + \frac{\partial W}{\partial x}\frac{\partial^2 W}{\partial x^2}\right) - B_{11}\left(\frac{\partial^3 W}{\partial x^3}\right) + E_{11}\left(\frac{\partial^3 W}{\partial x^3} - \frac{\partial^2 \varphi}{\partial x^2}\right)$$
$$= I_0 \frac{\partial^2 U}{\partial t^2} + (I_3 - I_1)\frac{\partial^3 W}{\partial x \partial t^2} - I_3 \frac{\partial^2 \varphi}{\partial t^2}, \tag{4–60}$$

$\delta W$ :

$$(B_{11} - E_{11})\left(\frac{\partial^3 U}{\partial x^3} + \left(\frac{\partial^2 W}{\partial x^2}\right)^2 + \frac{\partial W}{\partial x}\frac{\partial^3 W}{\partial x^3}\right) - (D_{11} - F_{11})\left(\frac{\partial^4 W}{\partial x^4}\right)$$
$$+ (F_{11} - H_{11})\left(\frac{\partial^4 W}{\partial x^4} - \frac{\partial^3 \varphi}{\partial x^3}\right) + L_{44}\left(\frac{\partial^2 W}{\partial x^2} - \frac{\partial \varphi}{\partial x}\right) + F_{Seismic} = I_0 \frac{\partial^2 W}{\partial t^2}$$
$$+ (I_1 - I_3)\frac{\partial^3 U}{\partial x \partial t^2} + (2I_4 - I_2 - I_5)\frac{\partial^4 W}{\partial x^2 \partial t^2} + (I_5 - I_4)\frac{\partial^3 \varphi}{\partial x \partial t^2}, \tag{4–61}$$

$\delta\varphi$ :

$$Q_x - E_{11}\left(\frac{\partial^2 U}{\partial x^2} + \frac{\partial W}{\partial x}\frac{\partial^2 W}{\partial x^2}\right) + F_{11}\left(\frac{\partial^{32} W}{\partial x^3}\right) - H_{11}\left(\frac{\partial^3 W}{\partial x^3} - \frac{\partial^2 \varphi}{\partial x^2}\right) =$$
$$I_5 \frac{\partial^2 \varphi}{\partial t^2} - I_3 \frac{\partial^2 U}{\partial t^2} + (I_4 - I_5)\frac{\partial^3 W}{\partial x \partial t^2}. \tag{4–62}$$

Also, the boundary conditions of the structure are considered as follows:

- **Clamped–clamped supported**

$$\begin{aligned} &W = U = \varphi = \frac{\partial W}{\partial x} = 0, && @ \quad x = 0 \\ &W = U = \varphi = \frac{\partial W}{\partial x} = 0. && @ \quad x = L \end{aligned} \tag{4–63}$$

- **Clamped–simply supported**

$$\begin{aligned} &W = U = \varphi = \frac{\partial W}{\partial x} = 0, && @ \quad x = 0 \\ &W = U = \varphi = M_x = 0. && @ \quad x = L \end{aligned} \tag{4–64}$$

- **Simply–simply supported**

$$\begin{aligned} &W = U = \varphi = M_x = 0, && @ \quad x = 0 \\ &W = U = \varphi = M_x = 0. && @ \quad x = L \end{aligned} \tag{4–65}$$

- **Clamped-free**

$$\begin{aligned} &W = U = \varphi = \frac{\partial W}{\partial x} = 0, && @ \quad x = 0 \\ &F_x = Q_x = N_x = M_x = 0. && @ \quad x = L \end{aligned} \tag{4–66}$$

## 4.5 NUMERICAL RESULTS

In this section, the effect of various parameters on the dynamic response of an NFRP-strengthened beam under seismic load is examined based on HDQM (presented in Section 2.2.2) and a MATLAB code in Appendix A. The outer radius and the inner radius of the beam are $R_0 = 205\,mm$ and $R_i = 56\,mm$, respectively, and the length of the beam is $L = 3\,m$. The elastic moduli of beam, epoxy resin, and carbon nanofiber are $E_c = 20\,GPa$, $E_f = 25\,GPa$, and $E_r = 1\ TPa$, respectively. In this chapter, the influences of the NFRP layer, the carbon nanofiber volume fraction, the geometric parameters, and the boundary conditions on the dynamic displacement of the structure are investigated. The earthquake acceleration, based on the Kobe earthquake that had a distribution of acceleration in 30 seconds, is shown in Figure 4.2.

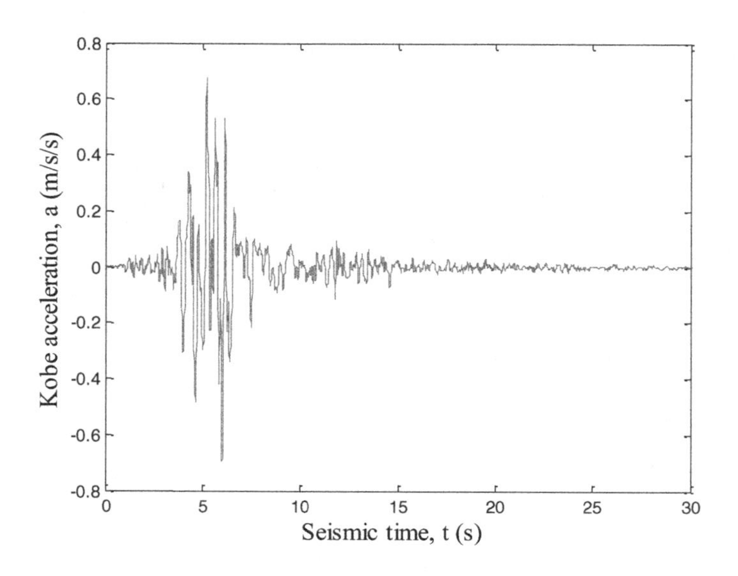


**FIGURE 4.2** Acceleration history of the Kobe earthquake.

### 4.5.1 Convergence of HDQM

Figure 4.3 shows the convergence of HDQM in evaluating the maximum deflection of the structure versus time. As can be seen, with increasing the number of grid points *N*, the maximum deflection of the structure decreases. For example, the maximum deflections of the structure for the number of grid points (*N*) of 7, 11, 15, and 17 are equal to 0.14, 0.06, 0.016, and 0015, respectively. It can be found that by increasing the number of grid points, the decay ratio of the dynamic deflection decreases as far as at $N = 17$ the dynamic deflection converges. It means that increasing the number of grid points does not affect the amount of dynamic displacement in the structure after $N = 17$. So the following results are based on 17 grid points for the HDQM solution.

### 4.5.2 Validation of Results

Given that no similar work has been done to validate the present study, an effort has been made to examine the results without considering the nonlinear terms of the governing equations and by comparing the linear dynamic response of the structure, which was obtained by two various solution methods. The results of the analytical and numerical (HDQ) methods are depicted in Figure 4.4. As can be observed, the results of numerical and analytical methods are identical, and therefore, the obtained results are accurate and acceptable.

### 4.5.3 Effect of an NFRP Layer on the Dynamic Response

Figures 4.5a–d illustrate the effect of an NFRP layer on the dynamic deflection versus time and various thicknesses of the NFRP layer. As can be observed, the

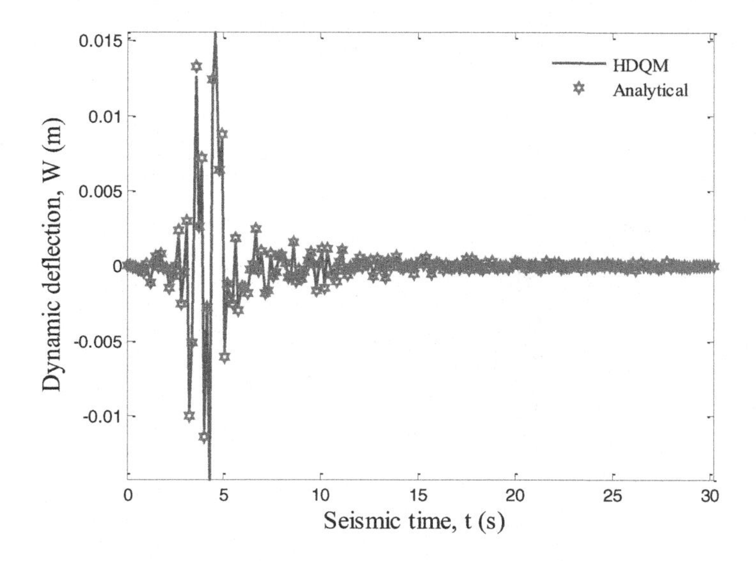


**FIGURE 4.3** Comparison of analytical and numerical results.

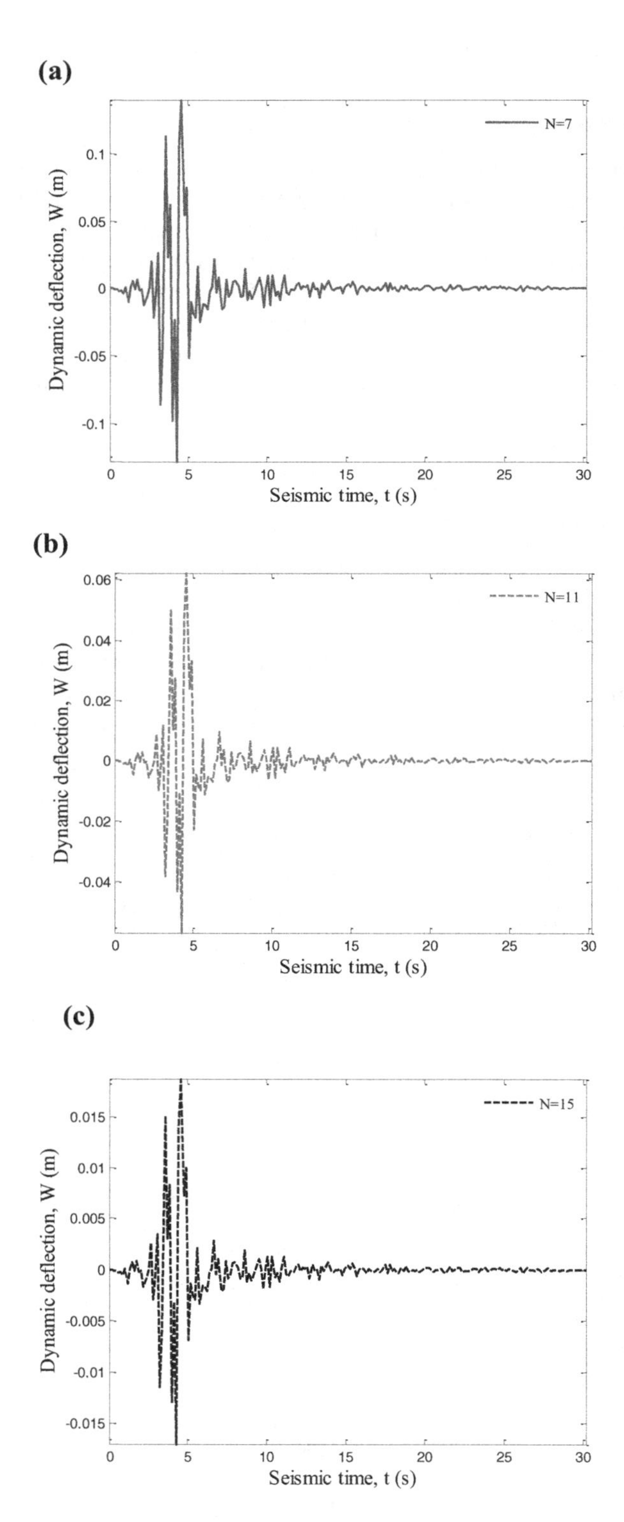


**FIGURE 4.4** Convergence and accuracy of HDQM.

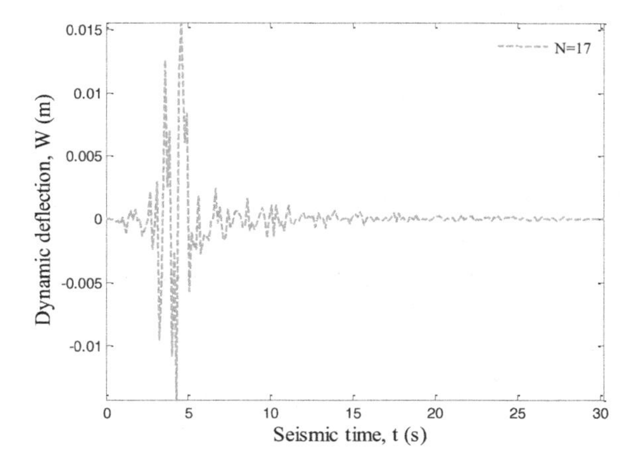


**FIGURE 4.4** (Continued)

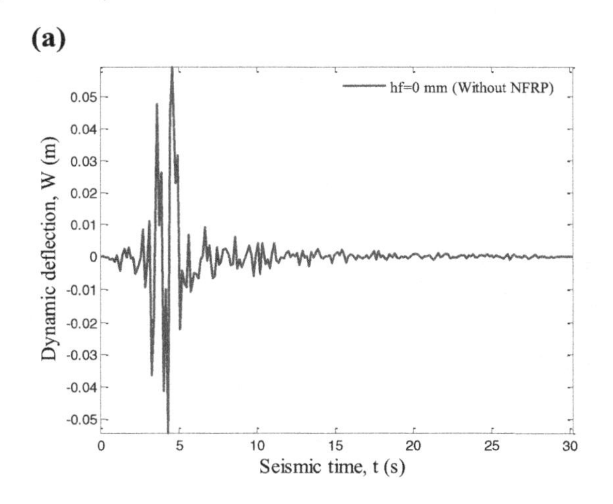


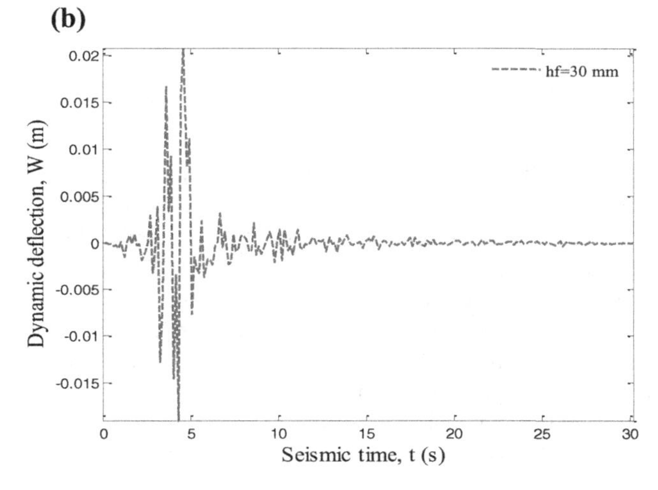


**FIGURE 4.5** The effect of an NFRP layer on the dynamic deflection of the structure.

(c)

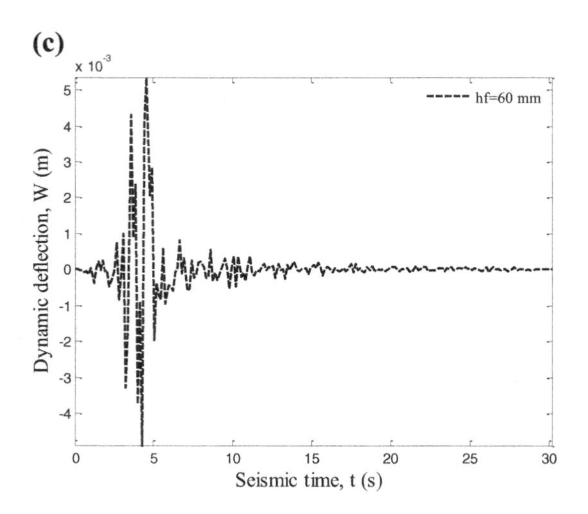


(d)

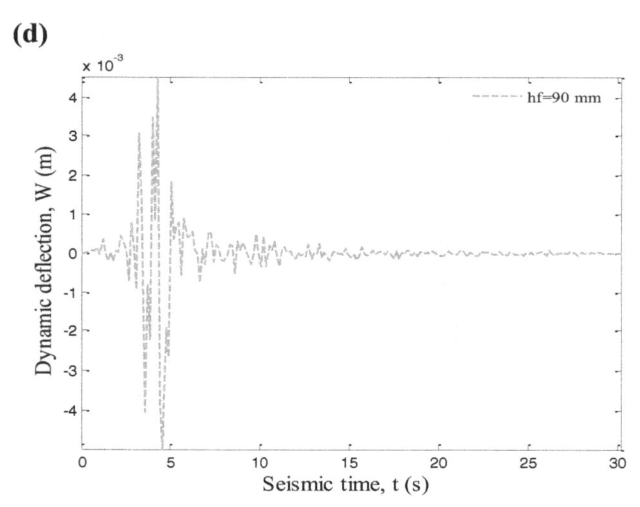


**FIGURE 4.5** (Continued)

structure without an NFRP layer has a greater dynamic deflection with respect to the beam covered with an NFRP layer. The reason is that the NFRP layer increases the stiffness of the structure. Figure 4.5a shows the maximum dynamic deflection of the structure without an NFRP layer was equal to 0.059 while by applying an NFRP layer with a thickness of 30, 60, and 90 mm, the maximum dynamic displacement of the structure is 0.0207, 0.0053, and 0.0045, respectively. By comparing the results, we can say that using an NFRP layer with a thickness of 30, 60, and 90 mm decreases the maximum dynamic displacement of the structure up to 64.9, 91.02, and 92.37 percent, respectively, which is a remarkable result in the dynamic designing of the structures. Also, it should be noted that excessively increasing the NFRP layer increases costs while it does not have a noticeable effect on the dynamic response of the structure. Hence, an NFRP layer with a thickness of 60 mm is the best choice for the present structure.

### 4.5.4 Effect of Carbon Nanofibers on the Dynamic Response

As mentioned in the previous sections, the NFRP is reinforced by carbon nanofibers instead of macro-fibers. In this section, the effect of the nanofiber volume percent on the dynamic response of the structure is studied. Figures 4.6a–d shows the dynamic deflection of the structure versus time for different values of the nanofiber volume fraction as $c_r = 0$, $c_r = 0.1$, $c_r = 0.2$, and $c_r = 0.3$, respectively. It is apparent

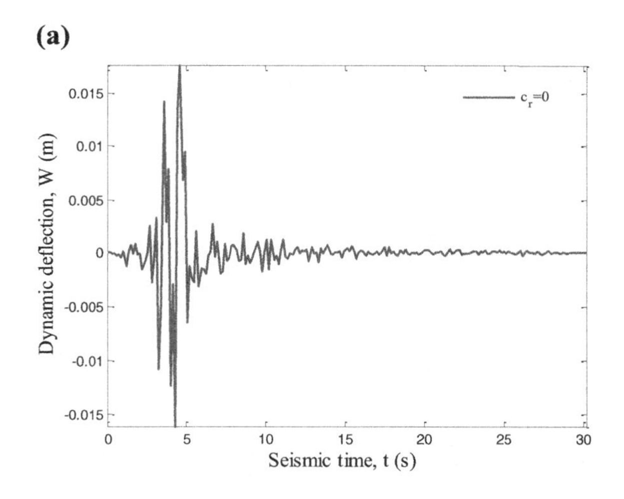


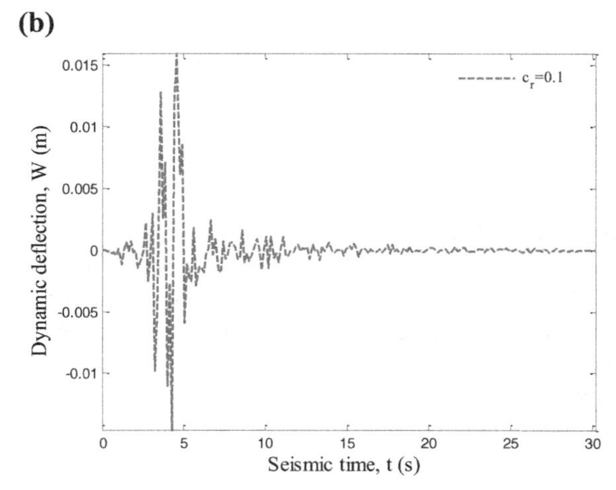


**FIGURE 4.6** The effect of nanofiber volume percent on the dynamic deflection of the structure.

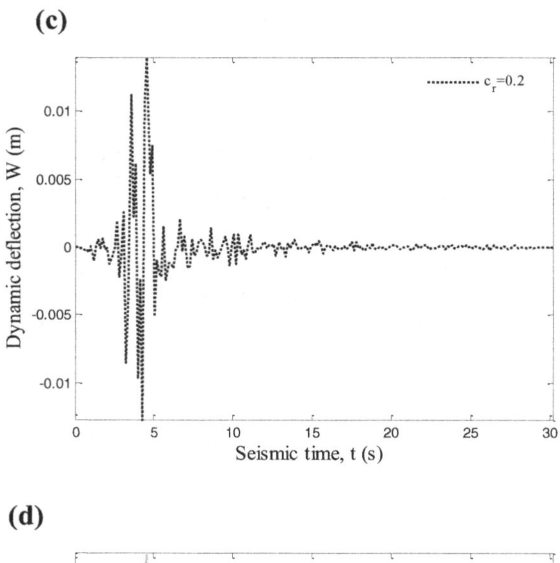


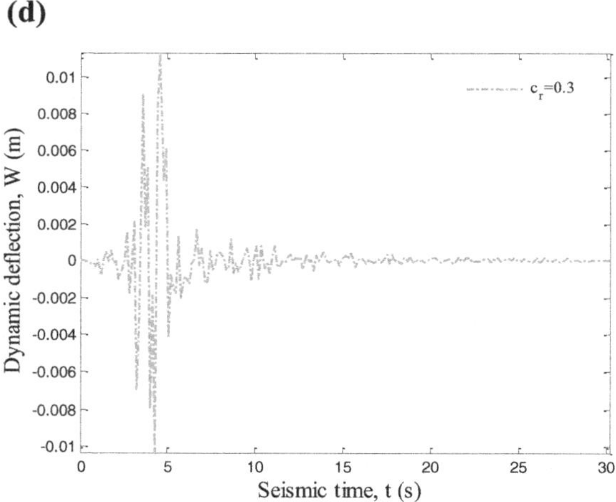


**FIGURE 4.6** (Continued)

that the maximum dynamic displacement of the structure is equal to 0.0176 for the case of $c_r = 0$ (without nanofibers). By applying nanofibers in volume fractions of 0.1, 0.2, and 0.3, the amount of maximum dynamic displacement is 0.0159, 0.0139, and 0.0113, respectively. Therefore, using nanofibers in volume fractions of 0.1, 0.2, and 0.3 increases the stiffness of the structure and reduces the maximum displacement of structure by 9.65, 21.02, and 35.79 percent, respectively. So it can be concluded that with increasing the volume fraction of nanofibers, the dynamic deflection of the system decreases, and this is because of the increase in the stiffness of the structure.

The agglomeration effect of nanofibers on the dynamic deflection of the structure versus time is illustrated in Figures 4.7a–d.

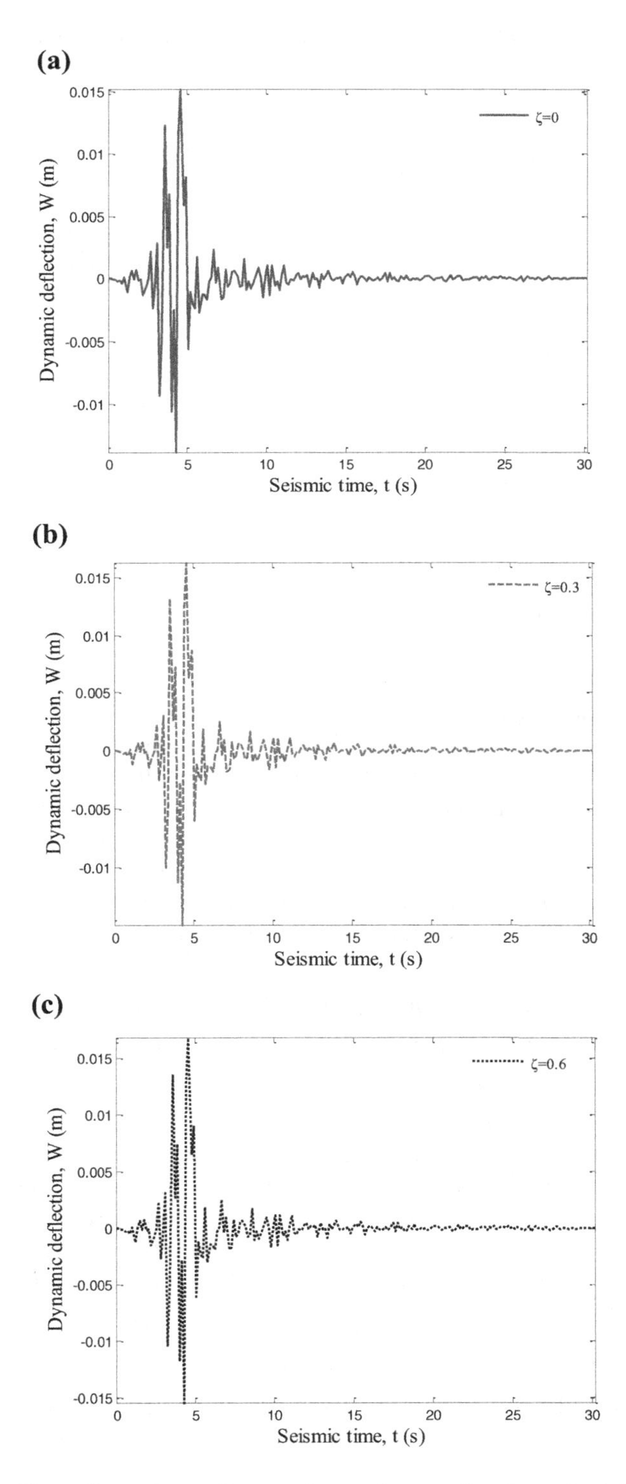


**FIGURE 4.7** The effect of nanofiber agglomeration on the dynamic deflection of the structure.

**(d)**

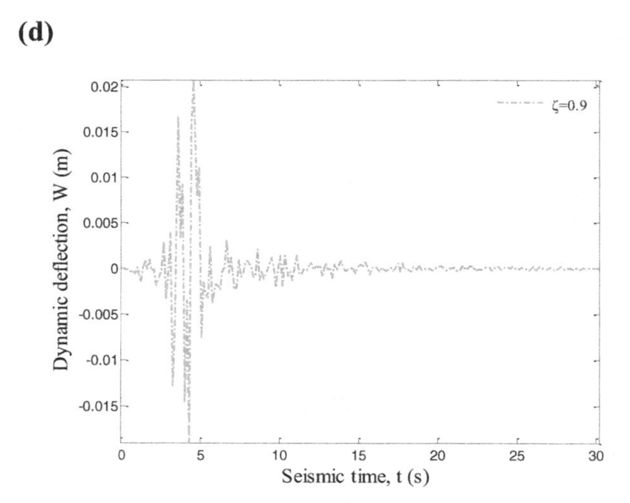


**FIGURE 4.7** (Continued)

As can be observed, by considering the agglomeration effect, the stiffness of the structure reduces while the dynamic displacement increases. For example, in the absence of the agglomeration effect ($\xi = 0$), the maximum dynamic deflection of the structure is 0.0152 while for $\xi = 0.9$, the maximum dynamic deflection is 0.0207. The results reveal that the existence of the agglomeration changes the maximum dynamic displacement of the structure up to 36.18 percent. Since during the process of nanocomposite manufacturing, the uniform distribution for nanofibers in the matrix is impossible, so the results of this figure can be very remarkable.

### 4.5.5 Effect of Geometric Parameters of a Beam on the Dynamic Response

The effect of the outer radius to the inner radius ratio of the beam on the dynamic response versus time is shown in Figures 4.8a–d.

It can be seen that with an increase in the outer-to-inner-radius ratio of the beam, the structure becomes softer, and the dynamic deflection of the system increases. The maximum dynamic displacements for the outer-to-inner-radius ratio of 2, 4, 6, and 8 are 0.00053, 0.0192, 0.0407, and 0.0585, respectively. For example, by increasing the outer-to-inner-radius ratio from 6 to 8, the maximum displacement increases up to 43.76 percent.

Figures 4.9a–d presents the effect of the beam length on the dynamic deflection of the structure versus time. It can be found that by increasing the length, the displacement of the structure increases. This is because of the reduction in the stiffness of the system when the beam becomes longer. For instance, an increase in the length of the beam from 2m to 3m leads to an increase in the maximum displacement of the structure up to 52.21 percent.

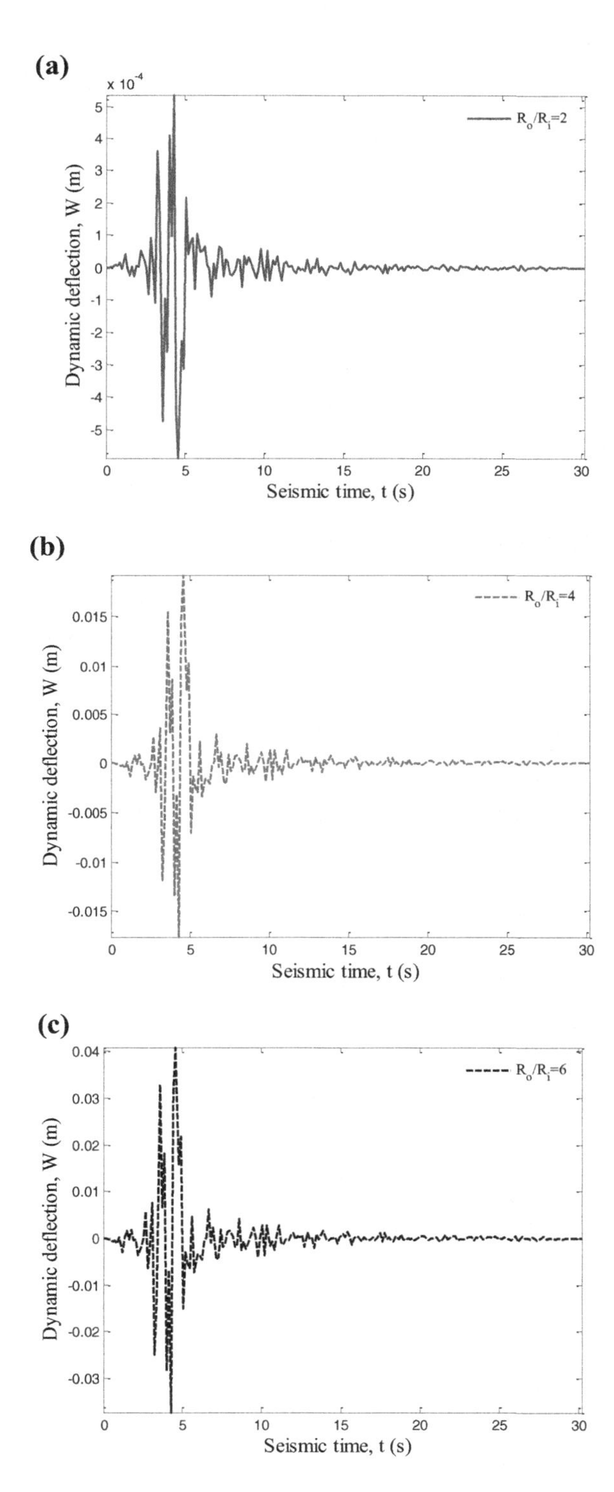


**FIGURE 4.8** The effect of the outer-to-inner-radius ratio of the beam on the dynamic deflection.

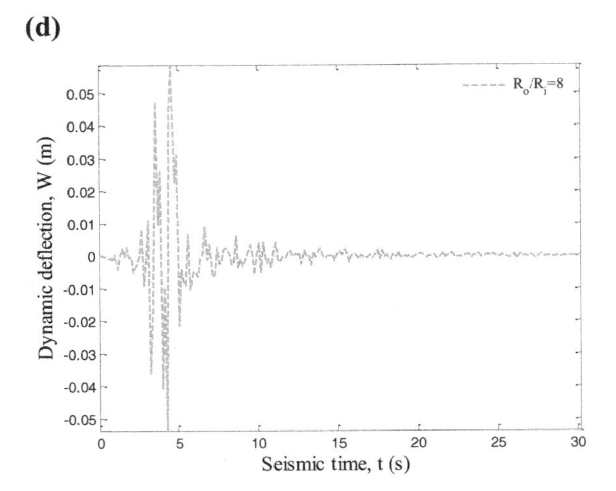


**FIGURE 4.8** (Continued)

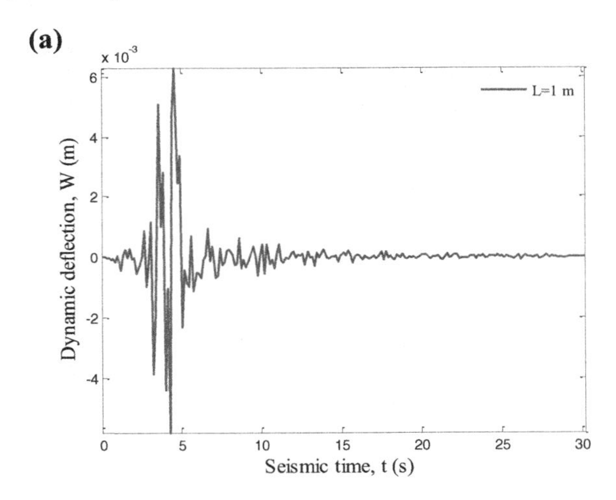


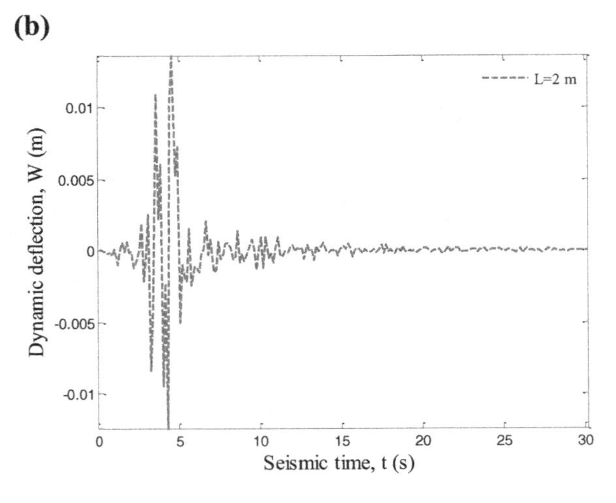


**FIGURE 4.9** The effect of beam length on the dynamic deflection of the structure.

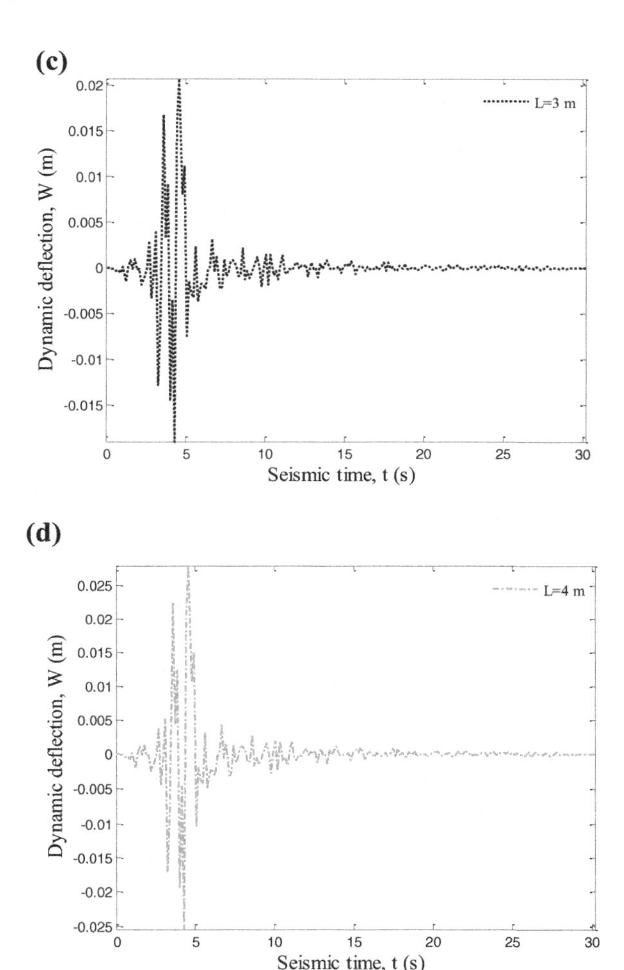


**FIGURE 4.9** (Continued)

### 4.5.6 Effect of Boundary Conditions on Dynamic Response

Figures 4.10a–d illustrate the effect of various boundary conditions on the dynamic response versus time. Four boundary conditions including clamped–clamped, clamped–simply, simply–simply, and free–simply supported are considered. The maximum dynamic deflections of the structure for clamped–clamped, clamped–simply supported, simply–simply supported, and free–simply supported boundary conditions are 0.0042, 0.0072, 0.0155, and 0.0311, respectively. As can be observed, boundary conditions have a significant effect on the dynamic response of the system so that the structure with a clamped–clamped boundary condition has the lowest displacement with respect to the other boundary conditions. It is because of the stronger constraint of the clamped boundary, which gives the structure a higher stiffness.

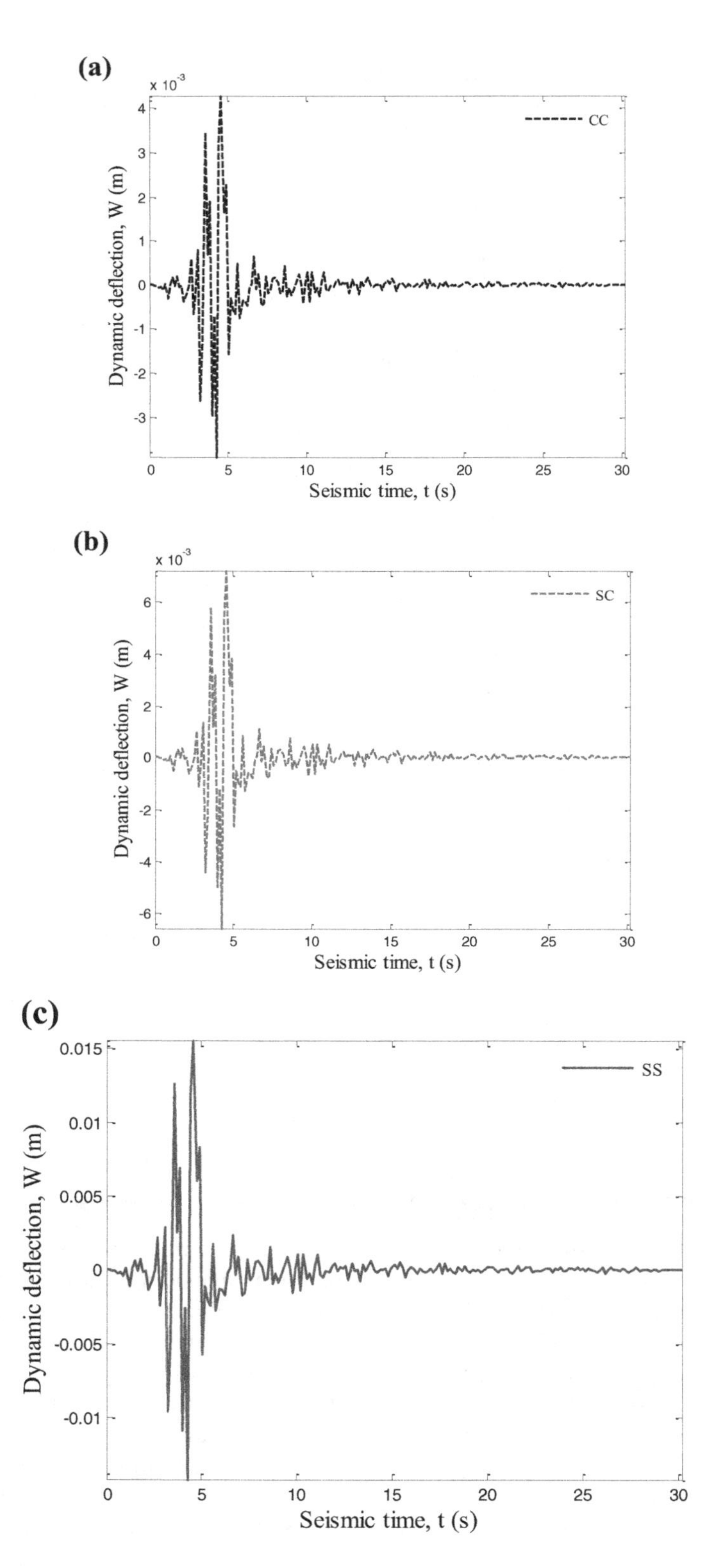


**FIGURE 4.10** Effect of boundary conditions on the dynamic deflection of the structure.

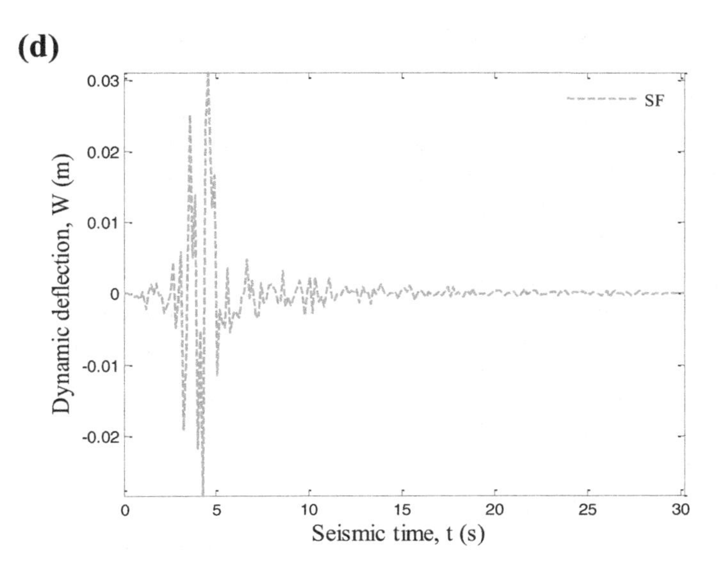


**FIGURE 4.10** (Continued)

**Acknowledgments:** This chapter is a slightly modified version of Ref. [20] and has been reproduced here with the permission of the copyright holder.

## REFERENCES

[1] Liang X, Parra-Montesinos GJ. Seismic behavior of reinforced beam-steel beam subassemblies and frame systems. *J Struct Eng* 2004;130:310–319.

[2] Cheng C, Chen C. Seismic behavior of steel beam and reinforced beam connections. *J Construct Steel Res* 2004;61:587–606.

[3] Changwang Y, Jinqing J, Ju Z. Seismic behavior of steel reinforced ultra high strength beam and reinforced beam connection. *Trans Tianjin Univ* 2010;16:309–316.

[4] Ji X, Zhang M, Kang H, Qian J, Hu H. Effect of cumulative seismic damage to steel tube-reinforced composite beams. *Eartq Struct* 2014;7:179–200.

[5] Cao VV, Ronagh HR. Reducing the potential seismic damage of reinforced frames using plastic hinge relocation by FRP. *Compos Part B: Eng* 2014;60:688–696.

[6] Choi SW, Yousok K, Park HS. Multi-objective seismic retrofit method for using FRP jackets in shear-critical reinforced frames. *Compos Part B: Eng* 2014;56:207–216.

[7] Liu ZQ, Xue JY, Zhao HT. Seismic behavior of steel reinforced special-shaped beam-beam joints. *Eartq Struct* 2016;11:665–680.

[8] Wuite J, Adali S. Deflection and stress behaviour of nanocomposite reinforced beams using a multiscale analysis. *Compos Struct* 2005;71:388–396.

[9] Matsuna H. Vibration and buckling of cross-ply laminated composite circular cylindrical shells according to a global higher-order theory. *Int J Mech Sci* 2007;49:1060–1075.

[10] Formica G, Lacarbonara W, Alessi R. Vibrations of carbon nanotube reinforced composites. *J Sound Vib* 2010;329:1875–1889.

[11] Liew KM, Lei ZX, Yu JL, Zhang LW. Postbuckling of carbon nanotube-reinforced functionally graded cylindrical panels under axial compression using a meshless approach. *Comput Meth Appl Mech Eng* 2014;268:1–17.

[12] Lei ZX, Zhang LW, Liew KM, Yu JL. Dynamic stability analysis of carbon nanotube-reinforced functionally graded cylindrical panels using the element-free kp-Ritz method. *Compos Struct* 2014;113:328–338.
[13] Ghorbanpour Arani A, Haghparast E, Khoddami Maraghi Z, Amir S. Static stress analysis of carbon nano-tube reinforced composite (CNTRC) cylinder under non-axisymmetric thermo-mechanical loads and uniform electro-magnetic fields. *Compos Part B: Eng* 2015;68:136–145.
[14] Kolahchi R, Rabani Bidgoli M, Beygipoor GH, Fakhar MH. A nonlocal nonlinear analysis for buckling in embedded FG-SWCNT-reinforced microplates subjected to magnetic field. *J Mech Sci Tech* 2013;5:2342–2355.
[15] Davar A, Khalili SMR, Malekzadeh Fard K. Dynamic response of functionally graded circular cylindrical shells subjected to radial impulse load. *Int J Mech Mater Des* 2013;9:65–81.
[16] Kolahchi R, Safari M, Esmailpour M. Dynamic stability analysis of temperature-dependent functionally graded CNT-reinforced visco-plates resting on orthotropic elastomeric medium. *Compos Struct* 2016;150:255–265.
[17] Jafarian Arani A, Kolahchi R. Buckling analysis of embedded beams armed with carbon nanotubes. *Comput* 2016;17:567–578.
[18] Shen HS, Yang DQ. Nonlinear vibration of anisotropic laminated cylindrical shells with piezoelectric fiber reinforced composite actuators. *Ocean Eng* 2014;80:36–49.
[19] Alibeigloo A. Thermoelastic analysis of functionally graded carbon nanotube reinforced composite cylindrical panel embedded in piezoelectric sensor and actuator layers. *Compos Part B: Eng* 2016;98:225–243.
[20] Rad SS, Bidgoli MR. Earthquake analysis of NFRP-reinforced-concrete beams using hyperbolic shear deformation theory. *Earthquak Struct* 2017;(13)3:241–253.
[21] Simsek M, Reddy JN. A unified higher order beam theory for buckling of a functionally graded microbeam embedded in elastic medium using modified couple stress theory. *Compos Struct* 2013;101:47–58.
[22] Mori T, Tanaka K. Average stress in matrix and average elastic energy of materials with misfitting inclusions. *Acta Metall Mater* 1973;21:571–574.
[23] Shu C, Xue H. Explicit computations of weighting coefficients in the harmonic differential quadrature. *J Sound Vib* 1997;204(3):549–555.

# 5 Buckling Analysis of Plates Reinforced with Graphene Platelets

## 5.1 INTRODUCTION

Graphene platelets (GPLs) have many applications in different industries due to their high hardness-to-weight and strength-to-weight ratios and other better properties compared to traditional isotropic ones. These structures can be used in aircraft, helicopters, missiles, launchers, satellites, and more. During the last five decades, the application of sandwich structures with light cores and two thin facesheets has been extensively investigated.

A new sinusoidal shear deformation theory was developed by Thai and Vo [1] for the bending, buckling, and vibration of functionally graded (FG) plates. A simple refined shear deformation theory was proposed by Huu-Tai Thai et al. [2] for the bending, buckling, and vibration of thick plates resting on an elastic foundation. The forced vibration response of a laminated composite and a sandwich shell was studied by Kumar et al. [3], using a two-dimensional (2D) finite element (FE) model based on the higher order zigzag theory. A study of composite and nanocomposite plates was presented by Duc et al. [4–9]. Duc et al. [10] investigated polymeric composite films using modified $TiO_2$ nanoparticles. A nonlocal dynamic buckling analysis of embedded microplates reinforced by single-walled carbon nanotubes was presented by Kolahchi and Cheraghbak [11]. Wang et al. [12] investigated the buckling of FG GPL-reinforced cylindrical shells consisting of multiple layers using the finite element method (FEM). A temperature-dependent buckling analysis of sandwich nanocomposite plates resting on an elastic medium subjected to a magnetic field was presented by Shokravi [13]. Li et al. [14] investigated the static linear elasticity, natural frequency, and buckling behavior of FG porous plates reinforced by GPLs. GPL-reinforced titanium composites were prepared by Liu et al. [15] using spark plasma sintering to evaluate a new type of structural material. Gao et al. [16] studied the free vibration of FG porous nanocomposite plates reinforced with a small amount of GPLs and supported by two-parameter elastic foundations with different boundary conditions. Polit et al. [17] investigated thick FG GPL-reinforced porous nanocomposite curved beams, considering the static bending and elastic stability analyses based on higher order shear deformation theory accounting for through-thickness stretching effect. A transient dynamic analysis of elastic wave propagation in an FG GPL-reinforced composite thick hollow cylinder was presented by Hosseini and Zhang [18]. The in-plane and out-of-plane forced vibration of a curved nanocomposite microbeam were considered by Allahkarami

DOI: 10.1201/9781003349525-5

et al. [19]. The vibration and nonlinear dynamic response of eccentrically stiffened FG composite truncated conical shells in thermal environments were presented by Chan et al. [20]. The nonlinear response and a buckling analysis of eccentrically stiffened functionally graded material (FGM) toroidal shell segments in a thermal environment were presented by Vuong and Duc [21]. The large amplitude vibration problem of laminated composite spherical shell panel under a combined temperature and moisture environment was analyzed by Mahapatra et al. [22]. The nonlinear free vibration behavior of a laminated composite spherical shell panel under an elevated hygrothermal environment was investigated by Mahapatra and Panda [23]. Mahapatra et al. [24] studied the geometrically nonlinear transverse bending behavior of a shear-deformable laminated composite spherical shell panel under hygro-thermo-mechanical loading. The nonlinear free vibration behavior of a laminated composite curved panel under a hygrothermal environment was investigated by Mahapatra et al. [25]. The flexural behavior of the laminated composite plate embedded with two different smart materials (piezoelectric and magnetostrictive [MS]) and the subsequent deflection suppression was investigated by Dutta et al. [26]. Suman et al. [27] studied the static bending and strength behavior of a laminated composite plate embedded with MS material numerically using a commercial FE tool. Javani et al. [28] investigated the buckling analyses of a composite plate reinforced with Graphene platelate (GPL). The effective material properties of the nano composite plate are determined using the Halphin-Tsai model. The modeling of the nano composite plate is carried out using the Third Order Shear Deformation Theory (TSDT).

In this chapter, buckling analyses of composite plates reinforced by GPLs are presented. The Halpin–Tsai model is used to obtain the effective material properties of the nanocomposite plate. The nanocomposite plate is modelled by third-order shear deformation theory (TSDT). The elastic medium is simulated using the Winkler model. Employing nonlinear strain displacement and stress strain, the energy equations of the plates are obtained, and using Hamilton's principle, the governing equations are derived. The governing equations are solved based on the Navier method. The effect of the GPL volume percent and the geometrical parameters of the plates and the elastic plates on the buckling load are investigated.

## 5.2 KINEMATICS OF DIFFERENT THEORIES

Figure 5.1 shows a nanocomposite plate reinforced by GPLs resting on an elastic medium. Based on the TSDT, the orthogonal components of the displacement vector are assumed based on Section 1.3.4.

The strain–stress relations for this structure are

$$\begin{bmatrix} \sigma_{xx} \\ \sigma_{yy} \\ \sigma_{zz} \\ \sigma_{zy} \\ \sigma_{xz} \\ \sigma_{zy} \end{bmatrix} = \begin{bmatrix} C_{11} & C_{12} & C_{13} & 0 & 0 & 0 \\ C_{12} & C_{22} & C_{23} & 0 & 0 & 0 \\ C_{13} & C_{23} & C_{33} & 0 & 0 & 0 \\ 0 & 0 & 0 & C_{44} & 0 & 0 \\ 0 & 0 & 0 & 0 & C_{55} & 0 \\ 0 & 0 & 0 & 0 & 0 & C_{66} \end{bmatrix} \begin{bmatrix} \varepsilon_{xx} \\ \varepsilon_{yy} \\ \varepsilon_{zz} \\ \gamma_{zy} \\ \gamma_{xz} \\ \gamma_{xy} \end{bmatrix}, \tag{5–1}$$

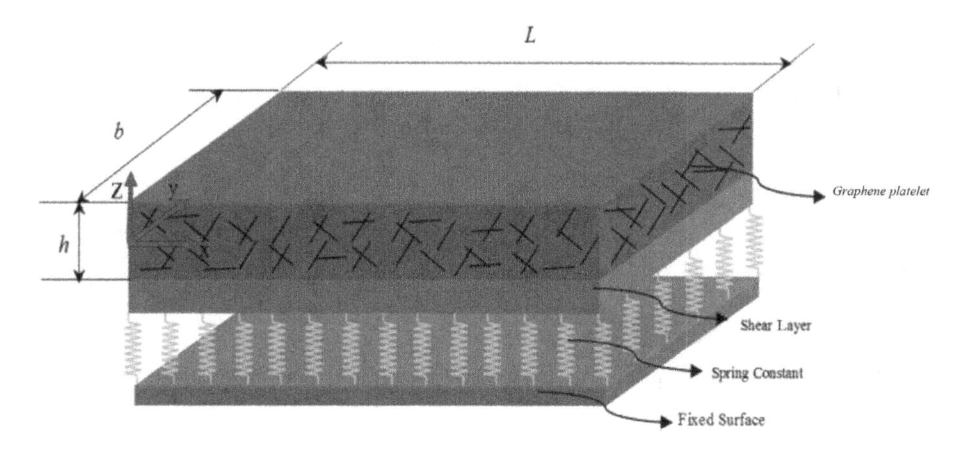


**FIGURE 5.1** A nanocomposite plate reinforced by GPLs resting on an elastic medium.

where parameters $C_{ij}$ are the elastic constant of a composite plate, which can be obtained by the Halpin–Tsai micromechanics model. Based on this model, we have [4]

$$E_c = \frac{3*(1+\xi_l V_{GPL})}{8*(1-\eta_L V_{GPL})} * E_m + \frac{5*(1+\xi_W \eta_W V_{GPL})}{8*(1-\eta_W V_{GPL})} * E_m, \tag{5–2}$$

where

$$\eta_L = \frac{\left(\frac{E_{GPL}}{E_M}\right) - 1}{\left(\frac{E_{GPL}}{E_M}\right) + \xi_L}, \tag{5–3}$$

$$\eta_W = \frac{\left(\frac{E_{GPL}}{E_M}\right) - 1}{\left(\frac{E_{GPL}}{E_M}\right) + \eta_L}, \tag{5–4}$$

and $E_c, E_m, E_{GPL}$ are the effective Young's moduli of the GPL/polymer nanocomposite, the polymer matrix, and GPLs, respectively. The effects of the geometry and size of the GPL reinforcements are described through parameters as

$$\xi_L = 2\left(\frac{l_{GPL}}{h_{GPL}}\right); \tag{5–5}$$

$$\xi_W = 2\left(\frac{w_{GPL}}{h_{GPL}}\right), \tag{5–6}$$

in which $l_{GPL}$, $w_{GPL}$, and $h_{GPL}$ denote the length, width, and thickness of the GPLs. The volume fraction of GPLs of the $i$th layer can be obtained from the GPL weight fraction $f_i$ and the mass densities of GPLs and polymer matrix, $\rho_{GPL}$ and $\rho_M$, by

$$V_i = \frac{f_i}{f_i + \left(\frac{\rho_{GPL}}{\rho_M}\right)(1-f_i)}. \tag{5–7}$$

## 5.3 MOTION EQUATION

For driving the motion equations, Hamilton's principle is used as follows:

$$\delta U - \delta W = 0, \tag{5–8}$$

where $\delta$ is variation, $\delta U$ is the variation of potential energy, and $\delta W$ is the variation of the external work.

The variation of potential energy for composite plate can be written as

$$U = \frac{1}{2}\int\left(\sigma_{xx}\varepsilon_{xx} + \sigma_{yy}\varepsilon_{yy} + \sigma_{xy}\gamma_{xy} + \sigma_{xz}\gamma_{xz} + \sigma_{yz}\gamma_{yz}\right)dV. \tag{5–8}$$

The variation of the external work, due to the elastic medium load simulated by the Pasternak model, can be express as

$$W_e = \int\int\left(-K_w w + K_g \nabla^2 w\right) w dA, \tag{5–9}$$

where $K_w$ and $K_g$ are the spring and shear constants of elastic medium, respectively. Using Hamilton's principle and a partial integral, the governing equations are computed as

$$\delta u : \frac{\partial N_{xx}}{\partial x} + \frac{\partial N_{xy}}{\partial y} = 0, \tag{5–10}$$

$$\delta v : \frac{\partial N_{xy}}{\partial x} + \frac{\partial N_{yy}}{\partial y} = 0, \tag{5–11}$$

$$\delta w : \frac{\partial Q_{xx}}{\partial x} + \frac{\partial Q_{yy}}{\partial y} + c_2\left(\frac{\partial K_{xx}}{\partial x} + \frac{\partial K_{yy}}{\partial y}\right) + N_{xx}\frac{\partial^2 w}{\partial x^2} + N_{yy}\frac{\partial^2 w}{\partial y^2}$$
$$- c_1\left(\frac{\partial^2 P_{xx}}{\partial x^2} + 2\frac{\partial^2 P_{xy}}{\partial x \partial y} + \frac{\partial^2 P_{yy}}{\partial y^2}\right) - K_w w + K_g \nabla^2 w = 0 \tag{5–12}$$

$$\delta\phi_x : \frac{\partial M_{xx}}{\partial x} + \frac{\partial M_{xy}}{\partial y} + c_1\left(\frac{\partial P_{xx}}{\partial x} + \frac{\partial P_{xy}}{\partial y}\right) - Q_{xx} - c_2 K_{xx} = 0, \tag{5–13}$$

$$\delta\phi_y : \frac{\partial M_{xy}}{\partial x} + \frac{\partial M_{yy}}{\partial y} + c_1\left(\frac{\partial P_{xy}}{\partial x} + \frac{\partial P_{yy}}{\partial y}\right) - Q_{yy} - c_2 K_{yy} = 0, \tag{5–14}$$

where the force and moment resultants can be defined as

$$\begin{Bmatrix} N_{xx} \\ N_{yy} \\ N_{xy} \end{Bmatrix} = \int_{-h/2}^{h/2} \begin{bmatrix} \sigma_{xx} \\ \sigma_{yy} \\ \sigma_{xy} \end{bmatrix} dz, \tag{5–15}$$

$$\begin{Bmatrix} M_{xx} \\ M_{yy} \\ M_{xy} \end{Bmatrix} = \int_{-h/2}^{h/2} \begin{bmatrix} \sigma_{xx} \\ \sigma_{yy} \\ \sigma_{xy} \end{bmatrix} zdz, \tag{5–16}$$

$$\begin{Bmatrix} P_{xx} \\ P_{yy} \\ P_{xy} \end{Bmatrix} = \int_{-h/2}^{h/2} \begin{bmatrix} \sigma_{xx} \\ \sigma_{yy} \\ \sigma_{xy} \end{bmatrix} z^3 dz, \tag{5–17}$$

$$\begin{bmatrix} Q_{xx} \\ Q_{yy} \end{bmatrix} = \int_{-h/2}^{h/2} \begin{bmatrix} \sigma_{xz} \\ \sigma_{yz} \end{bmatrix} dz, \tag{5–18}$$

$$\begin{bmatrix} K_{xx} \\ K_{yy} \end{bmatrix} = \int_{-h/2}^{h/2} \begin{bmatrix} \sigma_{xz} \\ \sigma_{yz} \end{bmatrix} z^2 dz, \tag{5–19}$$

$$I_i = \int_{-h/2}^{h/2} \rho z^i dz \qquad (i = 0, 1, \ldots, 6), \tag{5–20}$$

$$J_i = I_i - \frac{4}{3h^2} I_{i+2} \qquad (i = 1, 4), \tag{5–21}$$

$$K_2 = I_2 - \frac{8}{3h^2} I_4 + \left(\frac{4}{3h^2}\right)^2 I_6. \tag{5–22}$$

Therefore, the governing equations of the nanocomposite plate can be written as

$$
\begin{aligned}
&A_{11}\frac{\partial^2 u}{\partial x^2}+A_{12}\frac{\partial^2 v}{\partial x\partial y}+A_{16}\left(\frac{\partial^2 u}{\partial x\partial y}+\frac{\partial^2 v}{\partial x^2}\right)+B_{11}\frac{\partial^2 \varphi_x}{\partial x^2}+B_{12}\frac{\partial^2 \varphi_y}{\partial x\partial y}\\
&+B_{16}\left(\frac{\partial^2 \varphi_x}{\partial x\partial y}+\frac{\partial^2 \varphi_y}{\partial x^2}\right)+E_{11}c_1\left(\frac{\partial^2 \varphi_x}{\partial x^2}+\frac{\partial^3 w}{\partial x^3}\right)+E_{12}c_1\left(\frac{\partial^2 \varphi_y}{\partial x\partial y}+\frac{\partial^3 w}{\partial x\partial y^2}\right)\\
&+E_{16}c_1\left(\frac{\partial^2 \varphi_y}{\partial x^2}+\frac{\partial^2 \varphi_x}{\partial x\partial y}+2\frac{\partial^3 w}{\partial x^2\partial y}\right)+A_{16}\frac{\partial^2 u}{\partial x\partial y}+A_{26}\frac{\partial^2 v}{\partial y^2}\\
&+A_{66}\left(\frac{\partial^2 u}{\partial y^2}+\frac{\partial^2 v}{\partial x\partial y}\right)+B_{16}\frac{\partial^2 \varphi_x}{\partial x\partial y}+B_{26}\frac{\partial^2 \varphi_y}{\partial y^2}+B_{66}\left(\frac{\partial^2 \varphi_x}{\partial y^2}+\frac{\partial^2 \varphi_y}{\partial x\partial y}\right)\\
&+E_{16}c_1\left(\frac{\partial^2 \varphi_x}{\partial x\partial y}+\frac{\partial^3 w}{\partial y\partial x^2}\right)+E_{26}c_1\left(\frac{\partial^2 \varphi_y}{\partial y^2}+\frac{\partial^3 w}{\partial y^3}\right)\\
&+E_{66}c_1\left(\frac{\partial^2 \varphi_y}{\partial x\partial y}+\frac{\partial^2 \varphi_x}{\partial y^2}+2\frac{\partial^3 w}{\partial x\partial y^2}\right)=0,
\end{aligned}
\tag{5–23}
$$

$$
\begin{aligned}
&A_{16}\frac{\partial^2 u}{\partial x^2}+A_{26}\frac{\partial^2 v}{\partial x\partial y}+A_{66}\left(\frac{\partial^2 u}{\partial x\partial y}+\frac{\partial^2 v}{\partial x^2}\right)+B_{16}\frac{\partial^2 \varphi_x}{\partial x^2}+B_{26}\frac{\partial^2 \varphi_y}{\partial x\partial y}\\
&+B_{66}\left(\frac{\partial^2 \varphi_x}{\partial x\partial y}+\frac{\partial^2 \varphi_y}{\partial x^2}\right)+E_{16}c_1\left(\frac{\partial^2 \varphi_x}{\partial x^2}+\frac{\partial^3 w}{\partial x^3}\right)+E_{26}c_1\left(\frac{\partial^2 \varphi_y}{\partial x\partial y}+\frac{\partial^3 w}{\partial x\partial y^2}\right)\\
&+E_{66}c_1\left(\frac{\partial^2 \varphi_y}{\partial x^2}+\frac{\partial^2 \varphi_x}{\partial x\partial y}+2\frac{\partial^3 w}{\partial x^2\partial y}\right)+A_{21}\frac{\partial^2 u}{\partial x\partial y}+A_{22}\frac{\partial^2 v}{\partial y^2}\\
&+A_{26}\left(\frac{\partial^2 u}{\partial y^2}+\frac{\partial^2 v}{\partial x\partial y}\right)+B_{21}\frac{\partial^2 \varphi_x}{\partial x\partial y}+B_{22}\frac{\partial^2 \varphi_y}{\partial y^2}+B_{26}\left(\frac{\partial^2 \varphi_x}{\partial y^2}+\frac{\partial^2 \varphi_y}{\partial x\partial y}\right)\\
&+E_{21}c_1\left(\frac{\partial^2 \varphi_x}{\partial x\partial y}+\frac{\partial^3 w}{\partial y\partial x^2}\right)+E_{22}c_1\left(\frac{\partial^2 \varphi_y}{\partial y^2}+\frac{\partial^3 w}{\partial y^3}\right)\\
&+E_{26}c_1\left(\frac{\partial^2 \varphi_y}{\partial x\partial y}+\frac{\partial^2 \varphi_x}{\partial y^2}+2\frac{\partial^3 w}{\partial x\partial y^2}\right)=0,
\end{aligned}
\tag{5–24}
$$

$$
\begin{aligned}
&A_{55}\left(\frac{\partial^2 w}{\partial x^2}+\frac{\partial \varphi_x}{\partial x}\right)+A_{45}\left(\frac{\partial^2 w}{\partial x\partial y}+\frac{\partial \varphi_y}{\partial x}\right)+D_{55}c_2\left(\frac{\partial^2 w}{\partial x^2}+\frac{\partial \varphi_x}{\partial x}\right)\\
&+D_{45}c_2\left(\frac{\partial^2 w}{\partial x\partial y}+\frac{\partial \varphi_y}{\partial x}\right)+A_{45}\left(\frac{\partial^2 w}{\partial x\partial y}+\frac{\partial \varphi_x}{\partial y}\right)+A_{44}\left(\frac{\partial^2 w}{\partial y^2}+\frac{\partial \varphi_y}{\partial y}\right)
\end{aligned}
$$

$$+D_{45}c_2\left(\frac{\partial^2 w}{\partial x\partial y}+\frac{\partial \varphi_x}{\partial y}\right)+D_{44}c_2\left(\frac{\partial^2 w}{\partial y^2}+\frac{\partial \varphi_y}{\partial y}\right)$$

$$+c_2\left(\begin{array}{l} D_{55}\left(\frac{\partial^2 w}{\partial x^2}+\frac{\partial \varphi_x}{\partial x}\right)+D_{45}\left(\frac{\partial^2 w}{\partial x\partial y}+\frac{\partial \varphi_y}{\partial x}\right)+F_{55}c_2\left(\frac{\partial^2 w}{\partial x^2}+\frac{\partial \varphi_x}{\partial x}\right)+F_{45}c_2\left(\frac{\partial^2 w}{\partial x\partial y}+\frac{\partial \varphi_y}{\partial x}\right) \\ +D_{45}\left(\frac{\partial^2 w}{\partial x\partial y}+\frac{\partial \varphi_x}{\partial y}\right)+D_{44}\left(\frac{\partial^2 w}{\partial y^2}+\frac{\partial \varphi_y}{\partial y}\right)+F_{45}c_2\left(\frac{\partial^2 w}{\partial x\partial y}+\frac{\partial \varphi_x}{\partial y}\right)+F_{44}c_2\left(\frac{\partial^2 w}{\partial y^2}+\frac{\partial \varphi_y}{\partial y}\right) \end{array}\right)$$

$$+-K_w w+K_g\nabla^2 w-c_1\left(E_{11}\frac{\partial^3 u}{\partial x^3}+E_{12}\frac{\partial^3 v}{\partial x^2\partial y}+E_{16}\left(\frac{\partial^3 u}{\partial x^2\partial y}+\frac{\partial^3 v}{\partial x^3}\right)+F_{11}\frac{\partial^3 \varphi_x}{\partial x^3}+F_{12}\frac{\partial^3 \varphi_y}{\partial x^2\partial y}\right.$$

$$+F_{16}\left(\frac{\partial^3 \varphi_x}{\partial x^2\partial y}+\frac{\partial^3 \varphi_y}{\partial x^3}\right)+H_{11}c_1\left(\frac{\partial^3 \varphi_x}{\partial x^3}+\frac{\partial^4 w}{\partial x^4}\right)+H_{12}c_1\left(\frac{\partial^3 \varphi_y}{\partial x^2\partial y}+\frac{\partial^4 w}{\partial x^2\partial y^2}\right)$$

$$+H_{16}c_1\left(\frac{\partial^3 \varphi_y}{\partial x^3}+\frac{\partial^3 \varphi_x}{\partial x^2\partial y}+2\frac{\partial^4 w}{\partial x^3\partial y}\right)+E_{12}\frac{\partial^3 u}{\partial y^2\partial x}+E_{22}\frac{\partial^3 v}{\partial y^3}+E_{26}\left(\frac{\partial^3 u}{\partial y^3}+\frac{\partial^3 v}{\partial x\partial y^2}\right)$$

$$+F_{12}\frac{\partial^3 \varphi_x}{\partial x\partial y^2}+F_{22}\frac{\partial^3 \varphi_y}{\partial y^3}+F_{26}\left(\frac{\partial^3 \varphi_x}{\partial y^3}+\frac{\partial^3 \varphi_y}{\partial x\partial y^2}\right)+H_{12}c_1\left(\frac{\partial^3 \varphi_x}{\partial x\partial y^2}+\frac{\partial^4 w}{\partial x^2\partial y^2}\right)$$

$$+H_{22}c_1\left(\frac{\partial^3 \varphi_y}{\partial y^3}+\frac{\partial^4 w}{\partial y^4}\right)+H_{26}c_1\left(\frac{\partial^3 \varphi_y}{\partial x\partial y^2}+\frac{\partial^3 \varphi_x}{\partial y^3}+2\frac{\partial^4 w}{\partial x\partial y^3}\right)+2E_{16}\frac{\partial^3 u}{\partial y\partial x^2}$$

$$+2E_{26}\frac{\partial^3 v}{\partial y^2\partial x}+2E_{66}\left(\frac{\partial^3 u}{\partial y^2\partial x}+\frac{\partial^3 v}{\partial x^2\partial y}\right)+2F_{16}\frac{\partial^3 \varphi_x}{\partial x^2\partial y}+2F_{26}\frac{\partial^3 \varphi_y}{\partial y^2\partial x}$$

$$+2F_{66}\left(\frac{\partial^3 \varphi_x}{\partial y^2\partial x}+\frac{\partial^3 \varphi_y}{\partial x^2\partial y}\right)+2H_{16}c_1\left(\frac{\partial^3 \varphi_x}{\partial x^2\partial y}+\frac{\partial^4 w}{\partial x^3\partial y}\right)+2H_{26}c_1\left(\frac{\partial^3 \varphi_y}{\partial x\partial y^2}+\frac{\partial^4 w}{\partial y^3\partial x}\right)$$

$$\left.+2H_{66}c_1\left(\frac{\partial^3 \varphi_y}{\partial x^2\partial y}+\frac{\partial^3 \varphi_x}{\partial y^2\partial x}+2\frac{\partial^4 w}{\partial x^2\partial y^2}\right)\right)+N_{xx}\frac{\partial^2 w}{\partial x^2}+N_{yy}\frac{\partial^2 w}{\partial y^2}=0, \tag{5–25}$$

$$B_{11}\frac{\partial^2 u}{\partial x^2}+B_{12}\frac{\partial^2 v}{\partial x\partial y}+B_{16}\left(\frac{\partial^2 u}{\partial x\partial y}+\frac{\partial^2 v}{\partial x^2}\right)+D_{11}\frac{\partial^2 \varphi_x}{\partial x^2}+D_{12}\frac{\partial^2 \varphi_y}{\partial x\partial y}$$

$$+D_{16}\left(\frac{\partial^2 \varphi_x}{\partial x\partial y}+\frac{\partial^2 \varphi_y}{\partial x^2}\right)+F_{11}c_1\left(\frac{\partial^2 \varphi_x}{\partial x^2}+\frac{\partial^3 w}{\partial x^3}\right)+F_{12}c_1\left(\frac{\partial^2 \varphi_y}{\partial x\partial y}+\frac{\partial^3 w}{\partial x\partial y^2}\right)$$

$$+F_{16}c_1\left(\frac{\partial^2 \varphi_y}{\partial x^2}+\frac{\partial^2 \varphi_x}{\partial x\partial y}+2\frac{\partial^3 w}{\partial x^2\partial y}\right)+B_{16}\frac{\partial^2 u}{\partial x\partial y}+B_{26}\frac{\partial^2 v}{\partial y^2}$$

$$+B_{66}\left(\frac{\partial^2 u}{\partial y^2}+\frac{\partial^2 v}{\partial x\partial y}\right)+D_{16}\frac{\partial^2 \varphi_x}{\partial x\partial y}+D_{26}\frac{\partial^2 \varphi_y}{\partial y^2}+D_{66}\left(\frac{\partial^2 \varphi_x}{\partial y^2}+\frac{\partial^2 \varphi_y}{\partial x\partial y}\right)$$

$$
\begin{aligned}
&+F_{16}c_1\left(\frac{\partial^2\varphi_x}{\partial x\partial y}+\frac{\partial^3 w}{\partial y\partial x^2}\right)+F_{26}c_1\left(\frac{\partial^2\varphi_y}{\partial y^2}+\frac{\partial^3 w}{\partial y^3}\right)+F_{66}c_1\left(\frac{\partial^2\varphi_y}{\partial x\partial y}+\frac{\partial^2\varphi_x}{\partial y^2}+2\frac{\partial^3 w}{\partial x\partial y^2}\right)\\
&+c_1\left(E_{11}\frac{\partial^2 u}{\partial x^2}+E_{12}\frac{\partial^2 v}{\partial x\partial y}+E_{16}\left(\frac{\partial^2 u}{\partial x\partial y}+\frac{\partial^2 v}{\partial x^2}\right)+F_{11}\frac{\partial^2\varphi_x}{\partial x^2}+F_{12}\frac{\partial^2\varphi_y}{\partial x\partial y}\right.\\
&+F_{16}\left(\frac{\partial^2\varphi_x}{\partial x\partial y}+\frac{\partial^2\varphi_y}{\partial x^2}\right)+H_{11}c_1\left(\frac{\partial^2\varphi_x}{\partial x^2}+\frac{\partial^3 w}{\partial x^3}\right)+H_{12}c_1\left(\frac{\partial^2\varphi_y}{\partial x\partial y}+\frac{\partial^3 w}{\partial x\partial y^2}\right)\\
&+H_{16}c_1\left(\frac{\partial^2\varphi_y}{\partial x^2}+\frac{\partial^2\varphi_x}{\partial x\partial y}+2\frac{\partial^3 w}{\partial x^2\partial y}\right)+E_{16}\frac{\partial^2 u}{\partial x\partial y}+E_{26}\frac{\partial^2 v}{\partial y^2}+E_{66}\left(\frac{\partial^2 u}{\partial y^2}+\frac{\partial^2 v}{\partial x\partial y}\right)\\
&+F_{16}\frac{\partial^2\varphi_x}{\partial x\partial y}+F_{26}\frac{\partial^2\varphi_y}{\partial y^2}+F_{66}\left(\frac{\partial^2\varphi_x}{\partial y^2}+\frac{\partial^2\varphi_y}{\partial x\partial y}\right)+H_{16}c_1\left(\frac{\partial^2\varphi_x}{\partial x\partial y}+\frac{\partial^3 w}{\partial y\partial x^2}\right)\\
&+H_{26}c_1\left(\frac{\partial^2\varphi_y}{\partial y^2}+\frac{\partial^3 w}{\partial y^3}\right)+H_{66}c_1\left(\frac{\partial^2\varphi_y}{\partial x\partial y}+\frac{\partial^2\varphi_x}{\partial y^2}+2\frac{\partial^3 w}{\partial x\partial y^2}\right)-A_{55}\left(\frac{\partial w}{\partial x}+\varphi_x\right)\\
&-A_{45}\left(\frac{\partial w}{\partial y}+\varphi_y\right)-D_{55}c_2\left(\varphi_x+\frac{\partial w}{\partial x}\right)-D_{45}c_2\left(\frac{\partial w}{\partial y}+\varphi_y\right)+c_2\left(-D_{55}\left(\frac{\partial w}{\partial x}+\varphi_x\right)\right.\\
&\left.\left.-D_{45}\left(\frac{\partial w}{\partial y}+\varphi_y\right)-F_{55}c_2\left(\varphi_x+\frac{\partial w}{\partial x}\right)-F_{45}c_2\left(\frac{\partial w}{\partial y}+\varphi_y\right)\right)\right)=0,
\end{aligned}
\tag{5–26}
$$

$$
\begin{aligned}
&B_{16}\frac{\partial^2 u}{\partial x^2}+B_{26}\frac{\partial^2 v}{\partial x\partial y}+B_{66}\left(\frac{\partial^2 u}{\partial x\partial y}+\frac{\partial^2 v}{\partial x^2}\right)+D_{16}\frac{\partial^2\varphi_x}{\partial x^2}+D_{26}\frac{\partial^2\varphi_y}{\partial x\partial y}\\
&+D_{66}\left(\frac{\partial^2\varphi_x}{\partial x\partial y}+\frac{\partial^2\varphi_y}{\partial x^2}\right)+F_{16}c_1\left(\frac{\partial^2\varphi_x}{\partial x^2}+\frac{\partial^3 w}{\partial x^3}\right)+F_{26}c_1\left(\frac{\partial^2\varphi_y}{\partial x\partial y}+\frac{\partial^3 w}{\partial x\partial y^2}\right)\\
&+F_{66}c_1\left(\frac{\partial^2\varphi_y}{\partial x^2}+\frac{\partial^2\varphi_x}{\partial x\partial y}+2\frac{\partial^3 w}{\partial x^2\partial y}\right)+B_{12}\frac{\partial^2 u}{\partial x\partial y}+B_{22}\frac{\partial^2 v}{\partial y^2}\\
&+B_{26}\left(\frac{\partial^2 u}{\partial y^2}+\frac{\partial^2 v}{\partial x\partial y}\right)+D_{12}\frac{\partial^2\varphi_x}{\partial x\partial y}+D_{22}\frac{\partial^2\varphi_y}{\partial y^2}+D_{26}\left(\frac{\partial^2\varphi_x}{\partial y^2}+\frac{\partial^2\varphi_y}{\partial x\partial y}\right)\\
&+F_{12}c_1\left(\frac{\partial^2\varphi_x}{\partial x\partial y}+\frac{\partial^3 w}{\partial y\partial x^2}\right)+F_{22}c_1\left(\frac{\partial^2\varphi_y}{\partial y^2}+\frac{\partial^3 w}{\partial y^3}\right)+F_{26}c_1\left(\frac{\partial^2\varphi_y}{\partial x\partial y}+\frac{\partial^2\varphi_x}{\partial y^2}+2\frac{\partial^3 w}{\partial x\partial y^2}\right)\\
&+c_1\left(E_{16}\frac{\partial^2 u}{\partial x^2}+E_{26}\frac{\partial^2 v}{\partial x\partial y}+E_{66}\left(\frac{\partial^2 u}{\partial x\partial y}+\frac{\partial^2 v}{\partial x^2}\right)+F_{16}\frac{\partial^2\varphi_x}{\partial x^2}+F_{26}\frac{\partial^2\varphi_y}{\partial x\partial y}\right.\\
&+F_{66}\left(\frac{\partial^2\varphi_x}{\partial x\partial y}+\frac{\partial^2\varphi_y}{\partial x^2}\right)+H_{16}c_1\left(\frac{\partial^2\varphi_x}{\partial x^2}+\frac{\partial^3 w}{\partial x^3}\right)+H_{26}c_1\left(\frac{\partial^2\varphi_y}{\partial x\partial y}+\frac{\partial^3 w}{\partial x\partial y^2}\right)
\end{aligned}
$$

$$+H_{66}c_1\left(\frac{\partial^2\varphi_y}{\partial x^2}+\frac{\partial^2\varphi_x}{\partial x\partial y}+2\frac{\partial^3 w}{\partial x^2\partial y}\right)+E_{12}\frac{\partial^2 u}{\partial x\partial y}+E_{22}\frac{\partial^2 v}{\partial y^2}+E_{26}\left(\frac{\partial^2 u}{\partial y^2}+\frac{\partial^2 v}{\partial x\partial y}\right)$$

$$+F_{12}\frac{\partial^2\varphi_x}{\partial x\partial y}+F_{22}\frac{\partial^2\varphi_y}{\partial y^2}+F_{26}\left(\frac{\partial^2\varphi_x}{\partial y^2}+\frac{\partial^2\varphi_y}{\partial x\partial y}\right)+H_{12}c_1\left(\frac{\partial^2\varphi_x}{\partial x\partial y}+\frac{\partial^3 w}{\partial y\partial x^2}\right)$$

$$+H_{22}c_1\left(\frac{\partial^2\varphi_y}{\partial y^2}+\frac{\partial^3 w}{\partial y^3}\right)+H_{26}c_1\left(\frac{\partial^2\varphi_y}{\partial x\partial y}+\frac{\partial^2\varphi_x}{\partial y^2}+2\frac{\partial^3 w}{\partial x\partial y^2}\right)-A_{45}\left(\frac{\partial w}{\partial x}+\varphi_x\right)$$

$$-A_{44}\left(\frac{\partial w}{\partial y}+\varphi_y\right)-D_{45}c_2\left(\varphi_x+\frac{\partial w}{\partial x}\right)-D_{44}c_2\left(\frac{\partial w}{\partial y}+\varphi_y\right)+c_2\left(-D_{45}\left(\frac{\partial w}{\partial x}+\varphi_x\right)\right.$$

$$\left.-D_{44}\left(\frac{\partial w}{\partial y}+\varphi_y\right)-F_{45}c_2\left(\varphi_x+\frac{\partial w}{\partial x}\right)-F_{44}c_2\left(\frac{\partial w}{\partial y}+\varphi_y\right)\right)=0, \qquad (5\text{–}27)$$

where

$$A_{ij}=\int_{-h/2}^{h/2}Q_{ij}dz, \qquad (i,j=1,2,6) \qquad (5\text{–}28)$$

$$B_{ij}=\int_{-h/2}^{h/2}Q_{ij}zdz, \qquad (5\text{–}29)$$

$$D_{ij}=\int_{-h/2}^{h/2}Q_{ij}z^2dz, \qquad (5\text{–}29)$$

$$E_{ij}=\int_{-h/2}^{h/2}Q_{ij}z^3dz, \qquad (5\text{–}30)$$

$$F_{ij}=\int_{-h/2}^{h/2}Q_{ij}z^4dz, \qquad (5\text{–}31)$$

$$H_{ij}=\int_{-h/2}^{h/2}Q_{ij}z^6dz. \qquad (5\text{–}32)$$

## 5.4 NUMERICAL RESULT AND DISCUSSION

In this section, a parametric study is done for the effects of different parameters on the nonlinear buckling load of the composite structure based on the Navier method presented in Section 2.1.1.

Figures 5.2 and 5.3 show the effect of different transverse–to–axial lode ratios and plate widths on the buckling load versus mode number, respectively. As it is inferred, by increasing transverse–to–axial lode ratios and plate widths, the buckling load is reduced. It is because by increasing the transverse–to–axial lode ratios and plate widths, the stiffness of the system is decreased. In addition, by increasing the mode number, the buckling load is increased.

Figure 5.4 illustrates the effect of plate thickness on the buckling lode versus mode number. It can be concluded that as the plate thickness increases, the stiffness of the system is increased. It is because the buckling load is increased.

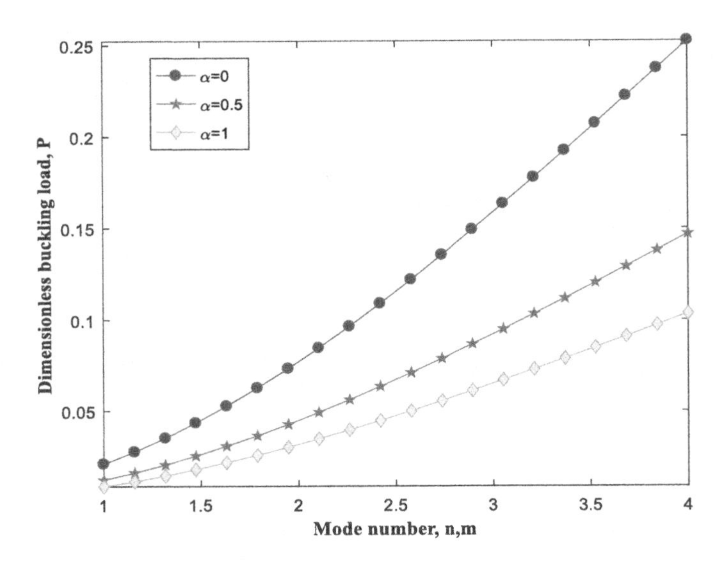


**FIGURE 5.2** Dimensionless buckling load versus the mode number for different transverse–to–axial lode ratios.

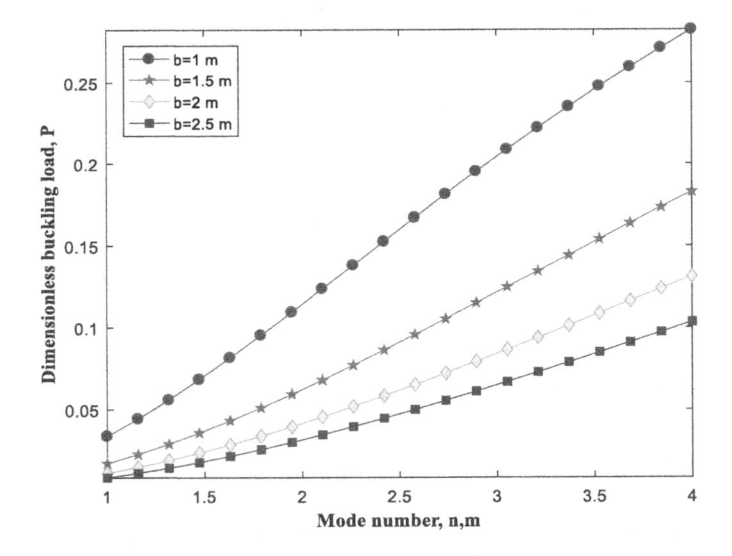


**FIGURE 5.3** Effect of plate width on the dimensionless buckling load versus the mode number.

Figure 5.5 indicates the effect of the spring constant of the elastic medium on the buckling load with respect to the mode number. It is observed that by increasing the spring constant of the elastic medium, the buckling load is increased. It is because the stiffness of the system is increased by the enhanced spring constant of the elastic medium.

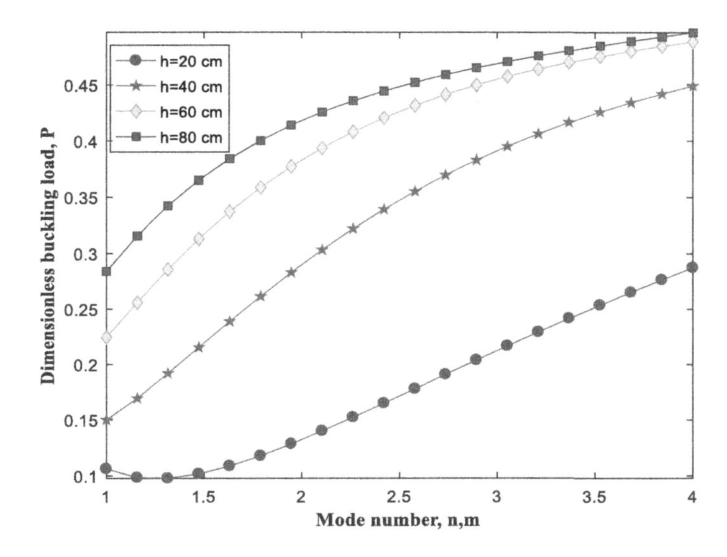


**FIGURE 5.4** Effect of plate thickness on the dimensionless buckling load versus the mode number.

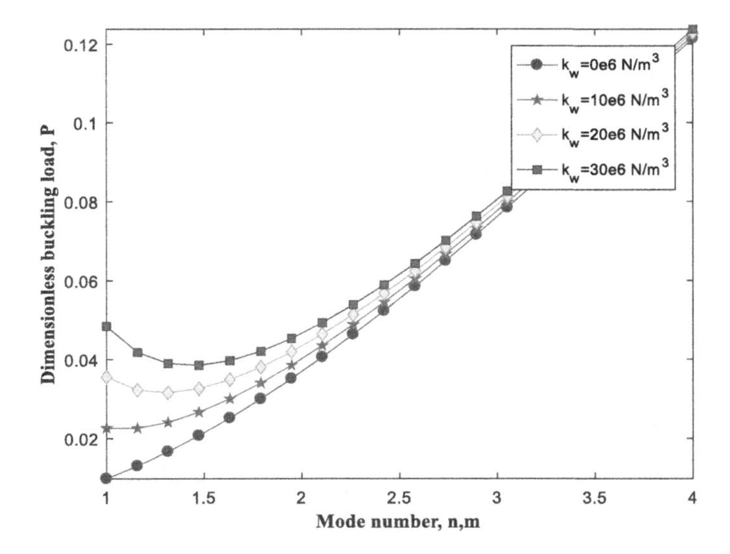


**FIGURE 5.5** Effect of the spring constant of the elastic medium on the dimensionless buckling load versus the mode number.

The effect of plate length on the buckling load as a function of the mode number is shown in Figure 5.6. By increasing the plate length, the buckling load decreases. This is because the stiffness of structure is decreased.

Figure 5.7 shows the buckling load versus the volume present of GPL for different traverse–to–axial lode ratios. As can be seen, the buckling load of a micro-composite

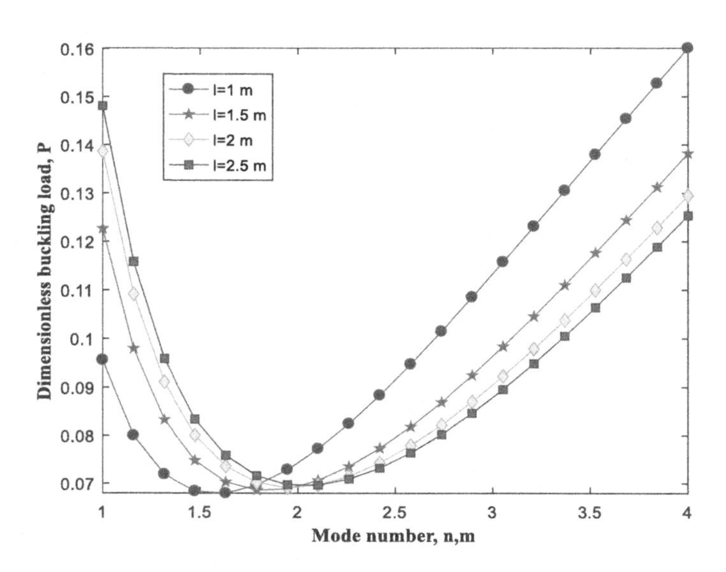


**FIGURE 5.6** Effect of plate length on the dimensionless buckling load versus the mode number.

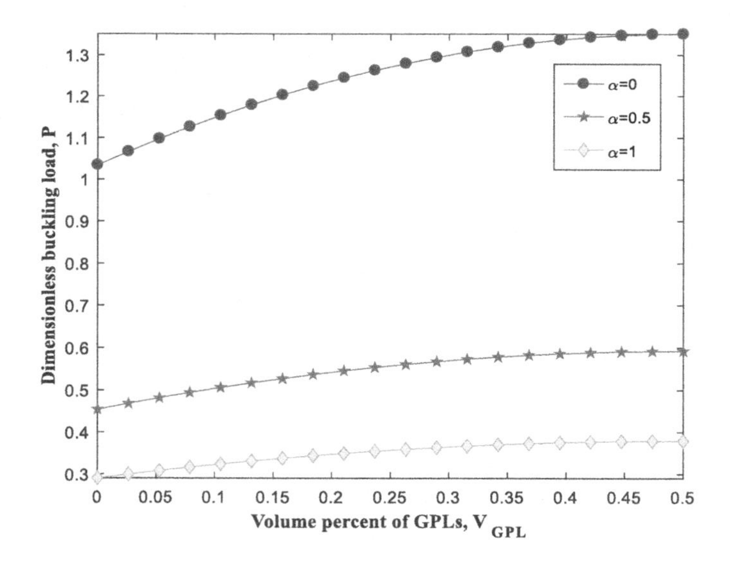


**FIGURE 5.7** Dimensionless buckling load versus volume percent of GPLs for different transverse–to–axial lode ratios.

structure with an increasing transverse–to–axial lode ratio is decreased. Furthermore, with an increase in the GPL volume percent, the buckling load is increased due to the increase in the bending rigidity of the structure.

The effect of the volume percent of GPLs on the buckling load is shown in Figure 5.8 for different length-to-thickness ratios of the GPLs ($\zeta$). It is found that

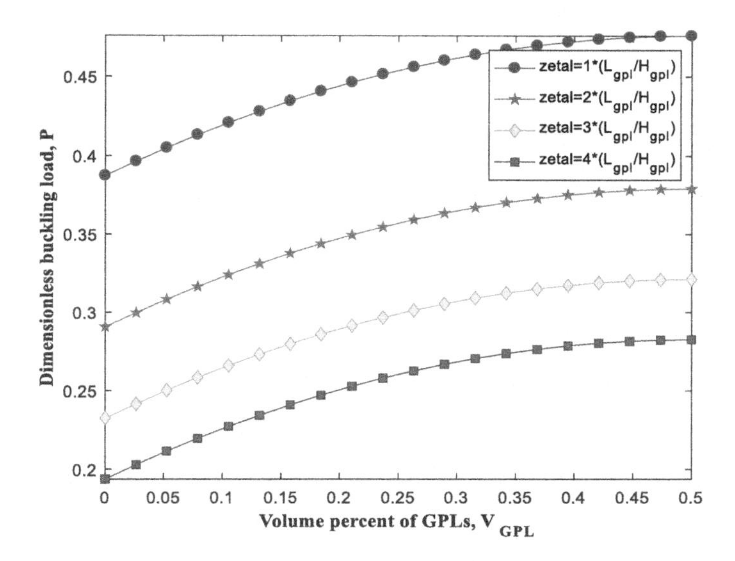


**FIGURE 5.8** Effect of the volume percent of GPLs on the dimensionless buckling load for different length-to-thickness ratios of the GPLs.

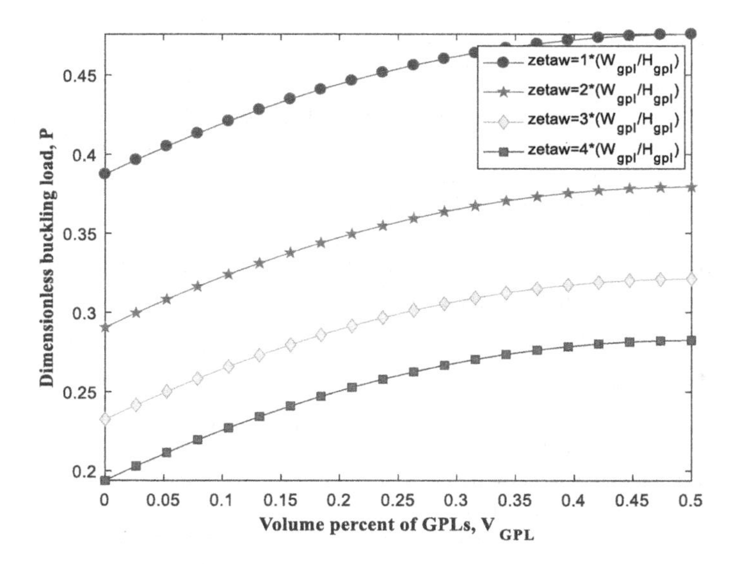


**FIGURE 5.9** Effect of the weight-to-thickness ratio of GPLs on the buckling load versus the volume percent of GPLs.

by increasing the zeta, the buckling load is decreased due to the enhanced stiffness of the structure.

Figure 5.9 shows the effect of the weight-to-thickness ratio of GPLs ($\zeta_w$) on the buckling load versus the volume percent of GPLs. As can be seen, by increasing the zetaw, the buckling load is decreased. It is because with an increase in the zetaw, the stiffness is decreased.

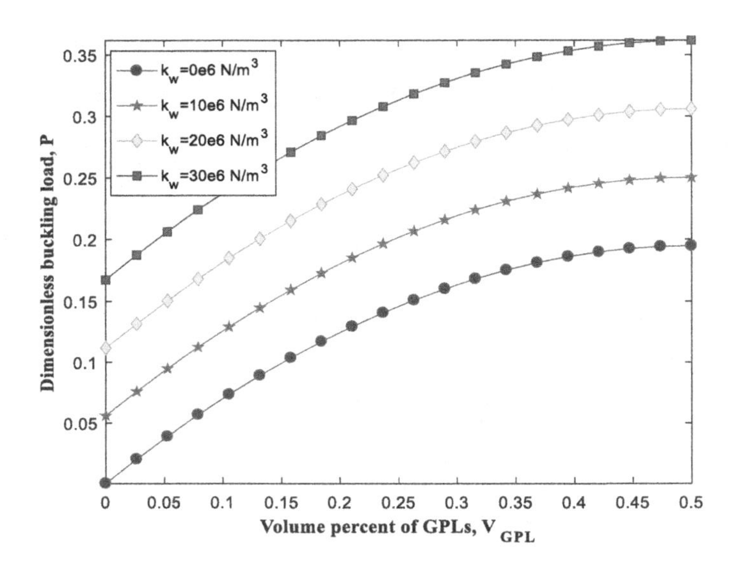


**FIGURE 5.10** Effect of the spring constant of the elastic medium on the buckling load versus the volume percent of GPLs.

Figure 5.10 indicates the effect of the spring constant of the elastic medium on the buckling load with respect to the volume percent of GPLs. It is observed that by increasing the spring constant of the elastic medium, the buckling load is increased. It is because the stiffness of the system is increased with the enhancement of the spring constant of the elastic medium.

**Acknowledgments:** This chapter is a slightly modified version of Ref. [28] and has been reproduced here with the permission of the copyright holder.

## REFERENCES

[1] Thai H, Vo T. A new sinusoidal shear deformation theory for bending, buckling and vibration of functionally graded plates. *Appl Math Model* 2013;37:3269–3281.
[2] Thai H, Park M, Choi, D. A simple refined theory for bending, buckling, and vibration of thick plates resting on elastic foundation. *Int J Mech Sci* 2013;73:40–52.
[3] Kumar A, Chakrabarti A, Bhargava P. Accurate dynamic response of laminated composites and sandwich shells using higher order zigzag theory. *Thin-Wall Struct* 2014;77:174–186.
[4] Duc ND, Minh DK. Bending analysis of three-phase polymer composite plates reinforced by glass fibers and Titanium oxide particles. *J Computat Mat Sci* 2010;49:194–198.
[5] Duc ND, Quan TQ, Nam D. Nonlinear stability analysis of imperfect three phase polymer composite plates. *J Mech Compos Mat* 2013;49:345–358.
[6] Duc ND. *Nonlinear static and dynamic stability of functionally graded plates and shells.* Vietnam National University Press, Hanoi, 2014.
[7] Duc ND. Nonlinear dynamic response of imperfect eccentrically stiffened FGM double curved shallow shells on elastic foundation. *J Compos Struct* 2014;102:306–314.

[8] Duc ND, Hadavinia H, Thu PV, Quan TQ. Vibration and nonlinear dynamic response of imperfect three-phase polymer nanocomposite panel resting on elastic foundations under hydrodynamic loads. *Compos Struct* 2015;131:229–237.

[9] Duc ND. Nonlinear thermal dynamic analysis of eccentrically stiffened S-FGM circular cylindrical shells surrounded on elastic foundations using the Reddy's third-order shear deformation shell theory. *Europ J Mech–A/Solids* 2016;58:10–30.

[10] Duc ND, Khoa ND, Thiem HT. Nonlinear thermo-mechanical response of eccentrically stiffened Sigmoid FGM circular cylindrical shells subjected to compressive and uniform radial loads using the Reddy's third-order shear deformation shell theory. *Mech Adv Mat Struct* 2018;25:1157–1167.

[11] Kolahchi R, Cheraghbak A. Agglomeration effects on the dynamic buckling of visco elastic micro plates reinforced with SWCNTs. *Non Dyn* 2017;90:479–492.

[12] Wang Y, Feng CH, Zhao ZH, Yang J. Eigenvalue buckling of functionally graded cylindrical shells reinforced with graphene platelets (GPL). *Compos Struct* 2018;20:238–246.

[13] Shokravi M. Buckling of sandwich plates with FG-CNT-reinforced layers resting on orthotropic elastic medium using Reddy plate theory. *Steel Compos Struct* 2017;23:623–631.

[14] Li K, Wu D, Chen X, Cheng J, Liu ZH, Gao W, Liu M. Isogeometric analysis of functionally graded porous plates reinforced by graphene platelets. *Compos Struct* 2018;204:114–130.

[15] Liu J, Wu M, Yang Y, Yang G, Yan H, Jiang K. Preparation and mechanical performance of graphene platelet reinforced titanium nanocomposites for high temperature applications. *J Alloy Compo* 2018;22:355–363.

[16] Gao K, Gao W, Chen D, Yang J. Nonlinear free vibration of functionally graded graphene platelets reinforced porous nano composite plates resting on elastic foundation. *Compos Struct* 2018;204:831–846.

[17] Polit O, Anant C, Anirudh B, Ganapathi M. Functionally graded graphene reinforced porous nanocomposite curved beams: Bending and elastic stability using a higher-order model with thickness stretch effect. *Compos Part B: Eng* 2019;166:310–327.

[18] Hosseini SM, Zhang Ch. Elastodynamic and wave propagation analysis in a FG Graphene platelets-reinforced nanocomposite cylinder using a modified nonlinear micromechanical model. *Steel Compos Struct* 2018;27:255–271.

[19] Allahkarami F, Nikkhah-bahrami M, Ghassabzadeh Saryazdi M. Nonlinear forced vibration of FG-CNTs-reinforced curved microbeam based on strain gradient theory considering out-of-plane motion. *Steel Compos Struct* 2018;22:673–691.

[20] Chan DQ, Anh VTT, Duc ND. Vibration and nonlinear dynamic response of eccentrically stiffened functionally graded composite truncated conical shells in thermal environments. *Acta Mech* 2018;230:157–178.

[21] Vuong PM, Duc ND. Nonlinear response and buckling analysis of eccentrically stiffened FGM toroidal shell segments in thermal environment. *Aerosp Sci Technol* 2018;79:383–398.

[22] Mahapatra TR, Panda SK. Nonlinear free vibration analysis of laminated composite spherical shell panel under elevated hygrothermal environment: A micromechanical approach. *Aerosp Sci Technol* 2016;49:276–288.

[23] Mahapatra TR, Panda SK, Kar VR. Nonlinear flexural analysis of laminated composite panel under hygro-thermo-mechanical loading—A micromechanical approach. *Int J Computat Meth* 2016;13:1650015.

[24] Mahapatra TR, Panda SK, Kar VR. Nonlinear hygro-thermo-elastic vibration analysis of doubly curved composite shell panel using finite element micromechanical model. *Mech Advan Mat Struct* 2016;23:1343–1359.

[25] Mahapatra TR, Panda SK, Kar VR. Geometrically nonlinear flexural analysis of hygro-thermo-elastic laminated composite doubly curved shell panel. *Int J Mech Mat Des* 2016;12:153–171.

[26] Dutta G, Panda SK, Mahapatra TR, Singh VK. Electro-magneto-elastic response of laminated composite plate: A finite element approach. *Int J Appl Computat Math* 2017;3:2573–2592.

[27] Suman SD, Hirwani CK, Chaturvedi A, Panda SK. Effect of magnetostrictive material layer on the stress and deformation behaviour of laminated structure. *IOP Conf Ser: Mat Sci Eng* 2017;178:012026.

[28] Javani R, Bidgoli MR, Kolahchi R. Buckling analysis of plates reinforced by Graphene platelet based on Halpin-Tsai and Reddy theories. *Steel Compos Struct* 2019;(31)4:419–427.

# 6 Vibration Analysis of Agglomerated Nanoparticle-Reinforced Plates

## 6.1 INTRODUCTION

A free vibrational analysis of a plate resting on a soil bed is of great importance for the design of many engineering problems, such as the footings of buildings, the pavement of roads, and the bases of machines. It is due to the vibrational effect on the performance of sensitive equipment located on plates. Engineers need to know the natural frequencies and the corresponding mode shapes of the plates so that they can design the system appropriately. Also, as it is the most usable material in the construction of plates, it has been required to improve its quality for reducing vibrations. Today, nanotechnology offers the possibility of great advances in construction materials. So, free vibrations can be reduced by improving the properties of plates by adding nanomaterial in order to increase its stiffness. On the other hand, for the accurate vibrational analysis of thick plates such as foundations, it is appropriate to use the high-order shear deformation theories. Because the results evaluated by using classical thin plate theory may not be reliable especially as the plate gets thicker. Therefore, a free vibration analysis of plates reinforced with silicon dioxide ($SiO_2$) nanoparticles using the shear deformation theory is done on this topic. Also, the Winker model is used for the simulation of a soil bed because of its simplicity and accuracy.

Investigating the vibration problems of plates on elastic plates has attracted the attention of many researchers working on structural foundation analysis and design. Lam et al. [1] applied Green's functions to achieve canonical exact solutions for the elastic bending, buckling, and vibration of Levy plates resting on two-parameter elastic foundations. The free and forced vibration analysis for a Reissner-Mindlin plate with four free edges resting on a Pasternak-type elastic foundation has been studied by Shen et al. [2]. By employing the Rayleigh–Ritz method, the three-dimensional vibration of rectangular thick plates on elastic foundations was investigated by Zhou et al. [3]. Zhong and Yin [4] investigated the free vibration behavior of plates on a Winkler foundation using the finite integral transform method. Ferreira et al. [5] used the radial basis function collocation method to study static deformation and free vibration of plates on a Pasternak foundation. A free vibration

DOI: 10.1201/9781003349525-6

analysis of moderately thick trapezoidal symmetrically laminated plates with various combinations of boundary conditions was studied by Zamani et al. [6]. Kumar and Lal [7] studied the vibration of nonhomogeneous orthotropic rectangular plates with a bilinear thickness variation resting on a Winkler foundation. A simple refined theory for the bending, buckling, and vibration of thick plates resting on an elastic foundation was introduced by Thai et al. [8]. An analytical solution of a refined plate theory is developed for free vibration analysis of functionally graded plates under various boundary conditions by Thai and Choi [9]. Nguyen-Thoi et al. presented an edge-based smoothed three-node Mindlin plate element for static and free vibration analyses of plates. An original first shear deformation theory for studying advanced composites on an elastic foundation was presented by Mantari and Granados [10]. Uğurlu [11] analyzed the vibration of elastic bottom plates of fluid storage tanks resting on a Pasternak foundation based on the boundary element method. A simplified first-order shear deformation theory for the bending, buckling, and free vibration of isotropic plates on elastic foundations was investigated by Park and Choi [12].

Furthermore, the mechanical behavior of structures containing nanoparticles has been investigated experimentally by a number of researchers, but there are few mathematical works in this field. Investigations into the development of a powder with $SiO_2$ nanoparticles have been made by Jo et al. [13]. Fathi et al. [14] investigated the mechanical and physical properties of an expanded polystyrene structure containing micro-silica and nano-silica. The effect of nano-silica on the compressive strength development and water absorption properties of cement paste containing fly ash was tested by Ehsani et al. [15]. In the field of mathematical modeling of structures, Jafarian Arani and Kolahchi [16] considered a buckling analysis of beams reinforced with carbon nanotubes by using the Euler–Bernoulli and Timoshenko beam models. The nonlinear buckling of a beam reinforced with $SiO_2$ nanoparticles was investigated by Zamanian et al. [17]. Also, Arbabi et al. [18] studied the buckling of beams reinforced with zinc oxide nanoparticles subjected to an electric field. Mechanical characteristics of a classical lightened concrete by the addition of treated straws were presented by Kammoun and Trabelsi [19]. Alijani and Bidgoli [20] explored the vibration analysis of a concrete foundation reinforced with $SiO_2$ nanoparticles and supported by a soil bed simulated with spring constants. The material properties of the nano-composite structure are determined using the Mori-Tanaka model, taking into account agglomeration effects.

It is worthy to note that this chapter is mainly concerned with the vibration of plates reinforced by $SiO_2$ nanoparticles resting on a soil medium. To the best of the authors' knowledge, the effects of using nanoparticles on the vibration of plates have not been investigated. In order to obtain the equivalent material properties of the nanocomposite structure, the Mori–Tanaka model is used. Also, higher order shear deformation theories (HSDTs) were proposed to avoid the use of shear correction factor and obtain a better prediction of the response of a thick plate. Applying the Reddy HSDT, the motion equations are obtained based on Hamilton's principle. Also, the Navier method is applied to obtain the frequency of the system. The effects of volume percent and agglomeration of $SiO_2$ nanoparticles, soil medium, and geometrical parameters of structure on the frequency of the system are discussed in detail.

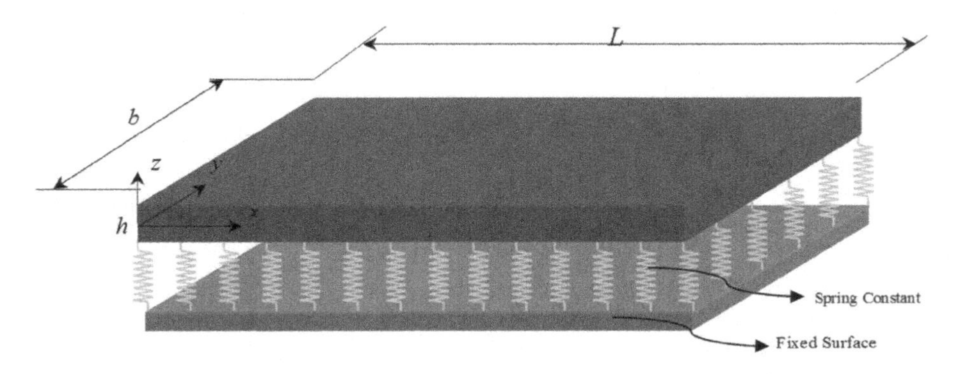


**FIGURE 6.1** Schematic of a plate reinforced with $SiO_2$ nanoparticles resting on a soil medium.

## 6.2 MATHEMATICAL MODELING

A plate reinforced with $SiO_2$ nanoparticles with length $L$, width $b$, and thickness $h$ is considered as shown in Figure 6.1.

### 6.2.1 Stress–Strain Relations

Based on the Reddy theory, the displacement field can be expressed based on Section 1.3.4 [21]. Based on Hook's law, we have

$$\sigma_{xx}^{c} = Q_{11}\varepsilon_{xx} + Q_{12}\varepsilon_{yy}, \tag{6–1}$$

$$\sigma_{yy}^{c} = Q_{12}\varepsilon_{xx} + Q_{22}\varepsilon_{yy}, \tag{6–2}$$

$$\tau_{yz}^{c} = Q_{44}\gamma_{yz}, \tag{6–3}$$

$$\tau_{xz}^{c} = Q_{55}\gamma_{zx}, \tag{6–4}$$

$$\tau_{xy}^{c} = Q_{66}\gamma_{xy}, \tag{6–5}$$

where $Q_{ij}$ are elastic constants that can be obtained by Mori–Tanaka model (this model is presented in Section 4.3).

### 6.2.2 Energy Method

The potential energy can be written as

$$U = \frac{1}{2}\int\left(\sigma_{xx}\varepsilon_{xx} + \sigma_{yy}\varepsilon_{yy} + \sigma_{xy}\gamma_{xy} + \sigma_{xz}\gamma_{xz} + \sigma_{yz}\gamma_{yz}\right)dV. \tag{6–6}$$

Combining Equations 1–33 and 1–34 with 6–6 yields

$$U = \frac{1}{2}\int \left( N_{xx}\left(\frac{\partial u}{\partial x}\right) + N_{yy}\left(\frac{\partial v}{\partial y}\right) + Q_{yy}\left(\frac{\partial w}{\partial y} + \phi_y\right) + Q_{xx}\left(\frac{\partial w}{\partial x} + \phi_x\right)\right.$$
$$+N_{xy}\left(\frac{\partial v}{\partial x} + \frac{\partial u}{\partial y}\right) + M_{xx}\frac{\partial \phi_x}{\partial x} + M_{yy}\frac{\partial \phi_y}{\partial y} + M_{xy}\left(\frac{\partial \phi_x}{\partial y} + \frac{\partial \phi_y}{\partial x}\right)$$
$$+K_{yy}\left(c_2\left(\phi_y + \frac{\partial w}{\partial y}\right)\right) + K_{xx}\left(c_2\left(\phi_x + \frac{\partial w}{\partial x}\right)\right) + P_{xx}\left(c_1\left(\frac{\partial \phi_x}{\partial x} + \frac{\partial^2 w}{\partial x^2}\right)\right)$$
$$\left.+P_{yy}\left(c_1\left(\frac{\partial \phi_y}{\partial y} + \frac{\partial^2 w}{\partial y^2}\right)\right) + P_{xy}\left(\frac{\partial \phi_y}{\partial x} + \frac{\partial \phi_x}{\partial y} + 2\frac{\partial^2 w}{\partial x \partial y}\right)\right) dA, \tag{6–7}$$

where the stress resultant–displacement relations can be written as

$$\begin{Bmatrix} N_{xx} \\ N_{yy} \\ N_{xy} \end{Bmatrix} = \int_{-h/2}^{h/2} \begin{bmatrix} \sigma_{xx} \\ \sigma_{yy} \\ \sigma_{xy} \end{bmatrix} dz, \tag{6–8}$$

$$\begin{Bmatrix} M_{xx} \\ M_{yy} \\ M_{xy} \end{Bmatrix} = \int_{-h/2}^{h/2} \begin{bmatrix} \sigma_{xx} \\ \sigma_{yy} \\ \sigma_{xy} \end{bmatrix} z dz, \tag{6–9}$$

$$\begin{Bmatrix} P_{xx} \\ P_{yy} \\ P_{xy} \end{Bmatrix} = \int_{-h/2}^{h/2} \begin{bmatrix} \sigma_{xx} \\ \sigma_{yy} \\ \sigma_{xy} \end{bmatrix} z^3 dz, \tag{6–10}$$

$$\begin{bmatrix} Q_{xx} \\ Q_{yy} \end{bmatrix} = \int_{-h/2}^{h/2} \begin{bmatrix} \sigma_{xz} \\ \sigma_{yz} \end{bmatrix} dz, \tag{6–11}$$

$$\begin{bmatrix} K_{xx} \\ K_{yy} \end{bmatrix} = \int_{-h/2}^{h/2} \begin{bmatrix} \sigma_{xz} \\ \sigma_{yz} \end{bmatrix} z^2 dz. \tag{6–12}$$

Substituting Equations 6–1 to 6–5 into Equations 6–8 through 6–12, the stress resultant–displacement relations can be obtained as follows:

$$N_{xx} = A_{11}\frac{\partial u}{\partial x} + A_{12}\frac{\partial v}{\partial y} + A_{16}\left(\frac{\partial u}{\partial y} + \frac{\partial v}{\partial x}\right) + B_{11}\frac{\partial \varphi_x}{\partial x} + B_{12}\frac{\partial \varphi_y}{\partial y} + B_{16}\left(\frac{\partial \varphi_x}{\partial y} + \frac{\partial \varphi_y}{\partial x}\right)$$
$$+E_{11}c_1\left(\frac{\partial \varphi_x}{\partial x} + \frac{\partial^2 w}{\partial x^2}\right) + E_{12}c_1\left(\frac{\partial \varphi_y}{\partial y} + \frac{\partial^2 w}{\partial y^2}\right) + E_{16}c_1\left(\frac{\partial \varphi_y}{\partial x} + \frac{\partial \varphi_x}{\partial y} + 2\frac{\partial^2 w}{\partial x \partial y}\right),$$

$$
\begin{aligned}
N_{yy} &= A_{12}\frac{\partial u}{\partial x} + A_{22}\frac{\partial v}{\partial y} + A_{26}\left(\frac{\partial u}{\partial y} + \frac{\partial v}{\partial x}\right) + B_{12}\frac{\partial \varphi_x}{\partial x} + B_{22}\frac{\partial \varphi_y}{\partial y} \\
&+ B_{26}\left(\frac{\partial \varphi_x}{\partial y} + \frac{\partial \varphi_y}{\partial x}\right) + E_{12}c_1\left(\frac{\partial \varphi_x}{\partial x} + \frac{\partial^2 w}{\partial x^2}\right) + E_{22}c_1\left(\frac{\partial \varphi_y}{\partial y} + \frac{\partial^2 w}{\partial y^2}\right) \\
&+ E_{26}c_1\left(\frac{\partial \varphi_y}{\partial x} + \frac{\partial \varphi_x}{\partial y} + 2\frac{\partial^2 w}{\partial x \partial y}\right), \\
N_{xy} &= A_{16}\frac{\partial u}{\partial x} + A_{26}\frac{\partial v}{\partial y} + A_{66}\left(\frac{\partial u}{\partial y} + \frac{\partial v}{\partial x}\right) + B_{16}\frac{\partial \varphi_x}{\partial x} + B_{26}\frac{\partial \varphi_y}{\partial y} \\
&+ B_{66}\left(\frac{\partial \varphi_x}{\partial y} + \frac{\partial \varphi_y}{\partial x}\right) + E_{16}c_1\left(\frac{\partial \varphi_x}{\partial x} + \frac{\partial^2 w}{\partial x^2}\right) + E_{26}c_1\left(\frac{\partial \varphi_y}{\partial y} + \frac{\partial^2 w}{\partial y^2}\right) \\
&+ E_{66}c_1\left(\frac{\partial \varphi_y}{\partial x} + \frac{\partial \varphi_x}{\partial y} + 2\frac{\partial^2 w}{\partial x \partial y}\right),
\end{aligned} \tag{6–13}
$$

$$
\begin{aligned}
M_{xx} &= B_{11}\frac{\partial u}{\partial x} + B_{12}\frac{\partial v}{\partial y} + B_{16}\left(\frac{\partial u}{\partial y} + \frac{\partial v}{\partial x}\right) + D_{11}\frac{\partial \varphi_x}{\partial x} + D_{12}\frac{\partial \varphi_y}{\partial y} \\
&+ D_{16}\left(\frac{\partial \varphi_x}{\partial y} + \frac{\partial \varphi_y}{\partial x}\right) + F_{11}c_1\left(\frac{\partial \varphi_x}{\partial x} + \frac{\partial^2 w}{\partial x^2}\right) + F_{12}c_1\left(\frac{\partial \varphi_y}{\partial y} + \frac{\partial^2 w}{\partial y^2}\right) \\
&+ F_{16}c_1\left(\frac{\partial \varphi_y}{\partial x} + \frac{\partial \varphi_x}{\partial y} + 2\frac{\partial^2 w}{\partial x \partial y}\right), \\
M_{yy} &= B_{12}\frac{\partial u}{\partial x} + B_{22}\frac{\partial v}{\partial y} + B_{26}\left(\frac{\partial u}{\partial y} + \frac{\partial v}{\partial x}\right) + D_{12}\frac{\partial \varphi_x}{\partial x} + D_{22}\frac{\partial \varphi_y}{\partial y} \\
&+ D_{26}\left(\frac{\partial \varphi_x}{\partial y} + \frac{\partial \varphi_y}{\partial x}\right) + F_{12}c_1\left(\frac{\partial \varphi_x}{\partial x} + \frac{\partial^2 w}{\partial x^2}\right) + F_{22}c_1\left(\frac{\partial \varphi_y}{\partial y} + \frac{\partial^2 w}{\partial y^2}\right) \\
&+ F_{26}c_1\left(\frac{\partial \varphi_y}{\partial x} + \frac{\partial \varphi_x}{\partial y} + 2\frac{\partial^2 w}{\partial x \partial y}\right), \\
M_{xy} &= B_{16}\frac{\partial u}{\partial x} + B_{26}\frac{\partial v}{\partial y} + B_{66}\left(\frac{\partial u}{\partial y} + \frac{\partial v}{\partial x}\right) + D_{16}\frac{\partial \varphi_x}{\partial x} + D_{26}\frac{\partial \varphi_y}{\partial y} \\
&+ D_{66}\left(\frac{\partial \varphi_x}{\partial y} + \frac{\partial \varphi_y}{\partial x}\right) + F_{16}c_1\left(\frac{\partial \varphi_x}{\partial x} + \frac{\partial^2 w}{\partial x^2}\right) + F_{26}c_1\left(\frac{\partial \varphi_y}{\partial y} + \frac{\partial^2 w}{\partial y^2}\right) \\
&+ F_{66}c_1\left(\frac{\partial \varphi_y}{\partial x} + \frac{\partial \varphi_x}{\partial y} + 2\frac{\partial^2 w}{\partial x \partial y}\right),
\end{aligned} \tag{6–14}
$$

$$
\begin{aligned}
P_{xx} &= E_{11}\frac{\partial u}{\partial x} + E_{12}\frac{\partial v}{\partial y} + E_{16}\left(\frac{\partial u}{\partial y} + \frac{\partial v}{\partial x}\right) + F_{11}\frac{\partial \varphi_x}{\partial x} + F_{12}\frac{\partial \varphi_y}{\partial y} \\
&\quad + F_{16}\left(\frac{\partial \varphi_x}{\partial y} + \frac{\partial \varphi_y}{\partial x}\right) + H_{11}c_1\left(\frac{\partial \varphi_x}{\partial x} + \frac{\partial^2 w}{\partial x^2}\right) + H_{12}c_1\left(\frac{\partial \varphi_y}{\partial y} + \frac{\partial^2 w}{\partial y^2}\right) \\
&\quad + H_{16}c_1\left(\frac{\partial \varphi_y}{\partial x} + \frac{\partial \varphi_x}{\partial y} + 2\frac{\partial^2 w}{\partial x \partial y}\right), \\
P_{yy} &= E_{12}\frac{\partial u}{\partial x} + E_{22}\frac{\partial v}{\partial y} + E_{26}\left(\frac{\partial u}{\partial y} + \frac{\partial v}{\partial x}\right) + F_{12}\frac{\partial \varphi_x}{\partial x} + F_{22}\frac{\partial \varphi_y}{\partial y} \\
&\quad + F_{26}\left(\frac{\partial \varphi_x}{\partial y} + \frac{\partial \varphi_y}{\partial x}\right) + H_{12}c_1\left(\frac{\partial \varphi_x}{\partial x} + \frac{\partial^2 w}{\partial x^2}\right) + H_{22}c_1\left(\frac{\partial \varphi_y}{\partial y} + \frac{\partial^2 w}{\partial y^2}\right) \\
&\quad + H_{26}c_1\left(\frac{\partial \varphi_y}{\partial x} + \frac{\partial \varphi_x}{\partial y} + 2\frac{\partial^2 w}{\partial x \partial y}\right), \\
P_{xy} &= E_{16}\frac{\partial u}{\partial x} + E_{26}\frac{\partial v}{\partial y} + E_{66}\left(\frac{\partial u}{\partial y} + \frac{\partial v}{\partial x}\right) + F_{16}\frac{\partial \varphi_x}{\partial x} + F_{26}\frac{\partial \varphi_y}{\partial y} \\
&\quad + F_{66}\left(\frac{\partial \varphi_x}{\partial y} + \frac{\partial \varphi_y}{\partial x}\right) + H_{16}c_1\left(\frac{\partial \varphi_x}{\partial x} + \frac{\partial^2 w}{\partial x^2}\right) + H_{26}c_1\left(\frac{\partial \varphi_y}{\partial y} + \frac{\partial^2 w}{\partial y^2}\right) \\
&\quad + H_{66}c_1\left(\frac{\partial \varphi_y}{\partial x} + \frac{\partial \varphi_x}{\partial y} + 2\frac{\partial^2 w}{\partial x \partial y}\right),
\end{aligned} \tag{6–15}
$$

$$
\begin{aligned}
Q_{xx} &= A_{55}\left(\frac{\partial w}{\partial x} + \varphi_x\right) + A_{45}\left(\frac{\partial w}{\partial y} + \varphi_y\right) + D_{55}c_2\left(\varphi_x + \frac{\partial w}{\partial x}\right) + D_{45}c_2\left(\frac{\partial w}{\partial y} + \varphi_y\right), \\
Q_{yy} &= A_{45}\left(\frac{\partial w}{\partial y} + \varphi_y\right) + A_{44}\left(\frac{\partial w}{\partial y} + \varphi_y\right) + D_{45}c_2\left(\varphi_y + \frac{\partial w}{\partial y}\right) + D_{44}c_2\left(\frac{\partial w}{\partial y} + \varphi_y\right),
\end{aligned} \tag{6–16}
$$

$$
\begin{aligned}
K_{xx} &= D_{55}\left(\frac{\partial w}{\partial x} + \varphi_x\right) + D_{45}\left(\frac{\partial w}{\partial y} + \varphi_y\right) + F_{55}c_2\left(\varphi_x + \frac{\partial w}{\partial x}\right) + F_{45}c_2\left(\frac{\partial w}{\partial y} + \varphi_y\right), \\
K_{yy} &= D_{45}\left(\frac{\partial w}{\partial y} + \varphi_y\right) + D_{44}\left(\frac{\partial w}{\partial y} + \varphi_y\right) + F_{45}c_2\left(\varphi_y + \frac{\partial w}{\partial y}\right) + F_{44}c_2\left(\frac{\partial w}{\partial y} + \varphi_y\right),
\end{aligned} \tag{6–17}
$$

where

$$A_{ij} = \int_{-h/2}^{h/2} Q_{ij}\,dz, \qquad (i, j = 1,2,6) \tag{6–18}$$

$$B_{ij} = \int_{-h/2}^{h/2} Q_{ij}z\,dz, \tag{6–19}$$

$$D_{ij} = \int_{-h/2}^{h/2} Q_{ij}z^2\,dz, \tag{6–20}$$

$$E_{ij} = \int_{-h/2}^{h/2} Q_{ij} z^3 dz, \tag{6–21}$$

$$F_{ij} = \int_{-h/2}^{h/2} Q_{ij} z^4 dz, \tag{6–22}$$

$$H_{ij} = \int_{-h/2}^{h/2} Q_{ij} z^6 dz. \tag{6–23}$$

The kinetic energy of the system may be written as

$$U = \frac{\rho}{2} \int \left( \dot{u}_1^{\,2} + \dot{u}_2^{\,2} + \dot{u}_3^{\,2} \right) dV. \tag{6–24}$$

The external work due to a Pasternak medium can be written as [22]

$$W_e = \int \int \left( -K_w w \right) w dA, \tag{6–25}$$

where $K_w$ is Winkler's spring modulus. The governing equations can be derived using Hamilton's principle as follows:

$$\int_0^t (\delta U - \delta K - \delta W_e) dt = 0. \tag{6–26}$$

Substituting Equations 6–7, 6–24, and 6–25 into Equation 6–26 yields the following governing equations:

$$\delta u : \frac{\partial N_{xx}}{\partial x} + \frac{\partial N_{xy}}{\partial y} = I_0 \frac{\partial^2 u}{\partial t^2} + J_1 \frac{\partial^2 \phi_x}{\partial t^2} - \frac{4 I_3}{h^2} \frac{\partial^3 w}{\partial t^2 \partial x}, \tag{6–27}$$

$$\delta v : \frac{\partial N_{xy}}{\partial x} + \frac{\partial N_{yy}}{\partial y} = I_0 \frac{\partial^2 v}{\partial t^2} + J_1 \frac{\partial^2 \phi_y}{\partial t^2} - \frac{4 I_3}{h^2} \frac{\partial^3 w}{\partial t^2 \partial y}, \tag{6–28}$$

$$\begin{aligned} \delta w : & \frac{\partial Q_{xx}}{\partial x} + \frac{\partial Q_{yy}}{\partial y} + c_2 \left( \frac{\partial K_{xx}}{\partial x} + \frac{\partial K_{yy}}{\partial y} \right) + N_{xx} \frac{\partial^2 w}{\partial x^2} + N_{yy} \frac{\partial^2 w}{\partial y^2} \\ & - c_1 \left( \frac{\partial^2 P_{xx}}{\partial x^2} + 2 \frac{\partial^2 P_{xy}}{\partial x \partial y} + \frac{\partial^2 P_{yy}}{\partial y^2} \right) - K_w w = I_0 \frac{\partial^2 w}{\partial t^2} - \left( \frac{4}{3h^2} \right)^2 I_6 \left( \frac{\partial^4 w}{\partial x^2 \partial t^2} + \frac{\partial^4 w}{\partial y^2 \partial t^2} \right) \\ & + \frac{4}{3h^2} \left( I_3 \frac{\partial^3 u}{\partial t^2 \partial x} + I_3 \frac{\partial^3 v}{\partial t^2 \partial y} + J_4 \left( \frac{\partial^3 \phi_x}{\partial t^2 \partial x} + \frac{\partial^3 \phi_y}{\partial t^2 \partial y} \right) \right), \end{aligned} \tag{6–29}$$

$$\begin{aligned} \delta \phi_x : & \frac{\partial M_{xx}}{\partial x} + \frac{\partial M_{xy}}{\partial y} + c_1 \left( \frac{\partial P_{xx}}{\partial x} + \frac{\partial P_{xy}}{\partial y} \right) - Q_{xx} - c_2 K_{xx} \\ & = J_1 \frac{\partial^2 u}{\partial t^2} + K_2 \frac{\partial^2 \phi_x}{\partial t^2} - \frac{4}{3h^2} J_4 \frac{\partial^3 w}{\partial t^2 \partial x}, \end{aligned} \tag{6–30}$$

$$\delta\phi_y : \frac{\partial M_{xy}}{\partial x} + \frac{\partial M_{yy}}{\partial y} + c_1\left(\frac{\partial P_{xy}}{\partial x} + \frac{\partial P_{yy}}{\partial y}\right) - Q_{yy} - c_2 K_{yy}$$
$$= J_1\frac{\partial^2 v}{\partial t^2} + K_2\frac{\partial^2 \phi_y}{\partial t^2} - \frac{4}{3h^2}J_4\frac{\partial^3 w}{\partial t^2 \partial y}, \qquad (6\text{–}31)$$

where

$$I_i = \int_{-h/2}^{h/2} \rho z^i dz \qquad (i = 0,1,\ldots,6), \qquad (6\text{–}32)$$

$$J_i = I_i - \frac{4}{3h^2}I_{i+2} \qquad (i = 1,4), \qquad (6\text{–}32)$$

$$K_2 = I_2 - \frac{8}{3h^2}I_4 + \left(\frac{4}{3h^2}\right)^2 I_6. \qquad (6\text{–}33)$$

Substituting Equations 6–14 through 6–17 into Equations 6–27 through 6–31, the governing equations can be written as follows:

$$A_{11}\frac{\partial^2 u}{\partial x^2} + A_{12}\frac{\partial^2 v}{\partial x \partial y} + A_{16}\left(\frac{\partial^2 u}{\partial x \partial y} + \frac{\partial^2 v}{\partial x^2}\right) + B_{11}\frac{\partial^2 \varphi_x}{\partial x^2} + B_{12}\frac{\partial^2 \varphi_y}{\partial x \partial y}$$
$$+B_{16}\left(\frac{\partial^2 \varphi_x}{\partial x \partial y} + \frac{\partial^2 \varphi_y}{\partial x^2}\right) + E_{11}c_1\left(\frac{\partial^2 \varphi_x}{\partial x^2} + \frac{\partial^3 w}{\partial x^3}\right) + E_{12}c_1\left(\frac{\partial^2 \varphi_y}{\partial x \partial y} + \frac{\partial^3 w}{\partial x \partial y^2}\right)$$
$$+E_{16}c_1\left(\frac{\partial^2 \varphi_y}{\partial x^2} + \frac{\partial^2 \varphi_x}{\partial x \partial y} + 2\frac{\partial^3 w}{\partial x^2 \partial y}\right) + A_{16}\frac{\partial^2 u}{\partial x \partial y} + A_{26}\frac{\partial^2 v}{\partial y^2} + A_{66}\left(\frac{\partial^2 u}{\partial y^2} + \frac{\partial^2 v}{\partial x \partial y}\right)$$
$$+B_{16}\frac{\partial^2 \varphi_x}{\partial x \partial y} + B_{26}\frac{\partial^2 \varphi_y}{\partial y^2} + B_{66}\left(\frac{\partial^2 \varphi_x}{\partial y^2} + \frac{\partial^2 \varphi_y}{\partial x \partial y}\right) + E_{16}c_1\left(\frac{\partial^2 \varphi_x}{\partial x \partial y} + \frac{\partial^3 w}{\partial y \partial x^2}\right)$$
$$+E_{26}c_1\left(\frac{\partial^2 \varphi_y}{\partial y^2} + \frac{\partial^3 w}{\partial y^3}\right) + E_{66}c_1\left(\frac{\partial^2 \varphi_y}{\partial x \partial y} + \frac{\partial^2 \varphi_x}{\partial y^2} + 2\frac{\partial^3 w}{\partial x \partial y^2}\right)$$
$$= I_0\frac{\partial^2 u}{\partial t^2} + J_1\frac{\partial^2 \phi_x}{\partial t^2} - \frac{4I_3}{h^2}\frac{\partial^3 w}{\partial t^2 \partial x}, \qquad (6\text{–}34)$$

$$A_{16}\frac{\partial^2 u}{\partial x^2} + A_{26}\frac{\partial^2 v}{\partial x \partial y} + A_{66}\left(\frac{\partial^2 u}{\partial x \partial y} + \frac{\partial^2 v}{\partial x^2}\right) + B_{16}\frac{\partial^2 \varphi_x}{\partial x^2} + B_{26}\frac{\partial^2 \varphi_y}{\partial x \partial y}$$
$$+B_{66}\left(\frac{\partial^2 \varphi_x}{\partial x \partial y} + \frac{\partial^2 \varphi_y}{\partial x^2}\right) + E_{16}c_1\left(\frac{\partial^2 \varphi_x}{\partial x^2} + \frac{\partial^3 w}{\partial x^3}\right) + E_{26}c_1\left(\frac{\partial^2 \varphi_y}{\partial x \partial y} + \frac{\partial^3 w}{\partial x \partial y^2}\right)$$
$$+E_{66}c_1\left(\frac{\partial^2 \varphi_y}{\partial x^2} + \frac{\partial^2 \varphi_x}{\partial x \partial y} + 2\frac{\partial^3 w}{\partial x^2 \partial y}\right) + A_{21}\frac{\partial^2 u}{\partial x \partial y} + A_{22}\frac{\partial^2 v}{\partial y^2} + A_{26}\left(\frac{\partial^2 u}{\partial y^2} + \frac{\partial^2 v}{\partial x \partial y}\right)$$

$$+B_{21}\frac{\partial^2\varphi_x}{\partial x\partial y}+B_{22}\frac{\partial^2\varphi_y}{\partial y^2}+B_{26}\left(\frac{\partial^2\varphi_x}{\partial y^2}+\frac{\partial^2\varphi_y}{\partial x\partial y}\right)+E_{21}c_1\left(\frac{\partial^2\varphi_x}{\partial x\partial y}+\frac{\partial^3 w}{\partial y\partial x^2}\right)$$
$$+E_{22}c_1\left(\frac{\partial^2\varphi_y}{\partial y^2}+\frac{\partial^3 w}{\partial y^3}\right)+E_{26}c_1\left(\frac{\partial^2\varphi_y}{\partial x\partial y}+\frac{\partial^2\varphi_x}{\partial y^2}+2\frac{\partial^3 w}{\partial x\partial y^2}\right)$$
$$=I_0\frac{\partial^2 v}{\partial t^2}+J_1\frac{\partial^2\phi_y}{\partial t^2}-\frac{4I_3}{h^2}\frac{\partial^3 w}{\partial t^2\partial y},\tag{6–35}$$

$$A_{55}\left(\frac{\partial^2 w}{\partial x^2}+\frac{\partial\varphi_x}{\partial x}\right)+A_{45}\left(\frac{\partial^2 w}{\partial x\partial y}+\frac{\partial\varphi_y}{\partial x}\right)+D_{55}c_2\left(\frac{\partial^2 w}{\partial x^2}+\frac{\partial\varphi_x}{\partial x}\right)+D_{45}c_2\left(\frac{\partial^2 w}{\partial x\partial y}+\frac{\partial\varphi_y}{\partial x}\right)$$
$$+A_{45}\left(\frac{\partial^2 w}{\partial x\partial y}+\frac{\partial\varphi_x}{\partial y}\right)+A_{44}\left(\frac{\partial^2 w}{\partial y^2}+\frac{\partial\varphi_y}{\partial y}\right)+D_{45}c_2\left(\frac{\partial^2 w}{\partial x\partial y}+\frac{\partial\varphi_x}{\partial y}\right)+D_{44}c_2\left(\frac{\partial^2 w}{\partial y^2}+\frac{\partial\varphi_y}{\partial y}\right)$$
$$+c_2\left(\begin{array}{l}D_{55}\left(\frac{\partial^2 w}{\partial x^2}+\frac{\partial\varphi_x}{\partial x}\right)+D_{45}\left(\frac{\partial^2 w}{\partial x\partial y}+\frac{\partial\varphi_y}{\partial x}\right)+F_{55}c_2\left(\frac{\partial^2 w}{\partial x^2}+\frac{\partial\varphi_x}{\partial x}\right)+F_{45}c_2\left(\frac{\partial^2 w}{\partial x\partial y}+\frac{\partial\varphi_y}{\partial x}\right)\\ +D_{45}\left(\frac{\partial^2 w}{\partial x\partial y}+\frac{\partial\varphi_x}{\partial y}\right)+D_{44}\left(\frac{\partial^2 w}{\partial y^2}+\frac{\partial\varphi_y}{\partial y}\right)+F_{45}c_2\left(\frac{\partial^2 w}{\partial x\partial y}+\frac{\partial\varphi_x}{\partial y}\right)+F_{44}c_2\left(\frac{\partial^2 w}{\partial y^2}+\frac{\partial\varphi_y}{\partial y}\right)\end{array}\right)$$
$$+N_{xx}\frac{\partial^2 w}{\partial x^2}+N_{yy}\frac{\partial^2 w}{\partial y^2}-c_1\left(E_{11}\frac{\partial^3 u}{\partial x^3}+E_{12}\frac{\partial^3 v}{\partial x^2\partial y}+E_{16}\left(\frac{\partial^3 u}{\partial x^2\partial y}+\frac{\partial^3 v}{\partial x^3}\right)+F_{11}\frac{\partial^3\varphi_x}{\partial x^3}\right.$$
$$+F_{12}\frac{\partial^3\varphi_y}{\partial x^2\partial y}+F_{16}\left(\frac{\partial^3\varphi_x}{\partial x^2\partial y}+\frac{\partial^3\varphi_y}{\partial x^3}\right)+H_{11}c_1\left(\frac{\partial^3\varphi_x}{\partial x^3}+\frac{\partial^4 w}{\partial x^4}\right)+H_{12}c_1\left(\frac{\partial^3\varphi_y}{\partial x^2\partial y}+\frac{\partial^4 w}{\partial x^2\partial y^2}\right)$$
$$+H_{16}c_1\left(\frac{\partial^3\varphi_y}{\partial x^3}+\frac{\partial^3\varphi_x}{\partial x^2\partial y}+2\frac{\partial^4 w}{\partial x^3\partial y}\right)+E_{12}\frac{\partial^3 u}{\partial y^2\partial x}+E_{22}\frac{\partial^3 v}{\partial y^3}+E_{26}\left(\frac{\partial^3 u}{\partial y^3}+\frac{\partial^3 v}{\partial x\partial y^2}\right)$$
$$+F_{12}\frac{\partial^3\varphi_x}{\partial x\partial y^2}+F_{22}\frac{\partial^3\varphi_y}{\partial y^3}+F_{26}\left(\frac{\partial^3\varphi_x}{\partial y^3}+\frac{\partial^3\varphi_y}{\partial x\partial y^2}\right)+H_{12}c_1\left(\frac{\partial^3\varphi_x}{\partial x\partial y^2}+\frac{\partial^4 w}{\partial x^2\partial y^2}\right)$$
$$+H_{22}c_1\left(\frac{\partial^3\varphi_y}{\partial y^3}+\frac{\partial^4 w}{\partial y^4}\right)+H_{26}c_1\left(\frac{\partial^3\varphi_y}{\partial x\partial y^2}+\frac{\partial^3\varphi_x}{\partial y^3}+2\frac{\partial^4 w}{\partial x\partial y^3}\right)+2E_{16}\frac{\partial^3 u}{\partial y\partial x^2}+2E_{26}\frac{\partial^3 v}{\partial y^2\partial x}$$
$$+2E_{66}\left(\frac{\partial^3 u}{\partial y^2\partial x}+\frac{\partial^3 v}{\partial x^2\partial y}\right)+2F_{16}\frac{\partial^3\varphi_x}{\partial x^2\partial y}+2F_{26}\frac{\partial^3\varphi_y}{\partial y^2\partial x}+2F_{66}\left(\frac{\partial^3\varphi_x}{\partial y^2\partial x}+\frac{\partial^3\varphi_y}{\partial x^2\partial y}\right)$$
$$+2H_{16}c_1\left(\frac{\partial^3\varphi_x}{\partial x^2\partial y}+\frac{\partial^4 w}{\partial x^3\partial y}\right)+2H_{26}c_1\left(\frac{\partial^3\varphi_y}{\partial x\partial y^2}+\frac{\partial^4 w}{\partial y^3\partial x}\right)$$
$$\left.+2H_{66}c_1\left(\frac{\partial^3\varphi_y}{\partial x^2\partial y}+\frac{\partial^3\varphi_x}{\partial y^2\partial x}+2\frac{\partial^4 w}{\partial x^2\partial y^2}\right)\right)-k_w w=I_0\frac{\partial^2 w}{\partial t^2}-\left(\frac{4}{3h^2}\right)^2 I_6\left(\frac{\partial^4 w}{\partial x^2\partial t^2}+\frac{\partial^4 w}{\partial y^2\partial t^2}\right)$$
$$+\frac{4}{3h^2}\left(I_3\frac{\partial^3 u}{\partial t^2\partial x}+I_3\frac{\partial^3 v}{\partial t^2\partial y}+J_4\left(\frac{\partial^3\phi_x}{\partial t^2\partial x}+\frac{\partial^3\phi_y}{\partial t^2\partial y}\right)\right),\tag{6–36}$$

$$B_{11}\frac{\partial^2 u}{\partial x^2}+B_{12}\frac{\partial^2 v}{\partial x\partial y}+B_{16}\left(\frac{\partial^2 u}{\partial x\partial y}+\frac{\partial^2 v}{\partial x^2}\right)+D_{11}\frac{\partial^2 \varphi_x}{\partial x^2}+D_{12}\frac{\partial^2 \varphi_y}{\partial x\partial y}+D_{16}\left(\frac{\partial^2 \varphi_x}{\partial x\partial y}+\frac{\partial^2 \varphi_y}{\partial x^2}\right)$$

$$+F_{11}c_1\left(\frac{\partial^2 \varphi_x}{\partial x^2}+\frac{\partial^3 w}{\partial x^3}\right)+F_{12}c_1\left(\frac{\partial^2 \varphi_y}{\partial x\partial y}+\frac{\partial^3 w}{\partial x\partial y^2}\right)+F_{16}c_1\left(\frac{\partial^2 \varphi_y}{\partial x^2}+\frac{\partial^2 \varphi_x}{\partial x\partial y}+2\frac{\partial^3 w}{\partial x^2\partial y}\right)$$

$$+B_{16}\frac{\partial^2 u}{\partial x\partial y}+B_{26}\frac{\partial^2 v}{\partial y^2}+B_{66}\left(\frac{\partial^2 u}{\partial y^2}+\frac{\partial^2 v}{\partial x\partial y}\right)+D_{16}\frac{\partial^2 \varphi_x}{\partial x\partial y}+D_{26}\frac{\partial^2 \varphi_y}{\partial y^2}$$

$$+D_{66}\left(\frac{\partial^2 \varphi_x}{\partial y^2}+\frac{\partial^2 \varphi_y}{\partial x\partial y}\right)+F_{16}c_1\left(\frac{\partial^2 \varphi_x}{\partial x\partial y}+\frac{\partial^3 w}{\partial y\partial x^2}\right)+F_{26}c_1\left(\frac{\partial^2 \varphi_y}{\partial y^2}+\frac{\partial^3 w}{\partial y^3}\right)$$

$$+F_{66}c_1\left(\frac{\partial^2 \varphi_y}{\partial x\partial y}+\frac{\partial^2 \varphi_x}{\partial y^2}+2\frac{\partial^3 w}{\partial x\partial y^2}\right)+c_1\left(E_{11}\frac{\partial^2 u}{\partial x^2}+E_{12}\frac{\partial^2 v}{\partial x\partial y}+E_{16}\left(\frac{\partial^2 u}{\partial x\partial y}+\frac{\partial^2 v}{\partial x^2}\right)\right.$$

$$+F_{11}\frac{\partial^2 \varphi_x}{\partial x^2}+F_{12}\frac{\partial^2 \varphi_y}{\partial x\partial y}+F_{16}\left(\frac{\partial^2 \varphi_x}{\partial x\partial y}+\frac{\partial^2 \varphi_y}{\partial x^2}\right)+H_{11}c_1\left(\frac{\partial^2 \varphi_x}{\partial x^2}+\frac{\partial^3 w}{\partial x^3}\right)$$

$$+H_{12}c_1\left(\frac{\partial^2 \varphi_y}{\partial x\partial y}+\frac{\partial^3 w}{\partial x\partial y^2}\right)+H_{16}c_1\left(\frac{\partial^2 \varphi_y}{\partial x^2}+\frac{\partial^2 \varphi_x}{\partial x\partial y}+2\frac{\partial^3 w}{\partial x^2\partial y}\right)+E_{16}\frac{\partial^2 u}{\partial x\partial y}+E_{26}\frac{\partial^2 v}{\partial y^2}$$

$$+E_{66}\left(\frac{\partial^2 u}{\partial y^2}+\frac{\partial^2 v}{\partial x\partial y}\right)+F_{16}\frac{\partial^2 \varphi_x}{\partial x\partial y}+F_{26}\frac{\partial^2 \varphi_y}{\partial y^2}+F_{66}\left(\frac{\partial^2 \varphi_x}{\partial y^2}+\frac{\partial^2 \varphi_y}{\partial x\partial y}\right)$$

$$\left.+H_{16}c_1\left(\frac{\partial^2 \varphi_x}{\partial x\partial y}+\frac{\partial^3 w}{\partial y\partial x^2}\right)+H_{26}c_1\left(\frac{\partial^2 \varphi_y}{\partial y^2}+\frac{\partial^3 w}{\partial y^3}\right)+H_{66}c_1\left(\frac{\partial^2 \varphi_y}{\partial x\partial y}+\frac{\partial^2 \varphi_x}{\partial y^2}+2\frac{\partial^3 w}{\partial x\partial y^2}\right)\right)$$

$$-A_{55}\left(\frac{\partial w}{\partial x}+\varphi_x\right)-A_{45}\left(\frac{\partial w}{\partial y}+\varphi_y\right)-D_{55}c_2\left(\varphi_x+\frac{\partial w}{\partial x}\right)-D_{45}c_2\left(\frac{\partial w}{\partial y}+\varphi_y\right)$$

$$+c_2\left(-D_{55}\left(\frac{\partial w}{\partial x}+\varphi_x\right)-D_{45}\left(\frac{\partial w}{\partial y}+\varphi_y\right)-F_{55}c_2\left(\varphi_x+\frac{\partial w}{\partial x}\right)-F_{45}c_2\left(\frac{\partial w}{\partial y}+\varphi_y\right)\right)$$

$$=J_1\frac{\partial^2 u}{\partial t^2}+K_2\frac{\partial^2 \phi_x}{\partial t^2}-\frac{4}{3h^2}J_4\frac{\partial^3 w}{\partial t^2\partial x}, \tag{6–37}$$

$$B_{16}\frac{\partial^2 u}{\partial x^2}+B_{26}\frac{\partial^2 v}{\partial x\partial y}+B_{66}\left(\frac{\partial^2 u}{\partial x\partial y}+\frac{\partial^2 v}{\partial x^2}\right)+D_{16}\frac{\partial^2 \varphi_x}{\partial x^2}+D_{26}\frac{\partial^2 \varphi_y}{\partial x\partial y}+D_{66}\left(\frac{\partial^2 \varphi_x}{\partial x\partial y}+\frac{\partial^2 \varphi_y}{\partial x^2}\right)$$

$$+F_{16}c_1\left(\frac{\partial^2 \varphi_x}{\partial x^2}+\frac{\partial^3 w}{\partial x^3}\right)+F_{26}c_1\left(\frac{\partial^2 \varphi_y}{\partial x\partial y}+\frac{\partial^3 w}{\partial x\partial y^2}\right)+F_{66}c_1\left(\frac{\partial^2 \varphi_y}{\partial x^2}+\frac{\partial^2 \varphi_x}{\partial x\partial y}+2\frac{\partial^3 w}{\partial x^2\partial y}\right)$$

$$+B_{12}\frac{\partial^2 u}{\partial x\partial y}+B_{22}\frac{\partial^2 v}{\partial y^2}+B_{26}\left(\frac{\partial^2 u}{\partial y^2}+\frac{\partial^2 v}{\partial x\partial y}\right)+D_{12}\frac{\partial^2 \varphi_x}{\partial x\partial y}+D_{22}\frac{\partial^2 \varphi_y}{\partial y^2}$$

$$+D_{26}\left(\frac{\partial^2 \varphi_x}{\partial y^2}+\frac{\partial^2 \varphi_y}{\partial x\partial y}\right)+F_{12}c_1\left(\frac{\partial^2 \varphi_x}{\partial x\partial y}+\frac{\partial^3 w}{\partial y\partial x^2}\right)+F_{22}c_1\left(\frac{\partial^2 \varphi_y}{\partial y^2}+\frac{\partial^3 w}{\partial y^3}\right)$$

$$+F_{26}c_1\left(\frac{\partial^2\varphi_y}{\partial x\partial y}+\frac{\partial^2\varphi_x}{\partial y^2}+2\frac{\partial^3 w}{\partial x\partial y^2}\right)+c_1\left(E_{16}\frac{\partial^2 u}{\partial x^2}+E_{26}\frac{\partial^2 v}{\partial x\partial y}+E_{66}\left(\frac{\partial^2 u}{\partial x\partial y}+\frac{\partial^2 v}{\partial x^2}\right)\right.$$

$$+F_{16}\frac{\partial^2\varphi_x}{\partial x^2}+F_{26}\frac{\partial^2\varphi_y}{\partial x\partial y}+F_{66}\left(\frac{\partial^2\varphi_x}{\partial x\partial y}+\frac{\partial^2\varphi_y}{\partial x^2}\right)+H_{16}c_1\left(\frac{\partial^2\varphi_x}{\partial x^2}+\frac{\partial^3 w}{\partial x^3}\right)$$

$$+H_{26}c_1\left(\frac{\partial^2\varphi_y}{\partial x\partial y}+\frac{\partial^3 w}{\partial x\partial y^2}\right)+H_{66}c_1\left(\frac{\partial^2\varphi_y}{\partial x^2}+\frac{\partial^2\varphi_x}{\partial x\partial y}+2\frac{\partial^3 w}{\partial x^2\partial y}\right)+E_{12}\frac{\partial^2 u}{\partial x\partial y}+E_{22}\frac{\partial^2 v}{\partial y^2}$$

$$+E_{26}\left(\frac{\partial^2 u}{\partial y^2}+\frac{\partial^2 v}{\partial x\partial y}\right)+F_{12}\frac{\partial^2\varphi_x}{\partial x\partial y}+F_{22}\frac{\partial^2\varphi_y}{\partial y^2}+F_{26}\left(\frac{\partial^2\varphi_x}{\partial y^2}+\frac{\partial^2\varphi_y}{\partial x\partial y}\right)$$

$$+H_{12}c_1\left(\frac{\partial^2\varphi_x}{\partial x\partial y}+\frac{\partial^3 w}{\partial y\partial x^2}\right)+H_{22}c_1\left(\frac{\partial^2\varphi_y}{\partial y^2}+\frac{\partial^3 w}{\partial y^3}\right)+H_{26}c_1\left(\frac{\partial^2\varphi_y}{\partial x\partial y}+\frac{\partial^2\varphi_x}{\partial y^2}+2\frac{\partial^3 w}{\partial x\partial y^2}\right)$$

$$-A_{45}\left(\frac{\partial w}{\partial x}+\varphi_x\right)-A_{44}\left(\frac{\partial w}{\partial y}+\varphi_y\right)-D_{45}c_2\left(\varphi_x+\frac{\partial w}{\partial x}\right)-D_{44}c_2\left(\frac{\partial w}{\partial y}+\varphi_y\right)$$

$$\left.+c_2\left(-D_{45}\left(\frac{\partial w}{\partial x}+\varphi_x\right)-D_{44}\left(\frac{\partial w}{\partial y}+\varphi_y\right)-F_{45}c_2\left(\varphi_x+\frac{\partial w}{\partial x}\right)-F_{44}c_2\left(\frac{\partial w}{\partial y}+\varphi_y\right)\right)\right)$$

$$=J_1\frac{\partial^2 v}{\partial t^2}+K_2\frac{\partial^2\phi_y}{\partial t^2}-\frac{4}{3h^2}J_4\frac{\partial^3 w}{\partial t^2\partial y}. \tag{6–38}$$

## 6.3 NUMERICAL RESULTS AND DISCUSSION

A computer program (see Appendix B) is prepared for the vibration of a plate reinforced with $SiO_2$ nanoparticles based on the Navier method presented in Section 2.2.1. $SiO_2$ nanoparticles have a Young's modulus of $E_r = 70\ GPa$ and a Poisson's ratio of $v_r = 0.2$.

### 6.3.1 Validation

In this chapter, to validate the results, the frequency of the structure is obtained by assuming the absence of a soil medium ($K_w = 0$). Therefore, all the mechanical properties and the type of loading are the same as Whitney [23]. So, the nondimensional frequency is considered as $\Omega=\sqrt{\rho h\omega^2 L^4/D_0}$ in which $D_0 = E_1 h^3/(12(1-v_{12}v_{21}))$. The results are compared with five references that have used different solution methods. Whitney [23] used an exact solution while Seçgin and Sarıgül [24] applied a discrete singular convolution approach. The numerical solution method used by Dai et al. [25] and Chen et al. [26] are mesh-free, finite element, and Ritz, respectively. As observed in Table 6.1, the results of the present work are in accordance with the mentioned references.

**TABLE 6.1**
**Validation of Present Study with the Other Works**

| Method | Mode number | | | |
|---|---|---|---|---|
| | 1 | 2 | 3 | 4 |
| Whitney [23] | 15.171 | 33.248 | 44.387 | 60.682 |
| Seçgin and Sarıgül [24] | 15.171 | 33.248 | 44.387 | 60.682 |
| Dai et al. [25] | 15.17 | 33.32 | 44.51 | 60.78 |
| Chen et al. [26] | 15.18 | 33.34 | 44.51 | 60.78 |
| Present | 15.169 | 33.241 | 44.382 | 60.674 |

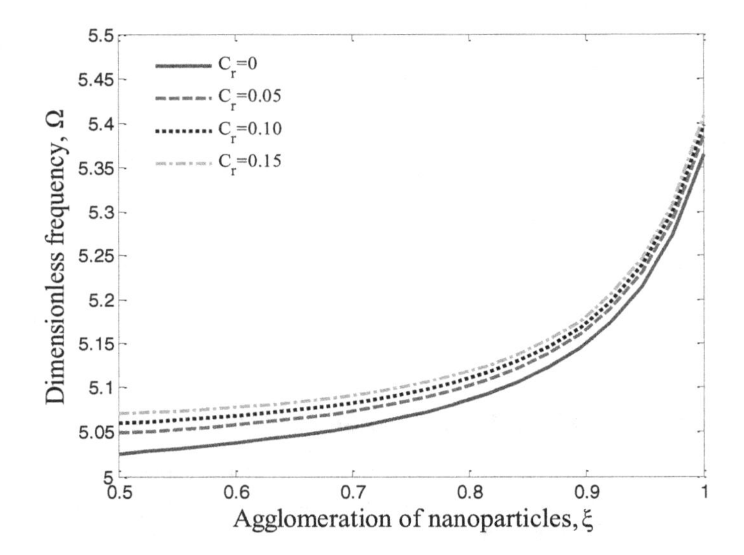


**FIGURE 6.2** Effects of the $SiO_2$ nanoparticle volume percent on the dimensionless frequency versus the agglomeration.

### 6.3.2 Effects of Different Parameters

Figure 6.2 illustrates the effect of the $SiO_2$ nanoparticle volume fraction on the dimensionless frequency of the structure ($\Omega = \omega L\sqrt{\rho_m / E_m}$) versus the agglomeration percent of the $SiO_2$ nanoparticles. It can be seen that by increasing the values of $SiO_2$ nanoparticle volume fraction, the frequency of the system is increased. This is due to the fact that an increase of the $SiO_2$ nanoparticles leads to a harder structure. In addition, by decreasing the agglomeration of the $SiO_2$ nanoparticles ($\xi \rightarrow 1$), the frequency is increased due to the plate having more stability.

The dimensionless frequency of the nanocomposite plate versus agglomeration percent of $SiO_2$ nanoparticles is demonstrated in Figure 6.3 for different soil mediums. In this figure, four cases of loose sand, dense sand, clayey medium-dense sand,

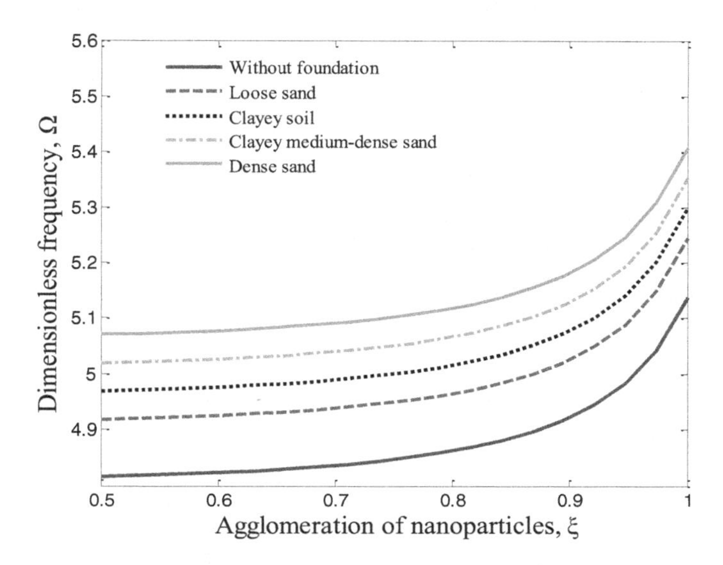


**FIGURE 6.3** Effect of soil medium on the dimensionless frequency versus the agglomeration.

**TABLE 6.2**
**Spring Constants of Different Soils**

| Soil | $K_w$ (N/m$^3$) |
|---|---|
| Loose sand | 4800–16000 |
| Dense sand | 64000–128000 |
| Clayey medium-dense sand | 32000–80000 |
| Clayey soil | 12000–24000 |

and clayey soil are considered with the spring constants of Table 6.2. As can be seen, considering soil medium increases the frequency of the structure. This is due to the fact that considering the soil medium leads to a stiffer structure. Furthermore, the frequency of the dense sand medium is higher than in the other cases since the spring constant of this medium is the maximum.

The effect of the length of the plate on the dimensionless frequency of the system versus the agglomeration percent of $SiO_2$ nanoparticles is depicted in Figure 6.4. As can be seen, the frequency of the structure decreases with an increase in the length of the plate. This is because increasing the length leads to a softer structure.

Figure 6.5 shows the dimensionless frequency of the structure versus the agglomeration percent of $SiO_2$ nanoparticles for different widths of the plate. It can be also found that the frequency of the structure decreases with an increase in the width, which is due to the higher stiffness of the system with a lower width of the plate.

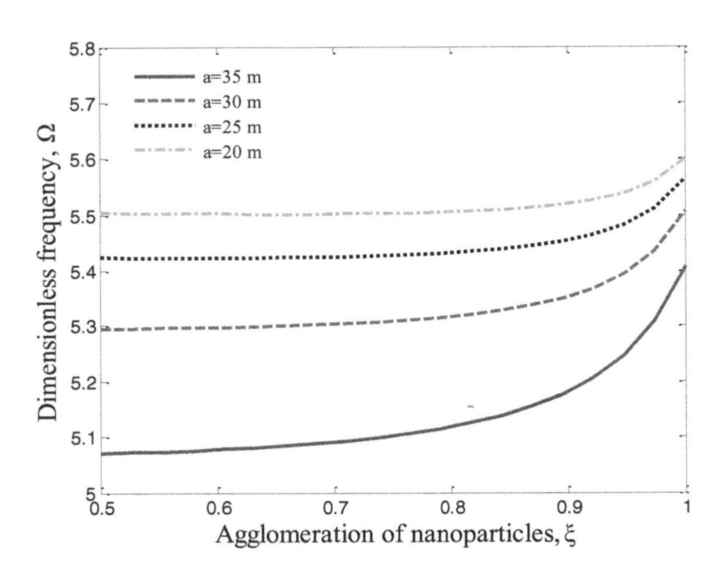


**FIGURE 6.4** Effect of plate length on the dimensionless frequency versus the agglomeration.

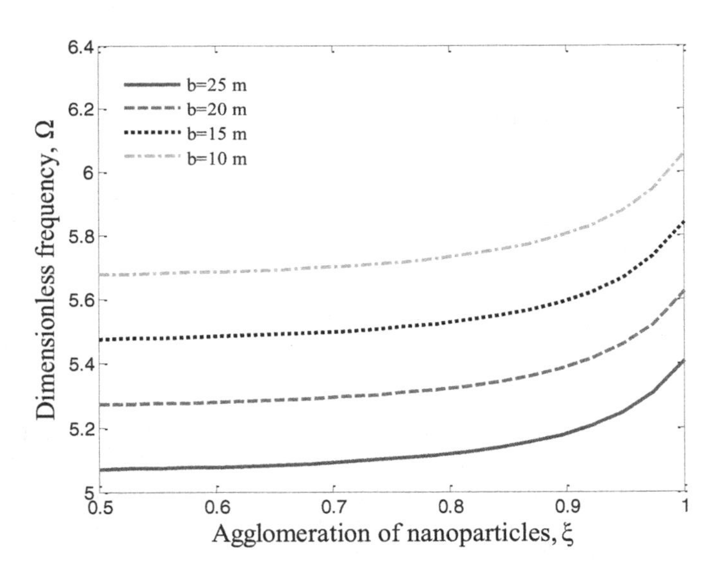


**FIGURE 6.5** Effect of plate width on the dimensionless frequency versus the agglomeration.

The effect of the thickness of the plate on the dimensionless frequency versus the agglomeration percent of $SiO_2$ nanoparticles is shown in Figure 6.6. It can be found that by increasing the thickness, the frequency of the structure is increased. This is because with an increase in the thickness, the stiffness of the structure will be improved.

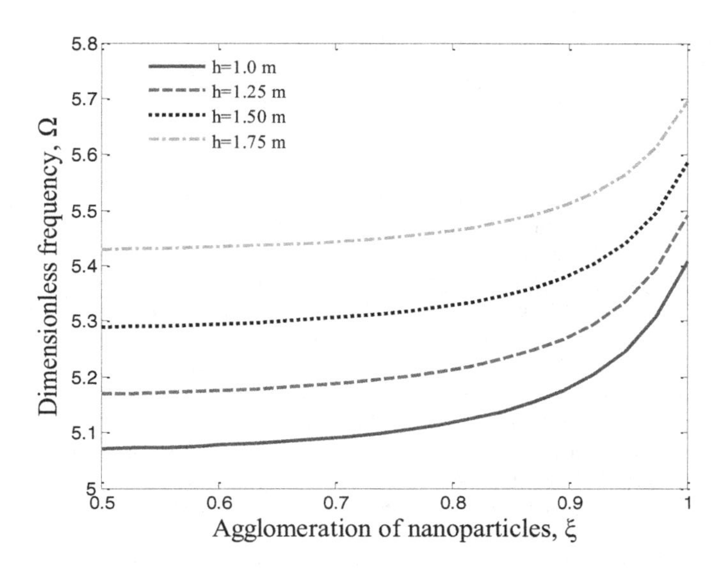


**FIGURE 6.6** Effect of plate thickness on the dimensionless frequency versus agglomeration.

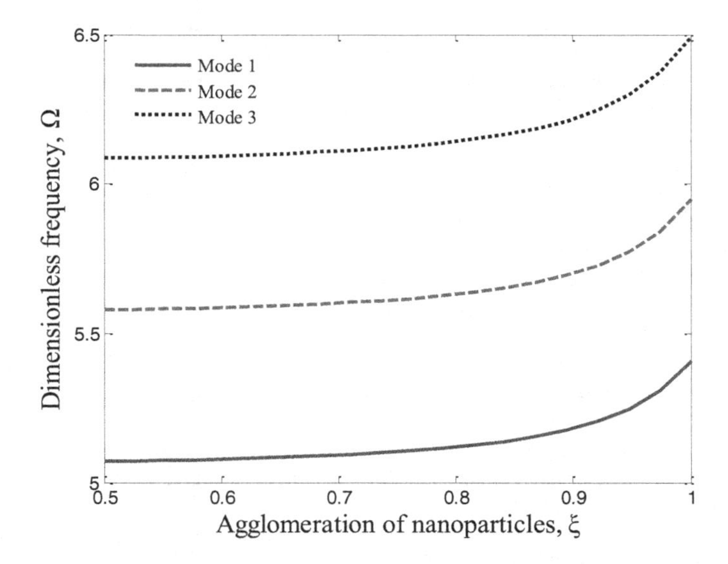


**FIGURE 6.7** Effect of the mode number on the dimensionless frequency versus the agglomeration.

The effect of mode numbers on the dimensionless frequency versus the agglomeration percent of $SiO_2$ nanoparticles of the system is plotted in Figure 6.7. As can be seen, by increasing the mode numbers, the frequency increases.

**Acknowledgments:** This chapter is a slightly modified version of Ref. [20] and has been reproduced here with the permission of the copyright holder.

## REFERENCES

[1] Lam KY, Wang CM, He XQ. Canonical exact solutions for levy-plates on two-parameter foundation using green's functions. *Eng Struct* 2000;22:364–378.

[2] Shen HS, Yang J, Zhang L. Free and forced vibration of Reissner-Mindlin plates with free edges resting on elastic foundations. *J Sound Vib* 2001;244:299–320.

[3] Zhou D, Cheung YK, Lo SH, Au FTK. Three-dimensional vibration analysis of rectangular thick plates on Pasternak foundation. *Int J Num Meth Eng* 2044;59:1313–1334.

[4] Zhong Y, Yin JH. Free vibration analysis of a plate on foundation with completely free boundary by finite integral transform method. *Mech Res Commun* 2008;35:268–275.

[5] Ferreira AJM, Roque CMC, Neves AMA, Jorge RMN, Soares CMM. Analysis of plates on Pasternak foundations by radial basis functions. *Comput Mech* 2010;46:791–803.

[6] Zamani M, Fallah A, Aghdam MM. Free vibration analysis of moderately thick trapezoidal symmetrically laminated plates with various combinations of boundary conditions. *Eur J Mech A/Solids* 2012;36:204–212.

[7] Kumar Y, Lal R. Vibrations of nonhomogeneous orthotropic rectangular plates with bilinear thickness variation resting on Winkler foundation. *Meccanica* 2012;47:893–915.

[8] Thai HT, Park M, Choi DH. A simple refined theory for bending, buckling, and vibration of thick plates resting on elastic foundation. *Int J Mech Sci* 2013;73:40–52.

[9] Thai HT, Choi DH. Levy Solution for free vibration analysis of functionally graded plates based on a refined plate theory. *KSCE J Civ Eng* 2014;18:1813–1824.

[10] Mantari JL, Granados EV. An original FSDT to study advanced composites on elastic foundation. *Thin Wall Struct* 2016;107:80–89.

[11] Uğurlu B. Boundary element method based vibration analysis of elastic bottom plates of fluid storage tanks resting on Pasternak foundation. *Eng Anal Bound Elem* 2016;62:163–176.

[12] Park M, Choi DH. A simplified first-order shear deformation theory for bending, buckling and free vibration analyses of isotropic plates on elastic foundations. *KSCE J Civ Eng* 2018;22:1235–1249.

[13] Jo BW, Kim CH, Lim JH. Investigations on the development of powder with Nano-SiO2 particles. *KSCE J Civ Eng* 2007;11:37–42.

[14] Fathi M, Yousefipour A, Hematpoury Farokhy E. Mechanical and physical properties of expanded polystyrene structural s containing Micro-silica and Nano-silica. *Constr Build Mater* 2017;136:590–597.

[15] Ehsani A, Nili M, Shaabani K. Effect of Nanosilica on the compressive strength development and water absorption properties of cement paste and containing fly ash. *KSCE J Civ Eng* 2017;21:1854–1865.

[16] Jafarian Arani A, Kolahchi R. Buckling analysis of embedded beams armed with carbon nanotubes. *Comput Concr* 2016;17:567–578.

[17] Zamanian M, Kolahchi R, Rabani Bidgoli M. Agglomeration effects on the buckling behavior of embedded beams reinforced with SiO2 nanoparticles. *Wind Struct* 2017;24:43–57.

[18] Arbabi A, Kolahchi R, Rabani Bidgoli M. Concrete columns reinforced with Zinc Oxide nanoparticles subjected to electric field: buckling analysis. *Wind Struct* 2017;24:431–446.

[19] Kammoun Z, Trabelsi A. Mechanical characteristics of a classical lightened by the addition of treated straws. *Adv Concr Constr* 2018;6:375–386.

[20] Alijani M, Bidgoli MR. Agglomerated $SiO_2$ nanoparticles reinforced-concrete foundations based on higher order shear deformation theory: Vibration analysis. *Adv Concrete Construct* 2018;(6)6:585–610.
[21] Reddy JN. *Mechanics of laminated composite plates and shells: Theory and analysis*. Second edition. CRC Press, Boca Raton, FL, 2002.
[22] Mori T, Tanaka K. Average stress in matrix and average elastic energy of materials with misfitting inclusions. *AcM&M* 1973;21:571–574.
[23] Shi DL, Feng XQ, Huang YY, Hwang KC, Gao H. The Effect of nanotube waviness and agglomeration on the elastic property of carbon nanotube-reinforced composites. *J Eng Mater Technol* 2004;126:250–270.
[24] Bowles JE. *Foundation analysis and design*. McGraw-Hill Inc, New York, 1988.
[23] Whitney JM. *Structural analysis of laminated anisotropic plates*. Technomic Publishing Company Inc., Lancaster, PA, 1987.
[24] Seçgin A, Sarıgül AS. Free vibration analysis of symmetrically laminated thin composite plates by using discrete singular convolution (DSC) approach: Algorithm and verification. *J Sound Vib* 2008;315:197–211.
[25] Dai KY, Liu GR, Lim KM, Chen XL. A mesh-free method for static and free vibration analysis of shear deformable laminated composite plates. *J Sound Vib* 2004;269:633–652.
[26] Chen XL, Liu GR, Lim SP. An element free Galerkin method for the free vibration analysis of composite laminates of complicated shape. *Compos Struct* 2003;59:279–289.

# 7 Vibration Analysis of Plates with an NFRP Layer

## 7.1 INTRODUCTION

Concrete plates modeled by a plate element located in a soil elastic foundation are one of the most used structural elements in civil application engineering, such as the footings of buildings, the paving of highways, and the bases of machines. With respect to the point that vibrations of plates create many problems in civil engineering, particularly for tools with high-level functionality and safety, study on this important topic is needed, and hence, vibrations may then be decreased.

There are various methods to decrease the vibration of plates, such as reinforcing with NFRP layers or refining their material properties. Persson et al. [1] presented numerical research on sinking concrete building vibrations using the improvement of concrete foundations. Research in recent years has established that imprisonment with an FRP cover leads to a considerable enhancement in the ductility, strength, and dynamical response of the concrete structure. Wei and Wu [2] presented a combined stress–strain theory of concrete for square, circular, and rectangular concrete columns narrowed by FRP jackets. A finite element analysis of sandstone panels reinforced by FRPs was presented by Grande et al. [3]. They discussed plans, on the basis of the finite element nonlinear analysis, regarding the optimal choice of concrete models and elements for reliable nonlinear analyses of sandstone structures strengthened with FRP strips. Hemmatnezhad et al. [4] presented a numerical, experimental, and analytical study on the free vibrational response of glass fiber-reinforced polymer (GFRP)-hardened cylindrical composite shells. The static and free vibration of strengthened carbon fiber-reinforced polymer (CFRP) concrete beams were examined by Capozucca and Bossoletti [5]. Pan and Wu [6] proposed an analytical method of bond behavior between a concrete plate and FRP layer. The CFRP composite reinforcing influence on the dynamic response of arch dams was examined by Altunisik et al. [7]. The effect of CFRP wraps on the strength of buried steel pipelines subjected to permanent ground deformations was considered by Mokhtari and Alai Nia [8]. The bond performance of FRP reinforced plates superficially attached over steel and concrete elements was measured numerically and experimentally by Ceroni et al. [9]. Seismic strengthening of infilled concrete frames reinforced by CFRP was investigated by Erol and Karadogan [10].

In contrast, recent progress in advanced and innovative materials has provided more emphasis on exploiting reinforcing elements with nanoparticles. One of the perfect uses for the strengthening phase of composite structures is carbon nanotubes (CNTs) due to their high Young's modulus of 1 Tpa and tensile strength of 150 Gpa [11]. Many researchers have studied the mechanical behaviors of CNT-reinforced composite (CNTRC) structures. The nonlinear vibration response of a reinforced composite plate

DOI: 10.1201/9781003349525-7

with CNTs subjected to combined force and parametric excitations were studied by Guo and Zhang [12]. An element-free evaluation of CNTRC plates with restrained elastically edges and column cares subjected to large twist was measured by Zhang et al. [13]. The vibration response of CNTRC laminated thick plates using higher order shear deformation theory (HSDT), as presented by Reddy, was studied by Zhang and Selim [14]. Zhang et al. [15] completed the first-known vibration analysis of functionally graded CNTRC triangular plates under in-plane stresses. Ahmadi et al. [11] explored the multiscale buckling, bending, and vibration of CNT-reinforced nanocomposite polymer plates with different cross shapes. Shen and Wang [16] examined the large- and small-amplitude vibration response of thermally and compressed postbuckled CNTRC polymeric plates located in elastic foundations. Kumar and Srinivas [17] studied the buckling, vibration, and bending behavior of functionally graded CNT–reinforced composite polymer plates on the basis of the layer-wise theory.

In addition, in recent years, various works have been reported on the vibration response of plates located on elastic foundations. Xiang et al. [18] analytically considered the vibration response of Mindlin rectangular plates with simply supported boundary conditions located on Pasternak elastic foundations. The finite element numerical method was applied by Omurtag et al. [19] to investigate the free vibration response of thin concrete plates on a Pasternak elastic foundation. Lam et al. [20] used Green's functions to get recognized exact solutions of elastic buckling, bending, and vibration for Levy plates located in two-parameter elastic foundations. By using the Rayleigh–Ritz analytical method, the 3D vibration response of thick rectangular plates on elastic foundations was examined by Zhou et al. [21]. Utilizing a 3D layer-wise finite element numerical method, evolving the differential quadrature method (DQM), and link a 3D discrete layer method with DQM, Malekzadeh et al. studied the vibration response of rectangular plates embedded in elastic foundations [22, 23]. Ferreira et al. [24] applied the radial basis collocation function solution method to show the free vibration and static deformation of polymeric plates located in Pasternak elastic medium. Kumar and Lal [25] investigated the vibration response of orthotropic nonhomogeneous rectangular plates with bilinear variation thickness on a Winkler medium with spring element. The element-free Galerkin method was applied by Bahmyari and Rahbar-Ranji [26] for vibration analysis of a variable thickness orthotropic polymeric plate on an elastic nonuniform medium. Bahmyari and khedmati [27] measured the vibration response of moderately thick nonhomogeneous plates with point wires located on elastic Pasternak medium based on element-free Galerkin numerical method. A vibrational analysis of advanced composite plates located on an elastic medium was offered by Mantari et al. [28]. They obtained the governing final equations for the functionally graded polymeric plate located in elastic medium by using Hamilton's principle. Nguyen-Thoi et al. [29] studied a smoothed edge-based three-node Mindlin plate theory for free vibration and static response. Uğurlu [30] examined the vibration response of the bottom elastic plates of fluid storage tanks located in elastic Pasternak medium using the element boundary solution method. A basic first order theory to model the composites on the elastic medium was offered by Mantari and Granados [31].

Recently, HSDTs are applied for the modeling of structures and thick plates. One of the HSDTs is sinusoidal shear deformation model. Neves et al. [32, 33] examined a quasi-3D sinusoidal theory for the free vibration and static response of functionally graded polymeric plates. Thai and Kim [34] used the theory of before paper for functionally graded polymeric plates as well. Their work showed that the gained outcomes with this theory are very accurate with respect to other HSDT with complex relations. A novel SSDT for the buckling, bending, and vibration of functionally graded polymeric plates was examined by Thai and Vo [35]. An isogeometric response of functionally graded polymeric plates based on HSTD was offered by Tran et al. [36]. The vibration response of a porous functionally graded cylindrical shell based on SSDT was calculated by Wang and Wu [37].

Dynamic analyses of concrete structures reinforced by nanoparticles using mathematical models are novel topics in the field of civil engineering. Zamani and Bidgoli [38] presented a mathematical model for the vibration analysis of a concrete foundation with a nano-fiber reinforced polymer (NFRP) layer resting on a soil medium. The NFRP layer incorporates nano-fibers made from carbon nanotubes (CNT). The effective material properties of the nano-composite structure are determined using the Mori-Tanaka model, taking into consideration agglomeration effects. They investigated the buckling response of CNT–reinforced concrete columns based on Timoshenko and Euler–Bernoulli beam theories. As far as the authors know, the mathematical model for concrete plates reinforced by a NFRP layer has not been reported yet. However, because of a lack of literature on this subject, the vibration response of concrete plates retrofitted by NFRP layer located in a soil elastic foundation is studied and presented in this chapter. The soil elastic foundation is modeled by spring elements of Winkler theory. To obtain the corresponding material characteristics of nanocomposite concrete structure, the Mori–Tanaka theory is utilized to study the agglomeration influences of nanofibers. Using SSDT, the motion equations are obtained on the basis of Hamilton's principle. In addition, the Navier solution method is used to get the frequency of the concrete structure. NFRP layers have long- and short-term influences on the lifetime and strength of the concrete structures respectively. However, the purpose of this investigation is to present the influences of an NFRP layer on the vibration analysis of concrete plates as well as showing the influences of the soil elastic foundation, agglomeration and volume percent of CNTs, the geometrical parameters and structural damping on the frequency response of the structure.

## 7.2 STRESS–STRAIN RELATIONS

Figure 7.1 shows a plate with NFRP layer with width $b$, length $L$, and thickness of $2h$, which is covered by an NFRP layer with a thickness $h_f$.

Using SSDT, the deflection field, which is presented in Section 1.4.4, may be applied in this chapter. On the basis of Hook's theory, the stress–strain equations for the plate may be given as

$$\begin{bmatrix} \sigma_{xx} \\ \sigma_{yy} \\ \sigma_{zz} \\ \sigma_{zy} \\ \sigma_{xz} \\ \sigma_{zy} \end{bmatrix} = \begin{bmatrix} C_{11} & C_{12} & C_{13} & 0 & 0 & 0 \\ C_{12} & C_{22} & C_{23} & 0 & 0 & 0 \\ C_{13} & C_{23} & C_{33} & 0 & 0 & 0 \\ 0 & 0 & 0 & C_{44} & 0 & 0 \\ 0 & 0 & 0 & 0 & C_{55} & 0 \\ 0 & 0 & 0 & 0 & 0 & C_{66} \end{bmatrix} \begin{bmatrix} \varepsilon_{xx} \\ \varepsilon_{yy} \\ \varepsilon_{zz} \\ \gamma_{zy} \\ \gamma_{xz} \\ \gamma_{xy} \end{bmatrix}, \tag{7–1}$$

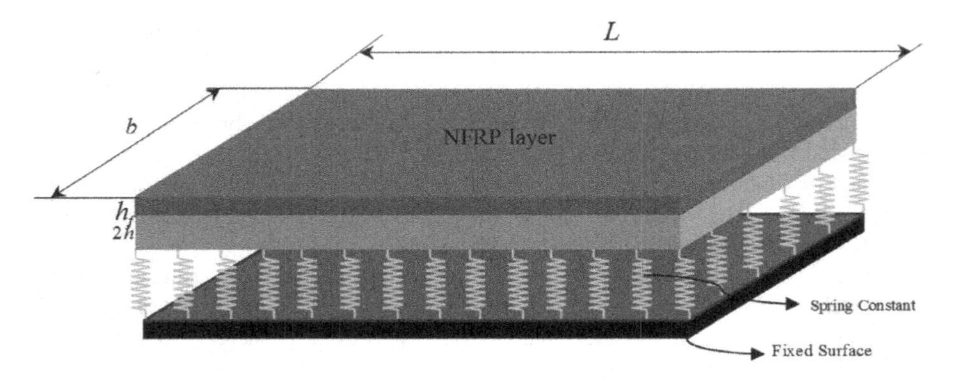


**FIGURE 7.1** Schematic of a concrete plate reinforced by NFRP layer. Reprinted with permission from Elsevier.

where $C_{ij}$ are elastic coefficients. Furthermore, the stress–strain equations of the NFRP layer are:

$$\begin{Bmatrix} \sigma_{11} \\ \sigma_{22} \\ \sigma_{33} \\ \sigma_{23} \\ \sigma_{13} \\ \sigma_{12} \end{Bmatrix} = \begin{bmatrix} \underbrace{k+m}_{Q_{11}} & \underbrace{l}_{Q_{12}} & \underbrace{k-m}_{Q_{13}} & 0 & 0 & 0 \\ \underbrace{l}_{Q_{12}} & \underbrace{n}_{Q_{22}} & \underbrace{l}_{Q_{23}} & 0 & 0 & 0 \\ \underbrace{k-m}_{Q_{31}} & \underbrace{l}_{Q_{32}} & \underbrace{k+m}_{Q_{33}} & 0 & 0 & 0 \\ 0 & 0 & 0 & \underbrace{p}_{Q_{44}} & 0 & 0 \\ 0 & 0 & 0 & 0 & \underbrace{m}_{Q_{55}} & 0 \\ 0 & 0 & 0 & 0 & 0 & \underbrace{p}_{Q_{66}} \end{bmatrix} \begin{Bmatrix} \varepsilon_{11} \\ \varepsilon_{22} \\ \varepsilon_{33} \\ \gamma_{23} \\ \gamma_{13} \\ \gamma_{12} \end{Bmatrix}, \tag{7–2}$$

Where $\sigma_{ij}, \varepsilon_{ij}, \gamma_{ij}, k, m, n, l, p$ are the stress, strain, and the stiffness constants, respectively. According to the Mori–Tanaka theory, the stiffness constants may be calculated (see Section 4.3).

## 7.3 ENERGY METHOD

The total potential energy for the structure may be given as

$$U = \frac{1}{2}\int_A \int_{-\frac{h}{2}}^{\frac{h}{2}} \left(\sigma_{xx}\varepsilon_{xx} + \sigma_{yy}\varepsilon_{yy} + \sigma_{xy}\gamma_{xy} + \sigma_{xz}\gamma_{xz} + \sigma_{yz}\gamma_{yz}\right) dzdA. \tag{7–3}$$

Combining Equations 1–36 and 1–37 with Equation 7–3 yields

$$U=\frac{1}{2}\int_A\left(N_{xx}\frac{\partial U}{\partial x}+N_{xy}\frac{\partial U}{\partial y}+N_{xy}\frac{\partial V}{\partial x}+N_{yy}\frac{\partial V}{\partial y}+Q_x\frac{\partial W_s}{\partial x}\right.$$
$$+Q_y\frac{\partial W_s}{\partial y}-M_{xxS}\frac{\partial^2 W_s}{\partial x^2}-M_{yyS}\frac{\partial^2 W_s}{\partial y^2}-2M_{xyS}\frac{\partial^2 W_s}{\partial y\partial x} \quad (7\text{–}4)$$
$$\left.-M_{xxB}\frac{\partial^2 W_b}{\partial x^2}-M_{yyB}\frac{\partial^2 W_b}{\partial y^2}-2M_{xyB}\frac{\partial^2 W_b}{\partial y\partial x}\right)\mathrm{dA},$$

in which the stress resultants in term of displacement may be given as

$$\begin{bmatrix}N_{xx}\\N_{yy}\\N_{xy}\end{bmatrix}=\int_{-h}^{h}\begin{bmatrix}\sigma_{xx}\\\sigma_{yy}\\\sigma_{xy}\end{bmatrix}dz+\int_{h}^{h+h_f}\begin{bmatrix}\sigma_{xx}^f\\\sigma_{yy}^f\\\sigma_{xy}^f\end{bmatrix}dz \quad (7\text{–}5)$$

$$\begin{bmatrix}M_{xxB}\\M_{yyB}\\M_{xyB}\end{bmatrix}=\int_{-h}^{h}\begin{bmatrix}\sigma_{xx}\\\sigma_{yy}\\\sigma_{xy}\end{bmatrix}zdz+\int_{h}^{h+h_f}\begin{bmatrix}\sigma_{xx}^f\\\sigma_{yy}^f\\\sigma_{xy}^f\end{bmatrix}zdz \quad (7\text{–}6)$$

$$\begin{bmatrix}M_{xxS}\\M_{yyS}\\M_{xyS}\end{bmatrix}=\int_{-h}^{h}\begin{bmatrix}\sigma_{xx}\\\sigma_{yy}\\\sigma_{xy}\end{bmatrix}fdz+\int_{h}^{h+h_f}\begin{bmatrix}\sigma_{xx}^f\\\sigma_{yy}^f\\\sigma_{xy}^f\end{bmatrix}fdz \quad (7\text{–}7)$$

$$\begin{bmatrix}Q_x\\Q_y\end{bmatrix}=\int_{-h}^{h}\begin{bmatrix}\sigma_{xx}\\\sigma_{yy}\end{bmatrix}pdz+\int_{-h}^{h+h_f}\begin{bmatrix}\sigma_{xx}\\\sigma_{yy}\end{bmatrix}pdz. \quad (7\text{–}8)$$

Substituting stress–strain equations into Equations 7–5 through 7–8, the stress resultant in terms of deflection may be given as:

$$N_{xx}=A_{11}\frac{\partial}{\partial x}U-A_{11z}\frac{\partial^2}{\partial x^2}W_b-A_{11f}\frac{\partial^2}{\partial x^2}W_s+A_{12}\frac{\partial}{\partial y}V$$
$$-A_{12z}\frac{\partial^2}{\partial y^2}W_b-A_{12f}\frac{\partial^2}{\partial y^2}W_s, \quad (7\text{–}9)$$

$$N_{yy}=A_{21}\frac{\partial}{\partial x}U-A_{21z}\frac{\partial^2}{\partial x^2}W_b-A_{21f}\frac{\partial^2}{\partial x^2}W_s+A_{22}\frac{\partial}{\partial y}V$$
$$-A_{22z}\frac{\partial^2}{\partial y^2}W_b-A_{22f}\frac{\partial^2}{\partial y^2}W_s, \quad (7\text{–}10)$$

$$N_{xy} = A_{44}\frac{\partial}{\partial y}U + A_{44}\frac{\partial}{\partial x}V - 2A_{44z}\frac{\partial^2}{\partial x\partial y}W_b - 2A_{44f}\frac{\partial^2}{\partial x\partial y}W, \tag{7–11}$$

$$Q_x = A_{55g}\frac{\partial}{\partial x}W_s + GA_{55g}\frac{\partial^2}{\partial x\partial t}W_s, \tag{7–12}$$

$$Q_y = A_{66g}\frac{\partial}{\partial y}W_s + GA_{66g}\frac{\partial^2}{\partial y\partial t}W_s, \tag{7–13}$$

$$M_{xxB} = A_{11z}\frac{\partial}{\partial x}U - B_{11}\frac{\partial^2}{\partial x^2}W_b - A_{11zf}\frac{\partial^2}{\partial x^2}W_s + A_{12z}\frac{\partial}{\partial y}V - B_{12}\frac{\partial^2}{\partial y^2}W_b - A_{12zf}\frac{\partial^2}{\partial y^2}W_s, \tag{7–14}$$

$$M_{xxS} = A_{11f}\frac{\partial}{\partial x}U - A_{11zf}\frac{\partial^2}{\partial x^2}W_b - E_{11}\frac{\partial^2}{\partial x^2}W_s + A_{12f}\frac{\partial}{\partial y}V - A_{12zf}\frac{\partial^2}{\partial y^2}W_b - E_{12}\frac{\partial^2}{\partial y^2}W_s, \tag{7–15}$$

$$M_{yyB} = A_{21z}\frac{\partial}{\partial x}U - B_{21}\frac{\partial^2}{\partial x^2}W_b - A_{21zf}\frac{\partial^2}{\partial x^2}W_s + A_{22z}\frac{\partial}{\partial y}V - B_{22}\frac{\partial^2}{\partial y^2}W_b - A_{22zf}\frac{\partial^2}{\partial y^2}W_s, \tag{7–16}$$

$$M_{yyS} = A_{21f}\frac{\partial}{\partial x}U - A_{21zf}\frac{\partial^2}{\partial x^2}W_b - E_{21}\frac{\partial^2}{\partial x^2}W_s + A_{22f}\frac{\partial}{\partial y}V - A_{22zf}\frac{\partial^2}{\partial y^2}W_b - E_{22}\frac{\partial^2}{\partial y^2}W_s, \tag{7–17}$$

$$M_{xyB} = 2A_{44z}\frac{\partial}{\partial y}U + 2A_{44z}\frac{\partial}{\partial x}V - 2B_{44}\frac{\partial^2}{\partial x\partial y}W_b - 2A_{44zf}\frac{\partial^2}{\partial x\partial y}W_s, \tag{7–18}$$

$$M_{xyS} = 2A_{44f}\frac{\partial}{\partial y}U + 2A_{44f}\frac{\partial}{\partial x}V - 2A_{44zf}\frac{\partial^2}{\partial x\partial y}W_b - 2E_{44}\frac{\partial^2}{\partial x\partial y}W_s, \tag{7–19}$$

where

$$\left(A_{11}, A_{12}, A_{22}, A_{44}\right) = \int_{-h}^{h}\left(C_{11}, C_{12}, C_{22}, C_{44}\right)dz + \int_{h}^{h+h_f}\left(Q_{11}, Q_{12}, Q_{22}, Q_{44}\right)dz, \tag{7–20}$$

$$\left(A_{11z}, A_{12z}, A_{22z}, A_{44z}\right) = \int_{-h}^{h}\left(C_{11}, C_{12}, C_{22}, C_{44}\right)z\,dz + \int_{h}^{h+h_f}\left(Q_{11}, Q_{12}, Q_{22}, Q_{44}\right)z\,dz, \tag{7–21}$$

$$\left(A_{11f},A_{12f},A_{22f},A_{44f}\right)=\int_{-h}^{h}\left(C_{11},C_{12},C_{22},C_{44}\right)fdz+\int_{h}^{h+h_f}\left(Q_{11},Q_{12},Q_{22},Q_{44}\right)fdz, \tag{7–22}$$

$$\left(A_{11zf},A_{12zf},A_{22zf},A_{44zf}\right)=\int_{-h}^{h}\left(C_{11},C_{12},C_{22},C_{44}\right)zfdz+\int_{h}^{h+h_f}\left(Q_{11},Q_{12},Q_{22},Q_{44}\right)zfdz, \tag{7–23}$$

$$\left(A_{55g},A_{66g}\right)=\int_{-h}^{h}\left(C_{55},C_{66}\right)gdz+\int_{h}^{h+h_f}\left(Q_{55},Q_{66}\right)gdz, \tag{7–24}$$

$$\left(B_{11},B_{12},B_{22},B_{44}\right)=\int_{-h}^{h}\left(C_{11},C_{12},C_{22},C_{44}\right)z^2dz+\int_{h}^{h+h_f}\left(Q_{11},Q_{12},Q_{22},Q_{44}\right)z^2dz, \tag{7–25}$$

$$\left(E_{11},E_{12},E_{22},E_{44}\right)=\int_{-h}^{h}\left(C_{11},C_{12},C_{22},C_{44}\right)f^2dz+\int_{h}^{h+h_f}\left(Q_{11},Q_{12},Q_{22},Q_{44}\right)f^2dz. \tag{7–26}$$

The total kinetic energy for the structure may be presented as

$$K=\frac{1}{2}\rho\int_A\int_{-\frac{h}{2}}^{\frac{h}{2}}\left(\left(\frac{\partial u}{\partial t}\right)^2+\left(\frac{\partial v}{\partial t}\right)^2+\left(\frac{\partial w}{\partial t}\right)^2\right)\mathrm{dz\,dA}. \tag{7–27}$$

The external force work induced by Pasternak elastic foundation may be given as

$$W_e=\int\int\left(-K_w w\right)wdA, \tag{7–28}$$

in which $K_w$ is spring Winkler constant. The governing final equations may be calculated by Hamilton's principle as:

$$\int_0^t(\delta U-\delta K-\delta W_e)dt=0. \tag{7–29}$$

Substituting Equations 7–4, 7–27, and 7–28 into Equation 7–29 yields the below final governing equations:

$$\frac{\partial}{\partial x}N_{xx}+\frac{\partial}{\partial y}N_{xy}-I_0\frac{\partial^2 U}{\partial t^2}+I_1\frac{\partial^3 W_b}{\partial x\partial t^2}+J_1\frac{\partial^3 W_s}{\partial x\partial t^2}=0, \tag{7–30}$$

$$\frac{\partial}{\partial x}N_{xy}+\frac{\partial}{\partial y}N_{yy}-I_0\frac{\partial^2 V}{\partial t^2}+I_1\frac{\partial^3 W_b}{\partial y\partial t^2}+J_1\frac{\partial^3 W_s}{\partial y\partial t^2}=0, \tag{7–31}$$

$$\frac{\partial^2}{\partial x^2} M_{xxB} + 2\frac{\partial^2}{\partial x \partial y} M_{xyB} + \frac{\partial^2}{\partial y^2} M_{yyB} - K_w w - I_0\left(\frac{\partial^2 W_b}{\partial t^2} + \frac{\partial^2 W_s}{\partial t^2}\right)$$
$$-I_1\left(\frac{\partial^3 U}{\partial x \partial t^2} + \frac{\partial^3 V}{\partial y \partial t^2}\right) + I_2\left(\frac{\partial^4 W_b}{\partial x^2 \partial t^2} + \frac{\partial^4 W_b}{\partial y^2 \partial t^2}\right)$$
$$+J_2\left(\frac{\partial^4 W_s}{\partial x^2 \partial t^2} + \frac{\partial^4 W_s}{\partial y^2 \partial t^2}\right) = 0, \tag{7–32}$$

$$\frac{\partial^2}{\partial x^2} M_{xxS} + 2\frac{\partial^2}{\partial x \partial y} M_{xyS} + \frac{\partial^2}{\partial y^2} M_{yyS} + \frac{\partial}{\partial x} Q_x + \frac{\partial}{\partial y} Q_y$$
$$-K_w w - I_0\left(\frac{\partial^2 W_b}{\partial t^2} + \frac{\partial^2 W_s}{\partial t^2}\right) - J_1\left(\frac{\partial^3 U}{\partial x \partial t^2} + \frac{\partial^3 V}{\partial y \partial t^2}\right)$$
$$+J_2\left(\frac{\partial^4 W_b}{\partial x^2 \partial t^2} + \frac{\partial^4 W_b}{\partial y^2 \partial t^2}\right) + K_2\left(\frac{\partial^4 W_s}{\partial x^2 \partial t^2} + \frac{\partial^4 W_s}{\partial y^2 \partial t^2}\right) = 0, \tag{7–33}$$

where

$$\left(I_0, I_1, I_2, J_1, J_1, K_2\right) = \int_{-h}^{h} \rho\left(1, z, f, zf, z^2, f^2\right)$$
$$+\int_{h}^{h+h_f} \rho_f\left(1, z, f, zf, z^2, f^2\right) dz. \tag{7–34}$$

Substituting Equations 7–9 through 7–19 into Equations 7–30 through 7–33, the governing final equations may be presented as:

$$\frac{\partial}{\partial x}\left(A_{11}\frac{\partial}{\partial x}U - A_{11z}\frac{\partial^2}{\partial x^2}W_b - A_{11f}\frac{\partial^2}{\partial x^2}W_s + A_{12}\frac{\partial}{\partial y}V - A_{12z}\frac{\partial^2}{\partial y^2}W_b - A_{12f}\frac{\partial^2}{\partial y^2}W_s\right)$$
$$+\frac{\partial}{\partial y}\left(A_{44}\frac{\partial}{\partial y}U + A_{44}\frac{\partial}{\partial x}V - 2A_{44z}\frac{\partial^2}{\partial x \partial y}W_b - 2A_{44f}\frac{\partial^2}{\partial x \partial y}W_s\right)$$
$$= \left[I_0\frac{\partial^2 U}{\partial t^2} - I_1\frac{\partial^3 W_b}{\partial x \partial t^2} - J_1\frac{\partial^3 W_s}{\partial x \partial t^2}\right], \tag{7–35}$$

$$\frac{\partial}{\partial x}\left(A_{44}\frac{\partial}{\partial y}U + A_{44}\frac{\partial}{\partial x}V - 2A_{44z}\frac{\partial^2}{\partial x \partial y}W_b - 2A_{44f}\frac{\partial^2}{\partial x \partial y}W_s\right)$$
$$+\frac{\partial}{\partial y}\left(A_{21}\frac{\partial}{\partial x}U - A_{21z}\frac{\partial^2}{\partial x^2}W_b - A_{21f}\frac{\partial^2}{\partial x^2}W_s + A_{22}\frac{\partial}{\partial y}V - A_{22z}\frac{\partial^2}{\partial y^2}W_b - A_{22f}\frac{\partial^2}{\partial y^2}W_s\right)$$
$$= \left[I_0\frac{\partial^2 V}{\partial t^2} - I_1\frac{\partial^3 W_b}{\partial y \partial t^2} - J_1\frac{\partial^3 W_s}{\partial y \partial t^2}\right], \tag{7–36}$$

$$\frac{\partial^2}{\partial x^2}\left(A_{11z}\frac{\partial}{\partial x}U - B_{11}\frac{\partial^2}{\partial x^2}W_b - A_{11zf}\frac{\partial^2}{\partial x^2}W_s + A_{12z}\frac{\partial}{\partial y}V - B_{12}\frac{\partial^2}{\partial y^2}W_b - A_{12zf}\frac{\partial^2}{\partial y^2}W_s\right)$$
$$+2\frac{\partial^2}{\partial x\partial y}\left(2A_{44z}\frac{\partial}{\partial y}U + 2A_{44z}\frac{\partial}{\partial x}V - 2B_{44}\frac{\partial^2}{\partial x\partial y}W_b - 2A_{44zf}\frac{\partial^2}{\partial x\partial y}W_s\right)$$
$$+\frac{\partial^2}{\partial y^2}\left(A_{21z}\frac{\partial}{\partial x}U - B_{21}\frac{\partial^2}{\partial x^2}W_b - A_{21zf}\frac{\partial^2}{\partial x^2}W_s + A_{22z}\frac{\partial}{\partial y}V - B_{22}\frac{\partial^2}{\partial y^2}W_b - A_{22zf}\frac{\partial^2}{\partial y^2}W_s\right)$$
$$-K_w W - K_w W_s = I_0\left(\frac{\partial^2 W_b}{\partial t^2} + \frac{\partial^2 W_s}{\partial t^2}\right) + I_1\left(\frac{\partial^3 U}{\partial x\partial t^2} + \frac{\partial^3 V}{\partial y\partial t^2}\right)$$
$$-I_2\left(\frac{\partial^4 W_b}{\partial x^2\partial t^2} + \frac{\partial^4 W_b}{\partial y^2\partial t^2}\right) - J_2\left(\frac{\partial^4 W_s}{\partial x^2\partial t^2} + \frac{\partial^4 W_s}{\partial y^2\partial t^2}\right)\Bigg], \tag{7–37}$$

$$\frac{\partial^2}{\partial x^2}\left(A_{11f}\frac{\partial}{\partial x}U - A_{11zf}\frac{\partial^2}{\partial x^2}W_b - E_{11}\frac{\partial^2}{\partial x^2}W_s + A_{12f}\frac{\partial}{\partial y}V - A_{12zf}\frac{\partial^2}{\partial y^2}W_b - E_{12}\frac{\partial^2}{\partial y^2}W_s\right)$$
$$+2\frac{\partial^2}{\partial x\partial y}\left(2A_{44f}\frac{\partial}{\partial y}U + 2A_{44f}\frac{\partial}{\partial x}V - 2A_{44zf}\frac{\partial^2}{\partial x\partial y}W_b - 2E_{44}\frac{\partial^2}{\partial x\partial y}W\right)$$
$$+\frac{\partial^2}{\partial y^2}\left(A_{21f}\frac{\partial}{\partial x}U - A_{21zf}\frac{\partial^2}{\partial x^2}W_b - E_{21}\frac{\partial^2}{\partial x^2}W_s + A_{22f}\frac{\partial}{\partial y}V - A_{22zf}\frac{\partial^2}{\partial y^2}W_b - E_{22}\frac{\partial^2}{\partial y^2}W_s + \right)$$
$$+\frac{\partial}{\partial x}\left(A_{55g}\frac{\partial}{\partial x}W_{ss}\right) + \frac{\partial}{\partial y}\left(A_{66g}\frac{\partial}{\partial y}W_s\right) + -K_w W - K_w W_s = I_0\left(\frac{\partial^2 W_b}{\partial t^2} + \frac{\partial^2 W_s}{\partial t^2}\right)$$
$$+J_1\left(\frac{\partial^3 U}{\partial x\partial t^2} + \frac{\partial^3 V}{\partial y\partial t^2}\right) - J_2\left(\frac{\partial^4 W_b}{\partial x^2\partial t^2} + \frac{\partial^4 W_b}{\partial y^2\partial t^2}\right) - K_2\left(\frac{\partial^4 W_s}{\partial x^2\partial t^2} + \frac{\partial^4 W_s}{\partial y^2\partial t^2}\right)\Bigg]. \tag{7–38}$$

## 7.4 NUMERICAL RESULTS AND DISCUSSION

In order to show the influence of various parameters on the frequency response of the structure, a plate based on Navier method (please see Appendix C) with a Young's modulus of $E_c = 20\ GPa$ and a Poisson's ratio of $v_c = 0.2$ is considered. The Young's modulus and Poisson's ratio of polymer for the NFRP layer is $E_m = 18\ GPa$ and $v_m = 0.3$, respectively which is armed by CNTs with a Young's modulus and Poisson's ratio of $E_r = 1\ TPa$ and $v_r = 0.3$, respectively.

Figure 7.2 demonstrates the influence of the CNT's volume percent on the non-dimensional frequency ($\Omega = \omega L\sqrt{\rho_m / E_m}$) as a function of the NFRP–to–plate thickness ratio. It may be found that by enhancing the CNT's volume percent, the non-dimensional frequency is improved. This is because the increase in CNT's volume percent leads to a stiffer system. It should also be noted that by increasing the NFRP–to–plate thickness ratio, the non-dimensional frequency is reduced for $h_f / h < 0.005$ and amplified for $h_f / h > 0.005$. It is perhaps due to this fact that for

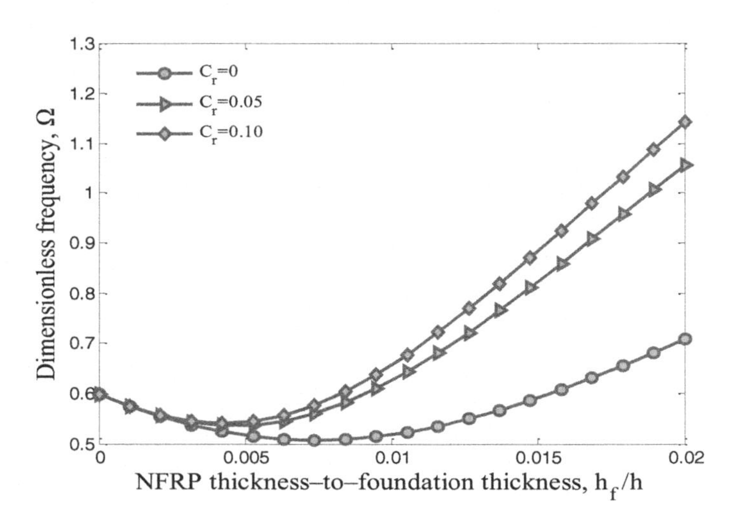


**FIGURE 7.2** Influence of CNT's volume percent on the non-dimensional frequency as a function of the NFRP layer–to–plate thickness ratio.

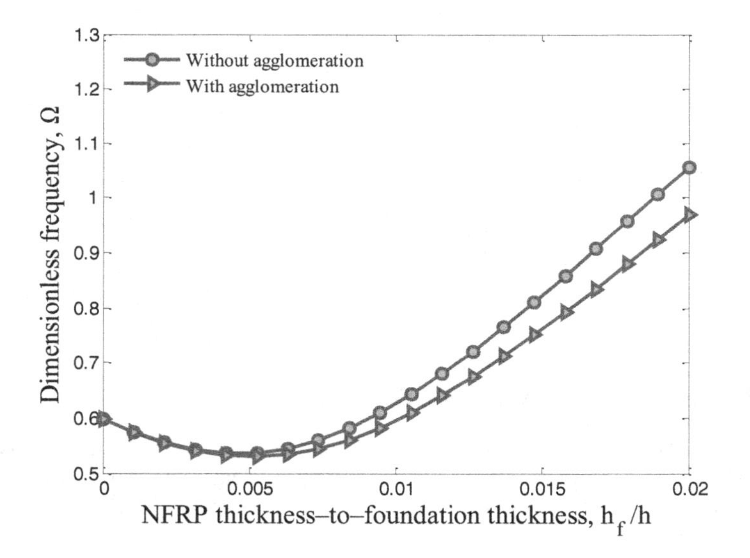


**FIGURE 7.3** Influence of CNT's agglomeration on the non-dimensional frequency as a function of the NFRP layer–to–plate thickness ratio.

$h_f / h < 0.005$, the negative influence of the CNT's agglomeration is greater than the influence of the NFRP layer, while for $h_f / h > 0.005$, this is converce (the NFRP layer displays a positive influence on the non-dimensional frequency).

Figure 7.3 shows the CNT's agglomeration on the non-dimensional frequency as a function of the NFRP–to–plate thickness ratio. It can be seen that assuming CNT's agglomeration leads to decrease of frequency due to the fact that the stiffness and stability of the structure falls. Also, the effect of the CNT's agglomeration on the non-dimensional frequency enhances with a rise in the NFRP–to–plate thickness ratio.

The non-dimensional frequency for the nanocomposite concrete plate as a function of the NFRP–to–plate thickness ratio for various soil elastic foundations is established in Figure 7.4.

In the figure four cases of dense sand, loose sand, clayey soil and clayey medium-dense sand are assumed with the spring coefficients listed in Table 7.1.

As may be observed, assuming soil elastic foundation enhances the frequency of the concrete nanocomposite plate. It is since assuming soil elastic foundation makes the structure stiffer. Additionally, the non-dimensional frequency of the structure with dense sand medium is greater than other cases due to high spring values of of this foundation.

The influence of the concrete plate length on the non-dimensional frequency of the concrete system as a function of the NFRP–to–plate thickness ratio is described

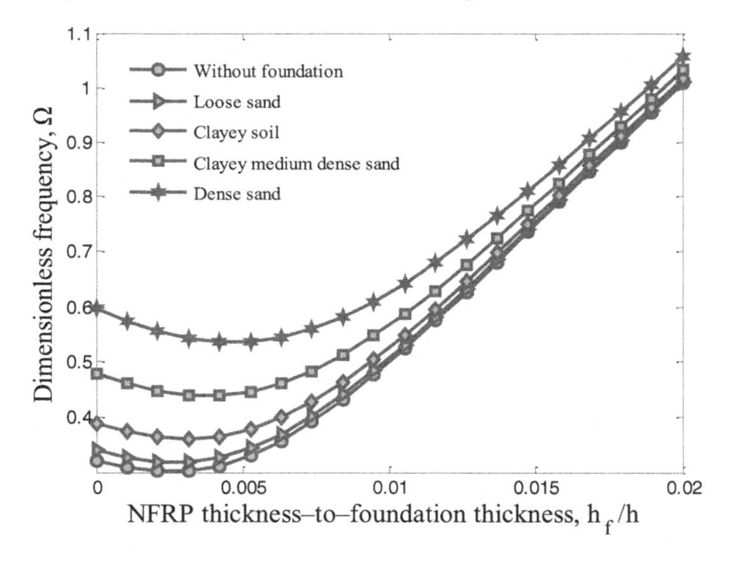


**FIGURE 7.4** Influence of the soil medium on the non-dimensional frequency as a function of the NFRP layer–to–plate thickness ratio.

**TABLE 7.1**
**Spring Constants of Soil Mediums**

| Soil | $K_w$ |
|---|---|
| Loose sand | 4800–16000 |
| Dense sand | 64000–128000 |
| Clayey medium-dense sand | 32000–80000 |
| Clayey soil | 12000–24000 |

in Figure 7.5. As can be seen, the non-dimensional frequency of the concrete structure reduces with an enhance in the length of plate. It is since enhancing the length of plate makes the structure softer. Moreover, the reducing rate of non-dimensional frequency is declined with the enhance of length.

Figure 7.6 demonstrations the non-dimensional frequency of the concrete structure as a function of NFRP–to–plate thickness ratio for the various plate widths. It can be seen that the non-dimensional frequency reduces with an increase in the plate widths, which is due to this fact that the structure has greater stiffness with worse plate width.

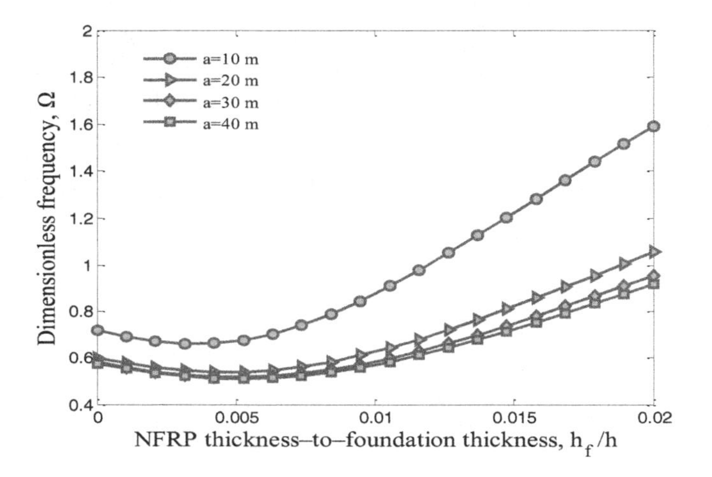


**FIGURE 7.5** Influence of the length of a plate on the non-dimensional frequency as a function of the NFRP–to–plate thickness ratio.

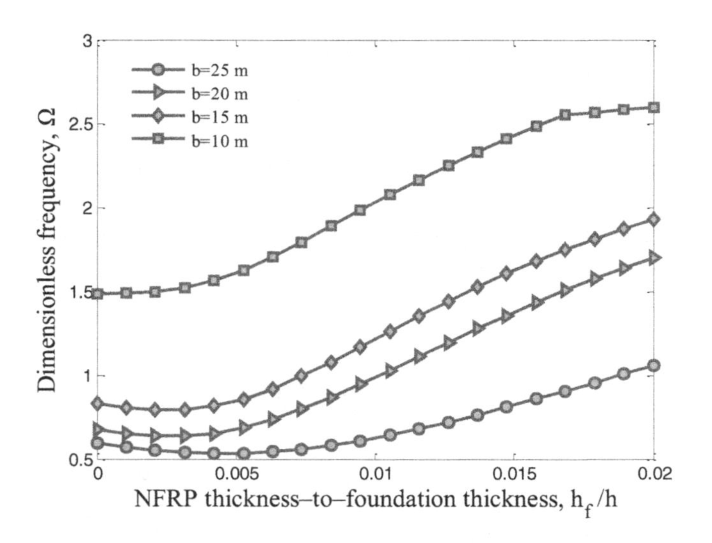


**FIGURE 7.6** Influence of the plate width on the non-dimensional frequency as a function of the NFRP–to–plate thickness ratio.

The influence of plate length on the thickness ratio on the non-dimensional frequency as a function of the NFRP–to–plate thickness ratio is shown in Figure 7.7. It may be seen that with an increase in the length-to-thickness ratio, the non-dimensional frequency is reduced. It is since with a growth in the plate length on the thickness ratio, the stiffness is reduced and decreased.

The influence of the mode numbers on the non-dimensional frequency as a function of the NFRP–to–plate thickness ratio is presented in Figure 7.8. As may be found, by enhancing the mode shape numbers, the non-dimensional frequency rises.

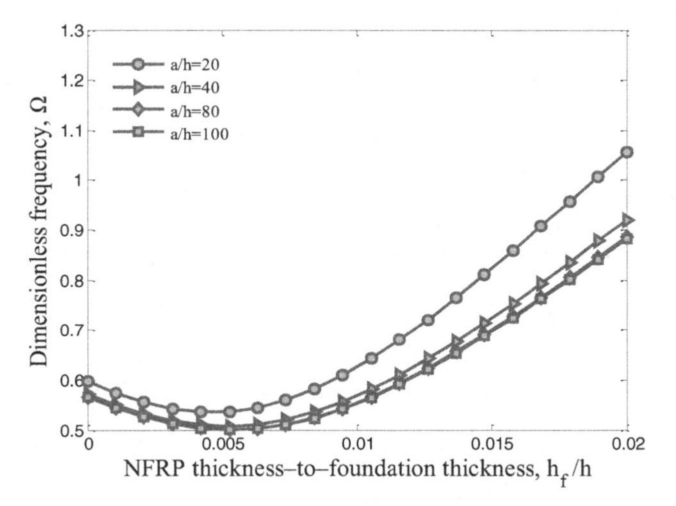


**FIGURE 7.7** Influence of plate length to thickness ratio on the non-dimensional frequency as a function of the NFRP–to–plate thickness ratio.

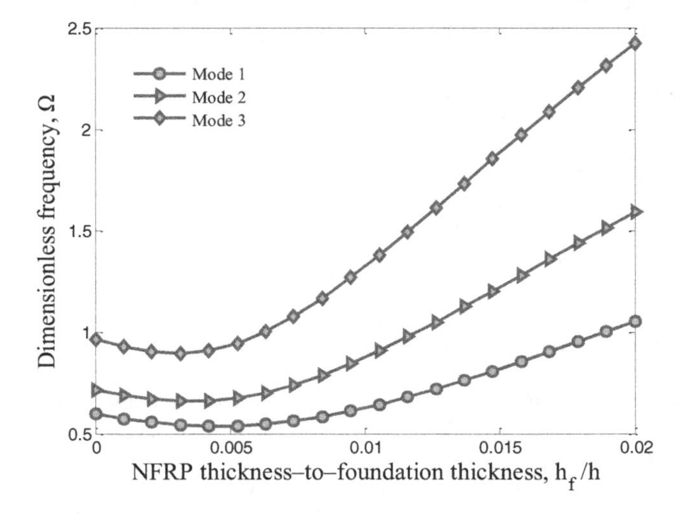


**FIGURE 7.8** Influence of the mode shape number on the non-dimensional frequency as a function of the NFRP layer–to–plate thickness ratio.

**Acknowledgments:** This chapter is a slightly modified version of Ref. [38] and has been reproduced here with the permission of the copyright holder.

## REFERENCES

[1] Persson P, Persson K, Sandberg G. Numerical study on reducing building vibrations by foundation improvement. *Eng Struct* 2016;124:361–375.
[2] Wei YY, Wu YF. Unified stress–strain model of concrete for FRP-confined columns. *Constr Build Mater* 2012;26:381–392.
[3] Grande E, Imbimbo M, Sacco E. Finite element analysis of masonry panels strengthened with FRPs. *Compos Part B-Eng* 2013;45:1296–1309.
[4] Hemmatnezhad M, Rahimi GH, Tajik M, Pellicano F. Experimental, numerical and analytical investigation of free vibrational behavior of GFRP-stiffened composite cylindrical shells. *Compos Struct* 2015;120:509–518.
[5] Capozucca R, Bossoletti S. Static and free vibration analysis of RC beams with NSM CFRP rectangular rods. *Compos Part B-Eng* 2014;67:95–110.
[6] Pan J, Wu YF. Analytical modeling of bond behavior between FRP plate and concrete. *Compos Part B-Eng* 2014;61:1–25.
[7] Altunisik AC, Gunaydin M, Sevim B, Bayraktar A, Adunar S. CFRP composite retrofitting effect on the dynamic characteristics of arch dams. *Soil Dyn Earthq Eng* 2015;74:1–9.
[8] Mokhtari M, Alai Nia A. The influence of using CFRP wraps on performance of buried steel pipelines under permanent ground deformations. *Soil Dyn Earthq Eng* 2015;73:29–41.
[9] Ceroni F, Ianniciello M, Pecce M. Bond behavior of FRP carbon plates externally bonded over steel and concrete elements: Experimental outcomes and numerical investigations. *Compos Part B-Eng* 2016;92:434–446.
[10] Erol G, Karadogan HF. Seismic strengthening of infilled reinforced concrete frames by CFRP. *Compos Part B-Eng* 2016;91:473–491.
[11] Ahmadi M, Ansari R, Rouhi H. Multi-scale bending, buckling and vibration analyses of carbonfiber/carbon nanotube-reinforced polymer nanocomposite plates with various shapes. *Physica E Low Dimens. Syst Nanostruct* 2017;93:17–25.
[12] Guo XY, Zhang W. Nonlinear vibrations of a reinforced composite plate with carbon nanotubes. *Compos Struct* 2015;135:96–108.
[13] Zhang LW, Liew KM, Jiang Z. An element-free analysis of CNT-reinforced composite plates with column supports and elastically restrained edges under large deformation. *Compos Part B-Eng* 2016;95:18–28.
[14] Zhang LW, Selim BA. Vibration analysis of CNT-reinforced thick laminated composite plates based on Reddy's higher-order shear deformation theory. *Compos Struct* 2017;160:689–705.
[15] Zhang LW, Zhang Y, Zhou GL, Liew KM. Free vibration analysis of triangular CNTreinforced composite plates subjected to in-plane stresses using FSDT element-free method. *Compos Struct* 2016;149:247–260.
[16] Shen HS, Wang H. Nonlinear vibration of compressed and thermally postbuckled nanotube-reinforced composite plates resting on elastic foundations. *Aerosp Sci Technol* 2017;64:63–74.
[17] Kumar P, Srinivas J. Vibration, buckling and bending behavior of functionally graded multi walled carbon nanotube reinforced polymer composite plates using the layer-wise formulation. *Compos Struct* 2017;177:158–170.
[18] Xiang Y, Wang CM, Kitipornchai S. Exact vibration solution for initially stressed Mindlin plates on Pasternak foundation. *Int J Mech Sci* 1994;36:311–316.

[19] Omurtag MH, Ozutok A, Akoz AY. Free vibration analysis of Kirchhoff plates resting on elastic foundation by mixedfinite element formulation based on gateaux differential. *Int J Numer Methods Eng* 1997;40:295–317.
[20] Lam KY, Wang CM, He XQ. Canonical exact solutions for levy-plates on two-parameter foundation using Green's functions. *Eng Struct* 2000;22:364–378.
[21] Zhou D, Cheung YK, Lo SH, Au FTK. Three-dimensional vibration analysis of rectangular thick plates on Pasternak foundation. *Int J Numer Methods Eng* 2004;59:1313–1334.
[22] Malekzadeh P, Karami G. Vibration of non-uniform thick plates on elastic foundation by differential quadrature method. *Eng Struct* 2004;26:1473–1482.
[23] Malekzadeh P. Three-dimensional free vibration analysis of thick functionally graded plates on elastic foundations. *Compos Struct* 2009;89:367–373.
[24] Ferreira AJM, Roque CMC, Neves AMA, Jorge RMN, Soares CMM. Analysis of plates on Pasternak foundations by radial basis functions. *Comput Mech* 2010;46:791–803.
[25] Kumar Y, Lal R. Vibrations of nonhomogeneous orthotropic rectangular plates with bilinear thickness variation resting on Winkler foundation. *Meccanica* 2012;47:893–915.
[26] Bahmyari E, Rahbar-Ranji A. Free vibration of orthotropic thin plates with variable thickness resting on nonuniform elastic foundation. *J Mech Sci Technol* 2012;26:2685–2694.
[27] Bahmyari E, Khedmati MR. Vibration analysis of nonhomogeneous moderately thick plates with point supports resting on Pasternak elastic foundation using element free Galerkin method. *Eng Anal Bound Elem* 2013;37:1212–1238.
[28] Mantari JL, Granados EV, Guedes Soares C. Vibrational analysis of advanced composite plates resting on elastic foundation. *Compos Part B Eng* 2014;66:407–419.
[29] Nguyen-Thoi T, Bui-Xuan T, Phung-Van P, Nguyen-Hoang S, Nguyen-Xuan H. An Edge-based smoothed three-node Mindlin plate element ((ES-MIN3)) for static and free vibration analyses of plates. *KSCE J Civ Eng* 2014;18:1072–1082.
[30] Uğurlu B. Boundary element method based vibration analysis of elastic bottom plates offluid storage tanks resting on Pasternak foundation. *Eng Anal Bound Elem* 2016;62:163–176.
[31] Mantari JL, Granados EV. An original FSDT to study advanced composites on elastic foundation. *Thin Wall Struct* 2016;107:80–89.
[32] Neves AMA, Ferreira AJM, Carrera E, Roque CMC, Cinefra M, Jorge RMN, et al. A quasi-3D sinusoidal shear deformation theory for the static and free vibration analysis of functionally graded plates. *Compos Part B-Eng* 2012;45:711–725.
[33] Neves AMA, Ferreira AJM, Carrera E, Roque CMC, Cinefra M, Jorge RMN, et al. A quasi-3D hyperbolic shear deformation theory for the static and free vibration analysis of functionally graded plates. *Compos Struct* 2012;45:1814–1825.
[34] Thai HT, Kim SE. A simple quasi-3D sinusoidal shear deformation theory for functionally graded plates. *Compos Struct* 2013;99:172–180.
[35] Thai HT, Vo TP. A new sinusoidal shear deformation theory for bending, buckling, and vibration of functionally graded plates. *Appl Math Model* 2013;37:3269–3281.
[36] Tran LV, Ferreira AJM, Nguyen-Xuan H. Isogeometric analysis of functionally graded plates using higher-order shear deformation theory. *Compos Part B Eng* 2013;51:368–383.
[37] Wang Y, Wu D. Free vibration of functionally graded porous cylindrical shell using a sinusoidal shear deformation theory. *Aerosp Sci Technol* 2017;66:83–91.
[38] Zamani A, Bidgoli MR. Vibration analysis of concrete foundations retrofit with NFRP layer resting on soil medium using sinusoidal shear deformation theory. *Soil Dynam Earthq Eng* 2017;103:141–150.

# 8 Vibration Analysis of Plates Reinforced with Nanoparticles and a Piezoelectric Layer

## 8.1 INTRODUCTION

Various external and internal sources generate vibration instabilities in buildings, such as motorway traffic, the existence of industrial equipment (for instance, compressors, crushers, sieve shakers, and dryers), or people walking inside the building. These vibrations also affect the performance of concrete foundations, and therefore, the vibrations may need to be reduced. Vibrations can be reduced by improving the properties of concrete plates or by using new technologies, such as using piezoelectric materials. A concrete foundation can be modified by being mixed with nanomaterial in order to increase its stiffness. In this subject, it can be mentioned research on reducing building vibrations through foundation improvement by Persson et al. [1].

The vibration behavior of plates on elastic foundations has attracted considerable attention in recent years. Lam et al. [2] used Green's functions to obtain canonical exact solutions of the elastic bending, buckling, and vibration for Levy plates resting on two-parameter elastic foundations. Buczkowski and Torbacki [3] presented a finite element method for the thick plates on a two-parameter elastic foundation. By employing the Rayleigh-Ritz method, the three-dimensional vibration of rectangular thick plates on elastic foundations was investigated by Zhou et al. [4]. The free vibrations of simply supported rectangular plates, resting on two different models of soils, were considered by De Rosa and Lippiello [5]. Ferreira et al. [6] used the radial basis function collocation method to study the static deformation and free vibration of plates on a Pasternak foundation. A nonlinear vibration analysis of laminated plates resting on nonlinear two-parameter elastic foundations was presented by Akgoz and Civalek [7]. Kumar and Lal [8] studied the vibration of nonhomogeneous orthotropic rectangular plates with bilinear thickness variation resting on a Winkler foundation. Bahmyari and Khedmati [9] considered the vibration analysis of nonhomogeneous moderately thick plates with point supports resting on a Pasternak elastic foundation using the element-free Galerkin method. A vibrational analysis of advanced composite plates resting on an elastic foundation was presented by Mantari et al. [10]. They derived the governing equations of a type of functionally graded (FG) plates resting on an elastic foundation by employing Hamilton's principle. An original first-order shear deformation theory to study advanced composites on an elastic foundation was presented by Mantari and Granados [11]. Ugurla [12] analyzed the vibration

DOI: 10.1201/9781003349525-8

of the elastic bottom plates of fluid storage tanks resting on a Pasternak foundation based on the boundary element method. Bounouara et al. [13] investigated a nonlocal zeroth-order shear deformation theory for the free vibration of FG nanoscale plates resting on elastic foundation. Also, a dimensionless parametric study for forced vibrations of foundation-soil systems was done by Chen et al. [14]. A nonpolynomial four-variable refined plate theory for the free vibration of FG thick rectangular plates on an elastic foundation was investigated by Meftah et al. [15].

None of the previously mentioned researchers have considered piezo-based nanocomposite structures. In recent years, theoretical and experimental studies have been conducted on nanocomposites subjected to an electric field. A static analysis of an FG carbon nanotube (CNT)–reinforced composite plate embedded in piezoelectric layers with three cases of CNT distribution based on the three-dimensional theory was discussed by Alibeigloo [16]. A piezo-based wireless sensor network for early-age concrete strength monitoring is planned by Chen et al. [17]. Sasmal et al. [18] investigated electrical conductivity and piezo-resistive characteristics of CNT- and carbon nanofiber (CNF)–incorporated cementitious nanocomposites under static and dynamic loading. In magneto-electro-elastic (MEE) composite materials, a coupling between mechanical, electric, and magnetic fields results in the ability to exchange energy among these three energy forms. These materials have direct applications in sensors and actuators, damping, and control of vibrations in structures. Xue et al. [19] studied the large deflection of a rectangular MEE thin plate for the first time based on the classical plate theory. Li and Zhang [20] investigated the free vibration of an MEE plate resting on a Pasternak foundation by using the Mindlin theory. Large amplitude free vibration of symmetrically laminated magneto-electro-elastic rectangular plates on a Pasternak-type foundation was investigated by Shooshtari and Razavi [21]. Ebrahimi et al. [22] proposed a four-variable shear deformation refined plate theory for a free vibration analysis of embedded smart plates made of porous MEE-FG materials resting on elastic foundations.

Furthermore, the mechanical behavior of concrete structures containing nanoparticles has been investigated experimentally by a number of researchers, but there is little mathematical control in this field. Nirmala and Dhanalakshmi [23] studied the influence of nanomaterials in a distressed retaining structure for crack filling. The influences of nanoparticles on the dynamic strength of ultra-high-performance concrete were tested by Su et al. [24]. Fathi et al. [25] investigated the mechanical and physical properties of expanded polystyrene structural concretes containing micro-silica and nano-silica. In the field of mathematical modeling of concrete structures, Jafarian Arani and Kolahchi [26] studied the buckling of concrete columns reinforced with CNTs using the Euler–Bernoulli and Timoshenko beam models. Zamanian et al. [27] investigated the nonlinear buckling of a concrete column reinforced with silicon dioxide ($SiO_2$) nanoparticles. Kargar and Bidgoli [28] studied the vibration and smart control analysis of a concrete foundation reinforced with $SiO_2$ nanoparticles and covered by a piezoelectric layer on a soil medium. The soil medium is represented using spring constants, and the Mori-Tanaka model is employed to determine the material properties of the nano-composite structure, taking into account agglomeration effects.

To the best of the authors' knowledge, the effects of using nanoparticles and a piezoelectric layer on the vibration of concrete plates have not been investigated. So this chapter is done to fill the gap in this area. The purpose of this chapter is to study the free vibration smart control of a concrete plate reinforced by $SiO_2$ nanoparticles embedded in a soil medium. The structure is covered by a piezoelectric layer subjected to an external voltage. In order to obtain the equivalent material properties of nanocomposite structure, the Mori–Tanaka model is used. Applying first-order shear deformation theory (FSDT), the motion equations are achieved based on Hamilton's principle. The Navier method is applied to obtain the frequency of the system. The effects of applied voltage, volume percent, and agglomeration of $SiO_2$ nanoparticles, soil medium, and geometrical parameters of structure on the frequency of the system are disused in detail.

## 8.2 CONSTITUTIVE EQUATIONS OF PIEZOELECTRIC MATERIAL

As shown in Figure 8.1, a concrete plate reinforced with $SiO_2$ nanoparticles and covered by a piezoelectric layer with length $L$, width $b$, concrete thickness $h$, and piezoelectric layer thickness $h_p$ is considered.

Based on FSDT theory, the displacement field can be expressed based on Section 1.4.2. In a piezoelectric material, applying an electric field to it will cause a strain proportional to the mechanical field strength and vice versa. The constitutive equation for stresses $\sigma$ and strains $\varepsilon$ matrix on the mechanical side, as well as flux density $D$ and field strength $E$ matrix on the electrostatic side, may be arbitrarily combined as follows [29]:

$$\begin{bmatrix} \sigma_{xx} \\ \sigma_{yy} \\ \sigma_{zz} \\ \tau_{yz} \\ \tau_{xz} \\ \tau_{xy} \end{bmatrix} = \begin{bmatrix} C_{11} & C_{12} & C_{13} & 0 & 0 & 0 \\ C_{12} & C_{22} & C_{23} & 0 & 0 & 0 \\ C_{13} & C_{23} & C_{33} & 0 & 0 & 0 \\ 0 & 0 & 0 & C_{44} & 0 & 0 \\ 0 & 0 & 0 & 0 & C_{55} & 0 \\ 0 & 0 & 0 & 0 & 0 & C_{66} \end{bmatrix} \left( \begin{Bmatrix} \varepsilon_{xx} \\ \varepsilon_{yy} \\ \varepsilon_{zz} \\ \gamma_{yz} \\ \gamma_{xz} \\ \gamma_{xy} \end{Bmatrix} \right) - \begin{bmatrix} 0 & 0 & e_{31} \\ 0 & 0 & e_{32} \\ 0 & 0 & e_{33} \\ 0 & e_{24} & 0 \\ e_{15} & 0 & 0 \\ 0 & 0 & 0 \end{bmatrix} \begin{Bmatrix} E_x \\ E_y \\ E_z \end{Bmatrix}, \tag{8–1}$$

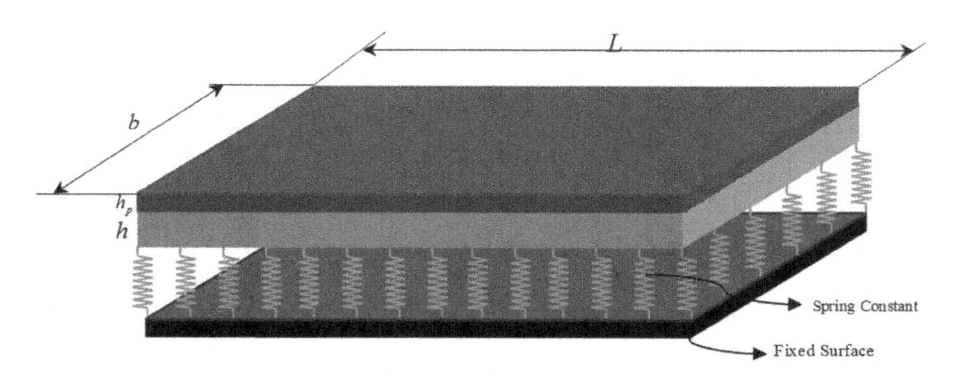


**FIGURE 8.1** Schematic of a concrete plate with piezoelectric layers reinforced with $SiO_2$ nanoparticles.

$$\begin{bmatrix} D_x \\ D_y \\ D_z \end{bmatrix} = \begin{bmatrix} 0 & 0 & 0 & 0 & e_{15} & 0 \\ 0 & 0 & 0 & e_{24} & 0 & 0 \\ e_{31} & e_{31} & e_{33} & 0 & 0 & 0 \end{bmatrix} \left( \begin{Bmatrix} \varepsilon_{xx} \\ \varepsilon_{yy} \\ \varepsilon_{zz} \\ \gamma_{yz} \\ \gamma_{xz} \\ \gamma_{xy} \end{Bmatrix} \right) + \begin{bmatrix} \in_{11} & 0 & 0 \\ 0 & \in_{22} & 0 \\ 0 & 0 & \in_{33} \end{bmatrix} \begin{Bmatrix} E_x \\ E_y \\ E_z \end{Bmatrix}, \tag{8–2}$$

where $\sigma_{ij}$, $\varepsilon_{ij}$, $D_{ii}$, and $E_{ii}$ are stress, strain, electric displacement, and electric field, respectively. Also, $C_{ij}$, $e_{ij}$, and $\in_{ij}$ denote the elastic, piezoelectric, and dielectric coefficients, respectively. Noted that $C_{ij}$ and $\alpha_{xx}$, $\alpha_{\theta\theta}$ may be obtained using Mori–Tanaka model (Section 4.3). The electric field in terms of electric potential ($\Phi$) is expressed as

$$E_k = -\nabla\Phi, \tag{8–3}$$

where the electric potential is assumed as the combination of a half-cosine and linear variation, which satisfies the Maxwell equation. It can be written as [29]

$$\Phi(x, y, z, t) = -\cos\left(\frac{\pi z}{h}\right)\varphi(x, y, t) + \frac{2V_0 z}{h}, \tag{8–4}$$

where $\varphi(x,\theta,t)$ is the time and spatial distribution of the electric potential that must satisfy the electric boundary conditions and $V_0$ is the external electric voltage. However, using Equations 8–1 and 8–2, the governing equations of piezoelectric material for FSDT may be written as

$$\sigma_{xx}^p = C_{11}\varepsilon_{xx} + C_{12}\varepsilon_{yy} + e_{31}\left(\frac{\pi}{h}\sin\left(\frac{\pi z}{h}\right)\varphi + \frac{2V_0}{h}\right), \tag{8–5}$$

$$\sigma_{yy}^p = C_{12}\varepsilon_{xx} + C_{22}\varepsilon_{yy} + e_{32}\left(\frac{\pi}{h}\sin\left(\frac{\pi z}{h}\right)\varphi + \frac{2V_0}{h}\right), \tag{8–6}$$

$$\tau_{yz}^p = C_{44}\gamma_{yz} - e_{15}\left(\cos\left(\frac{\pi z}{h}\right)\frac{\partial\varphi}{\partial y}\right), \tag{8–7}$$

$$\tau_{xz}^p = C_{55}\gamma_{zx} - e_{24}\left(\cos\left(\frac{\pi z}{h}\right)\frac{\partial\varphi}{\partial x}\right), \tag{8–8}$$

$$\tau_{xy}^p = C_{66}\gamma_{xy}, \tag{8–9}$$

$$D_x = e_{15}\gamma_{xz} + \in_{11}\left(\cos\left(\frac{\pi z}{h}\right)\frac{\partial\varphi}{\partial y}\right), \tag{8–10}$$

$$D_y = e_{24}\gamma_{zy} + \in_{22}\left(\cos\left(\frac{\pi z}{h}\right)\frac{\partial\varphi}{\partial y}\right), \tag{8–11}$$

$$D_z = e_{31}\varepsilon_{xx} + e_{32}\varepsilon_{yy} - \epsilon_{33}\left(\frac{\pi}{h}\sin\left(\frac{\pi z}{h}\right)\varphi + \frac{2V_0}{h}\right). \tag{8–12}$$

For the plate, for neglecting the piezoelectric properties, we have

$$\sigma_{xx}^{c} = Q_{11}\varepsilon_{xx} + Q_{12}\varepsilon_{yy}, \tag{8–13}$$

$$\sigma_{yy}^{c} = Q_{12}\varepsilon_{xx} + Q_{22}\varepsilon_{yy}, \tag{8–14}$$

$$\tau_{yz}^{c} = Q_{44}\gamma_{yz}, \tag{8–15}$$

$$\tau_{xz}^{c} = Q_{55}\gamma_{zx}, \tag{8–16}$$

$$\tau_{xy}^{c} = Q_{66}\gamma_{xy}. \tag{8–17}$$

## 8.3 ENERGY METHOD

The total potential energy, $V$, of the system is the sum of potential energy, $U$, kinetic energy, $K$, and the work done by the elastic medium, $W$. The potential energy can be written as

$$U = \frac{1}{2}\int\begin{pmatrix} \sigma_{xx}^{c}\varepsilon_{xx} + \sigma_{yy}^{c}\varepsilon_{yy} + \tau_{xz}^{c}\gamma_{xz} + \tau_{yz}^{c}\gamma_{yz} + \tau_{xy}^{c}\gamma_{xy} \\ +\sigma_{xx}^{p}\varepsilon_{xx} + \sigma_{yy}^{p}\varepsilon_{yy} + \tau_{xz}^{p}\gamma_{xz} + \tau_{yz}^{p}\gamma_{yz} + \tau_{xy}^{p}\gamma_{xy} \\ -D_xE_x - D_yE_y - D_zE_z \end{pmatrix} dV. \tag{8–18}$$

Combining Equations 1–31 and 8–18 yields

$$\begin{aligned} U = \frac{1}{2}\int_0^{2\pi}\int_0^{L} \Bigg\{ & \left[N_{xx}\frac{\partial u}{\partial x} + M_{xx}\frac{\partial \phi_x}{\partial x}\right] + \left[N_{yy}\frac{\partial v}{\partial y} + M_{yy}\frac{\partial \phi_y}{\partial y}\right] + Q_x\left(\phi_x + \frac{\partial w}{\partial x}\right) \\ & + N_{xy}\left[\frac{\partial v}{\partial x} + \frac{\partial u}{\partial y}\right] + M_{xy}\left[\frac{\partial \phi_y}{\partial x} + \frac{\partial \phi_x}{\partial y}\right] Rxdy + Q_y\left[\frac{\partial w}{\partial y} + \phi_y\right] \\ & + \int_{-h/2}^{h/2+h_p}\int_0^{2\pi}\int_0^{L} -D_x\left[\cos\left(\frac{\pi z}{h}\right)\frac{\partial \varphi}{\partial x}\right] \\ & - D_\theta\left[\cos\left(\frac{\pi z}{h}\right)\frac{\partial \varphi}{\partial y}\right] - D_z\left[-\frac{\pi}{h}\sin\left(\frac{\pi z}{h}\right)\varphi - \frac{2V_0}{h}\right]\Bigg\} dxdydz, \end{aligned} \tag{8–19}$$

where the stress resultant–displacement relations can be written as

$$\begin{Bmatrix} N_{xx} \\ N_{yy} \\ N_{xy} \end{Bmatrix} = \int_{-\frac{h}{2}}^{\frac{h}{2}} \begin{Bmatrix} \sigma_{xx}^{c} \\ \sigma_{yy}^{c} \\ \tau_{xy}^{c} \end{Bmatrix} dz + \int_{\frac{h}{2}}^{\frac{h}{2}+h_p} \begin{Bmatrix} \sigma_{xx}^{p} \\ \sigma_{yy}^{p} \\ \tau_{xy}^{p} \end{Bmatrix} dz \tag{8–20}$$

$$\begin{Bmatrix} Q_x \\ Q_\theta \end{Bmatrix} = k' \int_{-\frac{h}{2}}^{\frac{h}{2}} \begin{Bmatrix} \tau^c_{xz} \\ \tau^c_{yz} \end{Bmatrix} dz + k' \int_{-\frac{h}{2}}^{\frac{h}{2}+h_p} \begin{Bmatrix} \tau^p_{xz} \\ \tau^p_{yz} \end{Bmatrix} dz, \tag{8–21}$$

$$\begin{Bmatrix} M_{xx} \\ M_{yy} \\ M_{xy} \end{Bmatrix} = \int_{-\frac{h}{2}}^{\frac{h}{2}} \begin{Bmatrix} \sigma^c_{xx} \\ \sigma^c_{yy} \\ \tau^c_{xy} \end{Bmatrix} zdz + \int_{\frac{h}{2}}^{\frac{h}{2}+h_p} \begin{Bmatrix} \sigma^p_{xx} \\ \sigma^p_{yy} \\ \tau^p_{xy} \end{Bmatrix} zdz, \tag{8–22}$$

in which $k'$ is the shear correction coefficient. Substituting Equations 8–5 through 8–17 into Equations 8–20 through 8–22, the stress resultant–displacement relations can be obtained as follows:

$$N_{xx} = A_{110}\frac{\partial u}{\partial x} + A_{111}\frac{\partial \phi_x}{\partial x} + A_{120}\frac{\partial v}{\partial y} + A_{121}\frac{\partial \phi_y}{\partial y} + E_{31}\varphi, \tag{8–23}$$

$$N_{yy} = A_{120}\frac{\partial u}{\partial x} + A_{121}\frac{\partial \phi_x}{\partial x} + A_{220}\frac{\partial v}{\partial y} + A_{221}\frac{\partial \phi_y}{\partial y} + E_{32}\varphi, \tag{8–24}$$

$$Q_y = k'A_{44}\left[\frac{\partial w}{\partial y} + \phi_y\right] + E_{15}\frac{\partial \varphi}{\partial y}, \tag{8–25}$$

$$Q_x = k'A_{55}\left(\frac{\partial w}{\partial x} + \phi_x\right) + E_{24}\frac{\partial \varphi}{\partial x}, \tag{8–26}$$

$$N_{xy} = A_{660}\left(\frac{\partial u}{\partial y} + \frac{\partial v}{\partial x}\right) + A_{661}\left(\frac{\partial \phi_x}{\partial y} + \frac{\partial \phi_y}{\partial x}\right), \tag{8–27}$$

$$M_{xx} = A_{111}\frac{\partial u}{\partial x} + A_{112}\frac{\partial \phi_x}{\partial x} + A_{121}\frac{\partial v}{\partial y} + A_{122}\frac{\partial \phi_y}{\partial y} + F_{31}\varphi, \tag{8–28}$$

$$M_{yy} = A_{121}\frac{\partial u}{\partial x} + A_{122}\frac{\partial \phi_x}{\partial x} + A_{221}\frac{\partial v}{\partial y} + A_{222}\frac{\partial \phi_y}{\partial y} + F_{32}\varphi, \tag{8–29}$$

$$M_{xy} = A_{661}\left(\frac{\partial u}{\partial y} + \frac{\partial v}{\partial x}\right) + A_{662}\left(\frac{\partial \phi_x}{\partial y} + \frac{\partial \phi_\theta}{\partial x}\right), \tag{8–30}$$

where

$$A_{11k} = \int_{-h/2}^{h/2} Q_{11} z^k dz + \int_{-h/2}^{h/2+h_p} C_{11} z^k dz, \qquad k = 0,1,2 \tag{8–31}$$

$$A_{12k} = \int_{-h/2}^{h/2} Q_{12} z^k dz + \int_{-h/2}^{h/2+h_p} C_{12} z^k dz, \qquad k = 0,1,2 \tag{8–32}$$

$$A_{22k} = \int_{-h/2}^{h/2} Q_{22} z^k dz + \int_{-h/2}^{h/2+h_p} C_{22} z^k dz, \qquad k = 0,1,2 \tag{8–33}$$

$$A_{66k}=\int_{-h/2}^{h/2}Q_{66}z^k dz+\int_{-h/2}^{h/2+h_p}C_{66}z^k dz,\qquad k=0,1,2 \tag{8–34}$$

$$A_{44}=\int_{-h/2}^{h/2}Q_{44}dz+\int_{-h/2}^{h/2+h_p}C_{44}dz, \tag{8–35}$$

$$A_{55}=\int_{-h/2}^{h/2}Q_{55}dz+\int_{-h/2}^{h/2+h_p}C_{55}dz, \tag{8–36}$$

$$\left(E_{31},E_{32}\right)=\frac{\pi}{h}\int_{-h/2}^{h/2}\left(e_{31},e_{32}\right)\sin\left(\frac{\pi z}{h}\right)dz, \tag{8–37}$$

$$\left(E_{24},E_{15}\right)=-\int_{-h/2}^{h/2}\left(e_{24},e_{15}\right)\cos\left(\frac{\pi z}{h}\right)dz, \tag{8–38}$$

$$\left(F_{31},F_{32}\right)=\frac{\pi}{h}\int_{-h/2}^{h/2}\left(e_{31},e_{32}\right)\sin\left(\frac{\pi z}{h}\right)zdz. \tag{8–39}$$

The kinetic energy of the system may be written as

$$K=\frac{\left(\rho^c+\rho^p\right)}{2}\int\left(\left(\frac{\partial u}{\partial t}+z\frac{\partial \phi_x}{\partial t}\right)^2+\left(\frac{\partial v}{\partial t}+z\frac{\partial \phi_y}{\partial t}\right)^2+\left(\frac{\partial w}{\partial t}\right)^2\right)dV. \tag{8–40}$$

Defining the moments of inertia as follows,

$$\begin{Bmatrix}I_0\\ I_1\\ I_2\end{Bmatrix}=\int_{-h/2}^{h/2}\begin{bmatrix}\rho^c\\ \rho^c z\\ \rho^c z^2\end{bmatrix}dz+\int_{h/2}^{h/2+h_p}\begin{bmatrix}\rho^p\\ \rho^p z\\ \rho^p z^2\end{bmatrix}dz, \tag{8–41}$$

the kinetic energy may be written as

$$K=\frac{1}{2}\int\left(I_0\left(\left(\frac{\partial u}{\partial t}\right)^2+\left(\frac{\partial v}{\partial t}\right)^2+\left(\frac{\partial w}{\partial t}\right)^2\right)+I_1\left(2\frac{\partial u}{\partial t}\frac{\partial \phi_x}{\partial t}+2\frac{\partial v}{\partial t}\frac{\partial \phi_y}{\partial t}\right)\right.$$
$$\left.+I_2\left(\left(\frac{\partial \phi_x}{\partial t}\right)^2+\left(\frac{\partial \phi_y}{\partial t}\right)^2\right)\right)dA. \tag{8–42}$$

The external work due to the soil medium can be written as

$$W_e=\int_0^{2\pi}\int_0^{L}\left(-K_w w\right)dxdy, \tag{8–43}$$

where $K_w$ is Winkler's spring modulus. In addition, the in-plane forces may be written as

$$W_f = -\frac{1}{2}\int\left[N_{xx}^f\left(\frac{\partial w}{\partial x}\right)^2 + N_{yy}^f\left(\frac{\partial w}{\partial y}\right)^2\right]dxdy. \tag{8–44}$$

where

$$N_{xx}^f = N_{xx}^M + N_{xx}^E = N_{xx}^M + 2V_0 e_{31}, \tag{8–45}$$

$$N_{yy}^f = N_{yy}^M + N_{yy}^E = N_{yy}^M + 2V_0 e_{32}. \tag{8–46}$$

The governing equations can be derived using Hamilton's principle as follows:

$$\int_0^t (\delta U - \delta K - \delta W_e - \delta W_f)dt = 0. \tag{8–47}$$

Substituting Equations 8–19, 8–42, 8–43, and 8–44 into Equation 8–47 yields the following governing equations:

$$\delta u: \frac{\partial N_{xx}}{\partial x} + \frac{\partial N_{xy}}{\partial y} = I_0\frac{\partial^2 u}{\partial t^2} + I_1\frac{\partial^2 \phi_x}{\partial t^2}, \tag{8–48}$$

$$\delta v: \frac{\partial N_{xy}}{\partial x} + \frac{\partial N_{yy}}{\partial y} = I_0\frac{\partial^2 v}{\partial t^2} + I_1\frac{\partial^2 \phi_y}{\partial t^2}, \tag{8–49}$$

$$\delta w: \frac{\partial Q_x}{\partial x} + \frac{\partial Q_y}{\partial y} + N_{xx}^f\frac{\partial^2 w}{\partial x^2} + N_{yy}^f\frac{\partial^2 w}{\partial y^2} - K_w w = I_0\frac{\partial^2 w}{\partial t^2}, \tag{8–50}$$

$$\delta\phi_x: \frac{\partial M_{xx}}{\partial x} + \frac{\partial M_{xy}}{\partial y} - Q_x = I_2\frac{\partial^2 \phi_x}{\partial t^2} + I_1\frac{\partial^2 u}{\partial t^2}, \tag{8–51}$$

$$\delta\varphi_y: \frac{\partial M_{xy}}{\partial x} + \frac{\partial M_{xx}}{\partial y} - Q_y = I_2\frac{\partial^2 \phi_y}{\partial t^2} + I_1\frac{\partial^2 v}{\partial t^2}, \tag{8–52}$$

$$\delta\varphi: \int_{-h/2}^{h/2}\left\{\left[\cos\left(\frac{\pi z}{h}\right)\frac{\partial D_x}{\partial x}\right] + \left[\cos\left(\frac{\pi z}{h}\right)\frac{\partial D_y}{\partial y}\right] + D_z\left[\frac{\pi}{h}\sin\left(\frac{\pi z}{h}\right)\right]\right\}dz = 0. \tag{8–53}$$

Substituting Equations 8–23 through 8–30 into Equations 8–48 through 8–53, the governing equations can be written as follow:

$$A_{110}\frac{\partial^2 u}{\partial x^2} + A_{111}\frac{\partial^2 \phi_x}{\partial x^2} + A_{120}\frac{\partial^2 v}{\partial y\partial x} + A_{121}\frac{\partial^2 \phi_y}{\partial y\partial x} + E_{31}\frac{\partial \varphi}{\partial x} + A_{120}\frac{\partial^2 u}{\partial x\partial y}$$
$$+A_{121}\frac{\partial^2 \phi_x}{\partial x\partial y} + A_{220}\frac{\partial^2 v}{\partial y^2} + A_{221}\frac{\partial^2 \phi_y}{\partial y^2} + E_{32}\frac{\partial \varphi}{\partial y} = I_0\frac{\partial^2 u}{\partial t^2} + I_1\frac{\partial^2 \phi_x}{\partial t^2}, \tag{8–54}$$

$$A_{120}\frac{\partial^2 u}{\partial x\partial y}+A_{121}\frac{\partial^2 \phi_x}{\partial x\partial y}+A_{220}\frac{\partial^2 v}{\partial y^2}+A_{221}\frac{\partial^2 \phi}{\partial y^2}+E_{32}\frac{\partial \varphi}{\partial y}+A_{660}\left(\frac{\partial^2 u}{\partial y\partial x}+\frac{\partial^2 v}{\partial x^2}\right)$$
$$+A_{661}\left(\frac{\partial^2 \phi_x}{\partial y\partial x}+\frac{\partial^2 \phi_y}{\partial x^2}\right)=I_0\frac{\partial^2 v}{\partial t^2}+I_1\frac{\partial^2 \phi_y}{\partial t^2}, \tag{8–55}$$

$$k'A_{44}\left[\frac{\partial w}{\partial y^2}+\frac{\partial \phi_y}{\partial y}\right]+E_{15}\frac{\partial \varphi}{\partial y^2}+k'A_{55}\left(\frac{\partial^2 w}{\partial x^2}+\frac{\partial \phi_x}{\partial x}\right)+E_{24}\frac{\partial^2 \varphi}{\partial x^2}-A_{120}\frac{\partial u}{\partial x}$$
$$-A_{121}\frac{\partial \phi_x}{\partial x}-A_{220}\left(\frac{\partial v}{\partial y}\right)-A_{221}\frac{\partial \phi}{\partial y}+N_{xx}^f\frac{\partial^2 w}{\partial x^2}+N_{yy}^f\frac{\partial^2 w}{\partial y^2}-K_w w=I_0\frac{\partial^2 w}{\partial t^2}, \tag{8–56}$$

$$A_{111}\frac{\partial^2 u}{\partial x^2}+A_{112}\frac{\partial^2 \phi_x}{\partial x^2}+A_{121}\left(\frac{\partial^2 v}{\partial y\partial x}+\frac{\partial w}{\partial x}\right)+A_{122}\frac{\partial^2 \phi_y}{\partial y}+F_{31}\frac{\partial \phi}{\partial x}+A_{661}\left(\frac{\partial^2 u}{\partial y^2}+\frac{\partial^2 v}{\partial x\partial y}\right)$$
$$+A_{662}\left(\frac{\partial^2 \phi_x}{\partial y^2}+\frac{\partial^2 \phi_y}{\partial x\partial y}\right)-k'A_{55}\left(\frac{\partial w}{\partial x}+\phi_x\right)-E_{24}\frac{\partial \varphi}{\partial x}=I_2\frac{\partial^2 \phi_x}{\partial t^2}+I_1\frac{\partial^2 u}{\partial t^2}, \tag{8–57}$$

$$A_{121}\frac{\partial^2 u}{\partial x\partial y}+A_{122}\frac{\partial^2 \phi_x}{\partial x\partial y}+A_{221}\left(\frac{\partial^2 v}{\partial y^2}+\frac{\partial w}{\partial y}\right)+A_{222}\frac{\partial^2 \phi_\theta}{\partial y^2}+A_{661}\left(\frac{\partial^2 u}{\partial y\partial x}+\frac{\partial^2 v}{\partial x^2}\right)$$
$$+F_{32}\frac{\partial \varphi}{\partial y}+A_{662}\left(\frac{\partial^2 \phi_x}{\partial y\partial x}+\frac{\partial^2 \phi_y}{\partial x^2}\right)-k'A_{44}\left[\frac{\partial w}{\partial y}+\phi_y\right]-E_{15}\frac{\partial \varphi}{\partial y}=I_2\frac{\partial^2 \phi_y}{\partial t^2}+I_1\frac{\partial^2 v}{\partial t^2}, \tag{8–58}$$

$$\delta\phi:\quad -E_{15}\left(\frac{\partial \phi_x}{\partial x}+\frac{\partial^2 w}{\partial x^2}\right)+\Xi_{11}\left(\frac{\partial^2 \varphi}{\partial x\partial y}\right)-E_{24}\left(\frac{\partial^2 w}{\partial y^2}+\frac{\partial \phi_y}{\partial y}\right)$$
$$+\Xi_{22}\left(\frac{\partial^2 \varphi}{\partial y^2}\right)+E_{31}\frac{\partial u}{\partial x}+F_{31}\frac{\partial \phi_x}{\partial x}+\frac{E_{32}}{R}\left(w+\frac{\partial v}{\partial y}\right)+F_{32}\frac{\partial \phi_y}{\partial y}-\Xi_{33}\varphi=0. \tag{8–59}$$

where

$$\left(\Xi_{11},\Xi_{22}\right)=\int_{-h/2}^{h/2}\left(\in_{11},\in_{22}\right)\cos^2\left(\frac{\pi z}{h}\right)dz, \tag{8–60}$$

$$\left(\Xi_{33}\right)=\frac{\pi^2}{h^2}\int_{-h/2}^{h/2}\left(\in_{33}\right)\sin^2\left(\frac{\pi z}{h}\right)dz. \tag{8–61}$$

## 8.4 NUMERICAL RESULTS AND DISCUSSION

A computer program (see Appendix D) is prepared based on the Navier method presented in Section 2.2.1 for the vibration smart control solution of a plate reinforced with $SiO_2$ nanoparticles and a piezoelectric layer. Here, polyvinylidene fluoride (PVDF) is selected for the piezoelectric layer with the material properties of Table 8.1 [29]. In addition, the $SiO_2$ nanoparticles have a Young's modulus of $E_r =$ 70 GPa and a Poisson's ratio of $v_r = 0.2$.

Figure 8.2 illustrates the effect of the $SiO_2$ nanoparticles volume fraction on the dimensionless frequency of structure ($\Omega = \omega L\sqrt{\rho_m / E_m}$). It can be seen that by increasing the values of the $SiO_2$ nanoparticle volume fraction, the frequency of the system is increased. This is due to the fact that the increase of $SiO_2$ nanoparticles leads to a harder structure. However, it may be concluded that using nanotechnology to reinforce a plate has an important role in improving the vibration behavior of a system.

Figure 8.3 shows the effect of $SiO_2$ nanoparticle agglomeration on the dimensionless frequency of structure versus the external applied voltage. As can be seen, considering the agglomeration of $SiO_2$ nanoparticles leads to a lower frequency. It is due to this point that the agglomeration of $SiO_2$ nanoparticles decreases the stability and homogeneity of the structure.

The dimensionless frequency of the nanocomposite plate is demonstrated in Figure 8.4 for different soil mediums. In this figure, four cases of loose sand, dense sand, clayey medium-dense sand, and clayey soil are considered.

**TABLE 8.1**
**Material Properties of PVDF**

| Properties | PVDF |
|---|---|
| $C_{11}$ | 238.24 (GPa) |
| $C_{12}$ | 3.98 (GPa) |
| $C_{22}$ | 23.6 (GPa) |
| $e_{11}$ | −0.135 ($C/m^2$) |
| $e_{12}$ | −0.145 ($C/m^2$) |
| $\varepsilon_{11}$ | 1.1e-8 ($C^2/Nm^2$) |

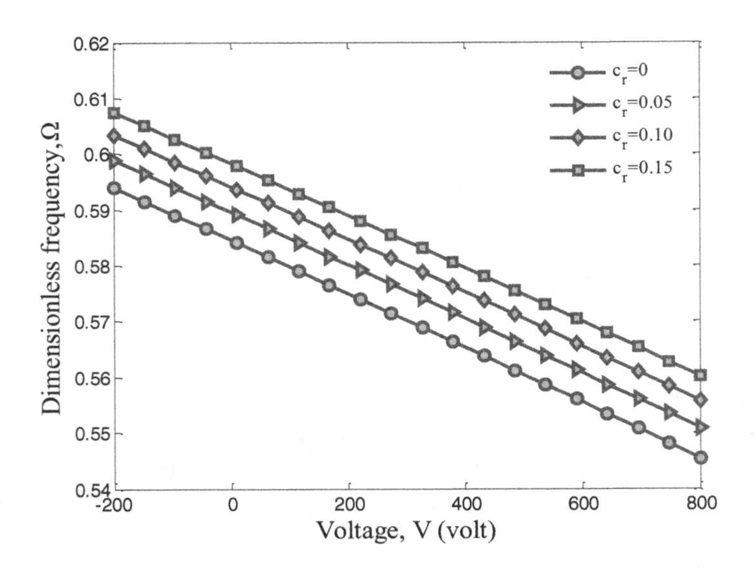


**FIGURE 8.2** Effect of $SiO_2$ nanoparticle volume percent on the dimension frequency versus the dimension external applied voltage.

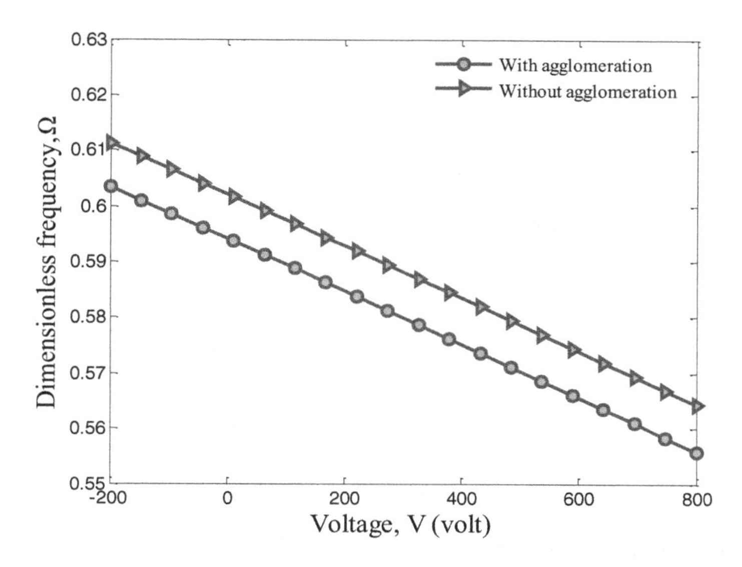


**FIGURE 8.3** Effect of $SiO_2$ nanoparticle agglomeration on the dimension frequency versus the dimension external applied voltage.

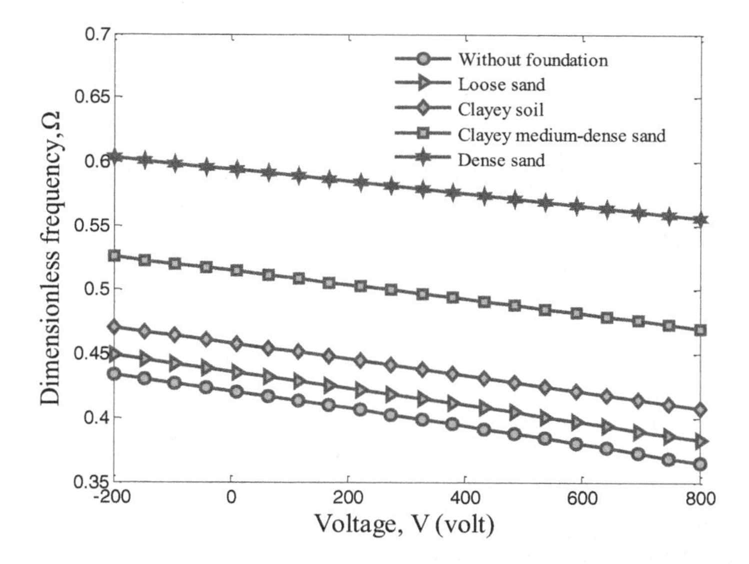


**FIGURE 8.4** Effect of soil medium on the dimension frequency versus the dimension external applied voltage.

As can be seen, considering the soil medium increases the frequency of the structure. This is due to the fact that considering soil medium leads to a stiffer structure. Furthermore, the frequency of the dense sand medium is higher than other cases since the spring constant is maximum.

The effect of the length-to-thickness ratio of plate on the dimensionless frequency of the system is depicted in Figure 8.5. As can be seen, the frequency of the structure

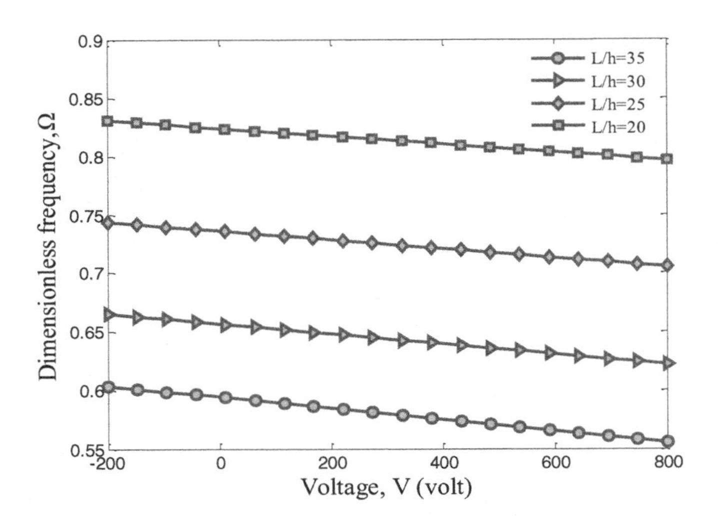


**FIGURE 8.5** Effect of length-to-thickness ratio of the plate on the dimension frequency versus the dimension external applied voltage.

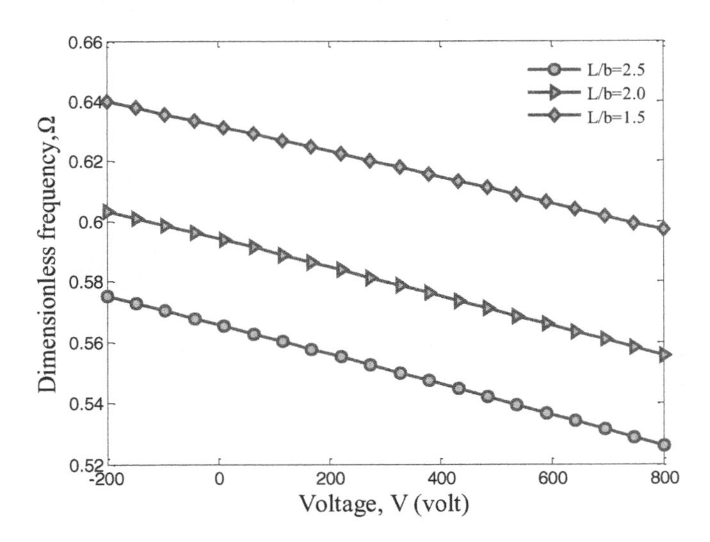


**FIGURE 8.6** Effect of the length-to-width ratio of the plate on the dimension frequency versus the dimension external applied voltage.

decreases with an increase in the length-to-thickness ratio. This is because increasing the length-to-thickness ratio leads to a softer structure.

Figure 8.6 shows the dimensionless frequency of the structure for different length-to-width ratios of the plate. It can be also found that the frequency of the structure decreases with an increase in the length-to-width ratio, which is due to the higher stiffness of the system with a lower length-to-width ratio.

The effect of piezoelectric layer thickness on the dimensionless frequency is shown in Figure 8.7. It can be found that by increasing the piezoelectric layer

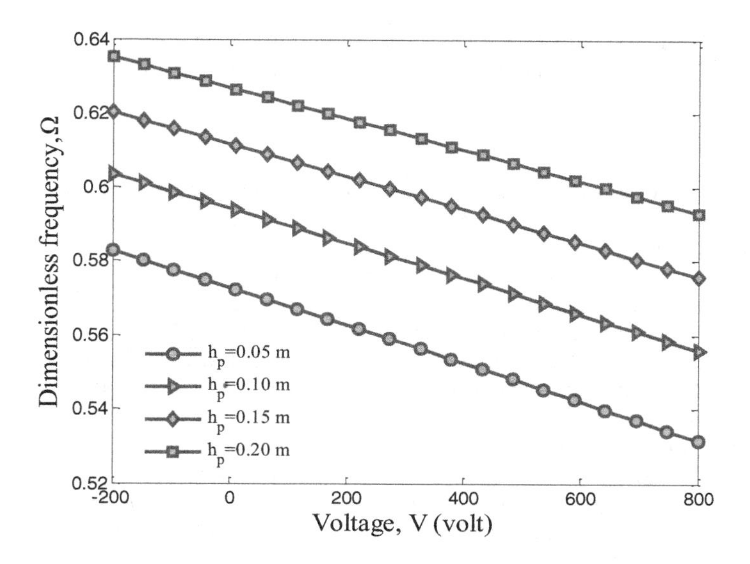


**FIGURE 8.7** Effect of piezoelectric layer thickness on the dimension frequency versus the dimension external applied voltage.

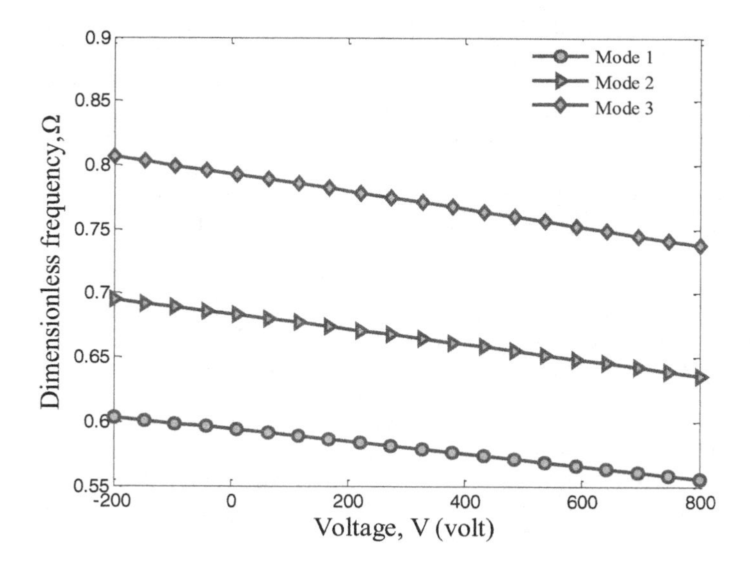


**FIGURE 8.8** Effect of mode number on the dimension frequency versus the dimension external applied voltage.

thickness, the frequency of the structure is increased. This is because by increasing the piezoelectric layer thickness, the stiffness of the structure will be improved.

The effect of mode numbers on the dimensionless frequency of the system against the external applied voltage is plotted in Figure 8.8. As can be seen, by increasing the mode numbers, the frequency increases.

**Acknowledgments:** This chapter is a slightly modified version of Ref. [28] and has been reproduced here with the permission of the copyright holder.

## REFERENCES

[1] Persson P, Persson K, Sandberg G. Numerical study on reducing building vibrations by foundation improvement. *Eng Struct* 2016;124:361–375.
[2] Lam KY, Wang CM, He XQ. Canonical exact solutions for levy-plates on two-parameter foundation using green's functions. *Eng Struct* 2000;22:364–378.
[3] Buczkowski R, Torbacki W. Finite element modelling of thick plates on two-parameter elastic foundation. *Int J Numer Anal Met* 2001;25:1409–1427.
[4] Zhou D, Cheung YK, Lo SH, Au FTK. Three-dimensional vibration analysis of rectangular thick plates on Pasternak foundation. *Int J Numer Meth Eng* 2004;59: 1313–1334.
[5] De Rosa MA, Lippiello M. Free vibrations of simply supported double plate on two models of elastic soils. *Num Anal Meth Geomech* 2008;33:331–353.
[6] Ferreira AJM, Roque CMC, Neves AMA, Jorge RMN, Soares CMM. Analysis of plates on Pasternak foundations by radial basis functions. *Comput Mecch* 2010;46:791–803.
[7] Akgoz B, Civalek O. Nonlinear vibration analysis of laminated plates resting on nonlinear two-parameters elastic foundations. *Steel Compos Struct* 2011;11:403–421.
[8] Kumar Y, Lal R. Vibrations of nonhomogeneous orthotropic rectangular plates with bilinear thickness variation resting on Winkler foundation. *Meccanica* 2012;47:893–915.
[9] Bahmyari E, Khedmati MR. Vibration analysis of nonhomogeneous moderately thick plates with point supports resting on Pasternak elastic foundation using element free [10] Galerkin method. *Eng Anal Bound Elem* 2013;37:1212–1238.
[10] Mantari JL, Granados EV, Guedes Soares C. Vibrational analysis of advanced composite plates resting on elastic foundation. *Compos Part B-Eng* 2014;66:407–419.
[11] Mantari JL, Granados EV. An original FSDT to study advanced composites on elastic foundation. *Thin Wall Struct* 2016;107:80–89.
[12] Uğurlu B. Boundary element method based vibration analysis of elastic bottom plates of fluid storage tanks resting on Pasternak foundation. *Eng Anal Bound Elem* 2016;62:163–176.
[13] Bounouara F, Benrahou KH, Belkorissat I, Tounsi A. A nonlocal zeroth-order shear deformation theory for free vibration of functionally graded nanoscale plates resting on elastic foundation. *Steel Compos Struct* 2016;20:227–249.
[14] Chen J, Li P, Song G, Ren Z. Piezo-based wireless sensor network for early-age concrete strength monitoring. *Optik* 2016;127:2983–2987.
[15] Meftah A, Bakora A, Zaoui FZ, Tounsi A, Adda Bedia EA. A non-polynomial four variable refined plate theory for free vibration of functionally graded thick rectangular plates on elastic foundation. *Steel Compos Struct* 2017;23:317–330.
[16] Alibeigloo A. Static analysis of functionally graded carbon nanotube-reinforced composite plate embedded in piezoelectric layers by using theory of elasticity. *Compos Struct* 2013;95:612–622.
[17] Chen SS, Liao KH, Shi JY. A dimensionless parametric study for forced vibrations of foundation-soil systems. *Comput Geotech* 2016;76:184–193.
[18] Sasmal S, Ravivarman N, Sindu BS, Vignesh K. Electrical conductivity and piezoresistive characteristics of CNT and CNF incorporated cementitious nanocomposites under static and dynamic loading. *Compos Part A Appl Sci* 2017;100:227–243.
[19] Xue CX, Pan E, Zhang SY, Chu HJ. Large deflection of a rectangular magnetoelectroelastic thin plate. *Mech Res Commun* 2011;38:518–523.

[20] Li Y, Zhang J. Free vibration analysis of magnetoelectroelastic plate resting on a Pasternak foundation. *Smart Mater Struct* 2014;23:025002.
[21] Shooshtari A, Razavi S. Large amplitude free vibration of symmetrically laminated magneto-electro-elastic rectangular plates on Pasternak type foundation. *Mech Res Commun* 2015;69:103–113.
[22] Ebrahimi F, Jafari A, Barati MR. Vibration analysis of magneto-electro-elastic heterogeneous porous material plates resting on elastic foundations. *Thin Wall Struct* 2017;119:33–46.
[23] Nirmala J, Dhanalakshmi G. Influence of nano materials in the distressed retaining structure for crack filling. *Constr Build Mater* 2015;88:225–231.
[24] Su Y, Li J, Wu C, Wu P, Li ZX. Influences of nano-particles on dynamic strength of ultra-high performance concrete. *Compos Part B-Eng* 2016;91:595–609.
[25] Fathi M, Yousefipour A, Hematpoury Farokhy E. Mechanical and physical properties of expanded polystyrene structural concretes containing Micro-silica and Nano-silica. *Constr Build Mater* 2017;136:590–597.
[26] Jafarian Arani A, Kolahchi R. Buckling analysis of embedded concrete columns armed with carbon nanotubes. *Comput Concrete* 2016;17:567–578.
[27] Zamanian M, Kolahchi R, Rabani Bidgo M. Agglomeration effects on the buckling behavior of embedded concrete columns reinforced with SiO2 nanoparticles. *Wind Struct* 2017;24:43–57.
[28] Kargar M, Bidgoli MR. Mathematical modeling of smart nanoparticles-reinforced concrete foundations: Vibration analysis. *Steel Compos Struct* 2018;(27)4:465–477.
[29] Kolahchi R, Hosseini H, Esmailpour M. Differential cubature and quadrature-Bolotin methods for dynamic stability of embedded piezoelectric nanoplates based on visco-nonlocal-peizoelasticity theories. *Compos Struct* 2016;157:174–186.

# 9 Forced Vibration Analysis of Plates Reinforced with Nanoparticles

## 9.1 INTRODUCTION

The vibration of plates means the effect of moving down and up due to several forces initiated by machinery, people, or the lower and upper building beams. This vibration can make people uncomfortable and create a fear of failure of structure due to it impacting people's sense of well-being and ability to complete tasks. When vibrating tools with a range of working speeds is operate on plates various problems can be produced. One of the most important vibration problems is resonance, which arises when using an external load on the plate at a specific frequency. For the correct design of the structure, it is vital to perform a precise engineering study to recognize its response to dynamic loads and particularly the resonance phenomenon. There are numerous approaches for decreasing the forced vibrations in structures, such as reinforcing the material parameters of the plate by nanoparticles, utilizing fiber-reinforced polymer layers, and so on. One of the best approaches is the usage of nanoparticles. Nanomaterials display particular characteristics with respect to conventional materials that create it probable to be useful in the field of building as an improved. Likewise, nanoparticles increase the bulk modulus of the common materials utilized in the plates. Consequently, this chapter emphases on expending nano-silica to decrease the forced vibration in the concrete plates.

Several papers have been presented on the field of forced vibration analysis of concrete plates recently. The forced vibration response of a two-layer pre-stretched plate on an elastic medium was studied by Akbarov et al. [1]. A numerical method for calculating vibration response induced by moving harmonic load on a floating discontinuous plate in a railway underground tunnel was studied by Hussein and Hunt [2]. The forced vibration induced by harmonic force and structural dynamics by earthquake load on a single-degree-of-freedom structure was presented by Rajasekaran [3]. A Mindlin–Reissner refined theory for the forced vibration of plates assuming shear deformable was presented by Batista [4]. They presented two shear deformable theories of vibration plate which can be used for the effect of the normal transverse stress. The vibration and buckling response of laminated and isotropic plates by radial basis solution functions was studied by Ferreira et al. [5]. Eftekhari and Jafari [6] investigated the application of two methods namely as differential quadrature (DQM) and Ritz methods to the vibration response of rectangular concrete plates. Lee et al. [7] applied the vertical global mode vibrations induced by human periodic group effort in a 39-story building system. Khan and Patel [8] measured the forced nonlinear vibration analysis of bimodular laminated material composite cross-ply plates under the

DOI: 10.1201/9781003349525-9

hatmonic excitation. The response was do on the basis of Bert's constitutive equations using first-order shear deformation theory on the basis of finite element numerical method (FEM), assuming geometric von Kármán nonlinearity. Ansari et al. [9] examined the forced nonlinear vibration analysis of nanocomposite plates armed by carbon nanotubes (CNTs) using numerical solution method. Song et al. [10] exposed active smart vibration control of functionally graded plates armed by CNT using higher order shear deformation model. The analytical nonlinear model for forced and free vibration analysis of composite concrete plates was offered by Gromysz [11]. The use of the surfaces sampling solution method for the 3D forced and free vibration response of composite laminated plates was studied by Kulikov et al. [12]. Soufeiani et al. [13] studied the influence of the fiber orientation and stacking laminate sequence on the dynamic analysis of fiber reinforced polymer (FRP) covered the composite plates subjected to external loads. A plate efficient element for the vibration response of composite concrete plates was gotten by Dey et al. [14]. Dynamic instability and nonlinear vibration response of thickness-tapered internally composite concrete plates subjected to external excitation were attained by Darabi and Ganesan [15]. A forced nonlinear vibration response of functionally graded axially nonuniform plates was investigated by Kumar et al. [16] on the basis of numerical method.

Besides, the mechanical response of concrete structures enclosing nanoparticles was studied numerically and experimentally by a various researchers, but work on the mathematical model applied for the dynamic response of the concrete structures with nanoparticles is novel topic. The influence of utilizing various types of nanomaterials on the mechanical parameters of high strength materials was presented by Amin and El-hassan [17]. Mohamed [18] studied the effect of nanomaterials on the compressive strength and flexural performance of the plates. Alrekabi et al. [19] calculated the mechanical parameters of novel composites with cement matrix organized with hybrid nano/micro fibers or nanofibers. In the field of mathematical model for the concrete plate, Jafarian Arani and Kolahchi [20] assumed the buckling analysis of concrete beams armed by CNTs based on the Timoshenko and Euler–Bernoulli beam theories. The large amplitude or nonlinear vibration response of a composite laminated spherical panel shell subjected to combined moisture and temperature loads was studied by Mahapatra et al. [21]. The free nonlinear vibration response of composite laminated spherical panel shell subjected to moisture and temperature loads was presented by Mahapatra and Panda [22]. Mahapatra et al. [23] presented the nonlinear geometrically transverse bending analysis of the laminated shear deformable composite spherical panel shell subjected to moisture and temperature loads. The free nonlinear vibration response of a composite laminated curved panel subjected to moisture and temperature loads was studied by Mahapatra et al. [24]. The buckling analysis of a beam armed by $SiO_2$ nanoparticles was studied by Zamanian et al. [25]. Moreover, Arbabi et al. [26] investigated the buckling analysis of concrete beams armed by zinc oxide nanoparticles under the electric load. The flexural response of the composite laminated plate covered by two various smart magnetostrictive and piezoelectric materials and the consequent displacement suppression were studied by Dutta et al. [27]. Jassas and Bidgoli [28] studied the forced vibration analysis of a concrete slab reinforced with $SiO_2$ nanoparticles is explored. The Mori-Tanaka model is utilized to derive the material properties of the nano-composite structure, taking into consideration agglomeration effects.

As far as the authors know, the influence of nanoparticles with agglomeration on the forced vibration response of the concrete plates are not studied. hence, this chapter emphases on gaining the forced vibration response of the concrete plates armed with $SiO_2$ nanoparticles. For obtaining the correspondent material parameters of the nanocomposite system, the Mori–Tanaka theory is applied. Using the Reddy model, the motion final equations are derived on the basis of Hamilton's principle. The FEM, harmonic differential quadrature method (HDQM), and Newmark numerical method are used to get the frequency curves of the structure. The influences of the boundary conditions, agglomeration and volume fraction of $SiO_2$ nanoparticles and geometrical constants of the structure on the frequency curves are studied.

## 9.2 MATHEMATICAL MODELING

A concrete plate reinforced with $SiO_2$ nanoparticles with width $b$, length $L$, and thickness $h$ subjected to harmonic load is shown in Figure 9.1. Using the Reddy higher order theory, the deflection field may be given (see Section 1.4.3).

Using Hook's law, the stress relations are

$$\sigma_{xx}^{c} = Q_{11}\varepsilon_{xx} + Q_{12}\varepsilon_{yy}, \tag{9–1}$$

$$\sigma_{yy}^{c} = Q_{12}\varepsilon_{xx} + Q_{22}\varepsilon_{yy}, \tag{9–2}$$

$$\tau_{yz}^{c} = Q_{44}\gamma_{yz}, \tag{9–3}$$

$$\tau_{xz}^{c} = Q_{55}\gamma_{zx}, \tag{9–4}$$

$$\tau_{xy}^{c} = Q_{66}\gamma_{xy}, \tag{9–5}$$

where $Q_{ij}$ are elastic coefficients that may be derived by Mori–Tanaka theory (see Section 4.3).

The total potential energy of the structure may be given as

$$U = \frac{1}{2}\int\left(\sigma_{xx}\varepsilon_{xx} + \sigma_{yy}\varepsilon_{yy} + \sigma_{xy}\gamma_{xy} + \sigma_{xz}\gamma_{xz} + \sigma_{yz}\gamma_{yz}\right)dV. \tag{9–6}$$

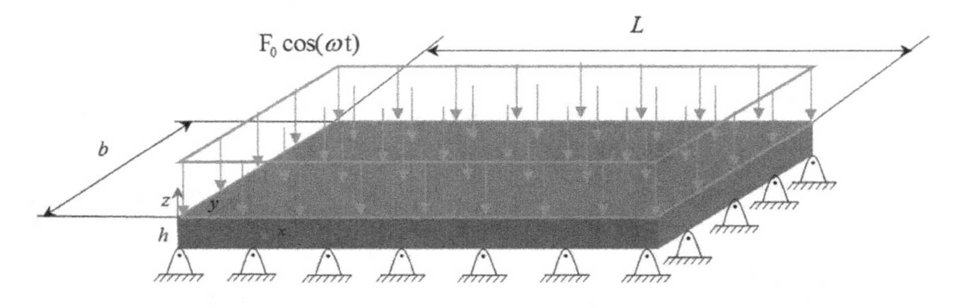


**FIGURE 9.1** Schematic for a plate armed by $SiO_2$ nanoparticles.

Combining Equations 1–33 and 1–34 with Equation 9–6 yields

$$U=\frac{1}{2}\int\left(N_{xx}\left(\frac{\partial u}{\partial x}+\frac{1}{2}\left(\frac{\partial w}{\partial x}\right)^2\right)+N_{yy}\left(\frac{\partial v}{\partial y}+\frac{1}{2}\left(\frac{\partial w}{\partial y}\right)^2\right)+Q_{yy}\left(\frac{\partial w}{\partial y}+\phi_y\right)\right.$$
$$+Q_{xx}\left(\frac{\partial w}{\partial x}+\phi_x\right)+N_{xy}\left(\frac{\partial v}{\partial x}+\frac{\partial u}{\partial y}+\frac{\partial w}{\partial x}\frac{\partial w}{\partial y}\right)+M_{xx}\frac{\partial \phi_x}{\partial x}+M_{yy}\frac{\partial \phi_y}{\partial y}$$
$$+M_{xy}\left(\frac{\partial \phi_x}{\partial y}+\frac{\partial \phi_y}{\partial x}\right)+K_{yy}\left(c_2\left(\phi_y+\frac{\partial w}{\partial y}\right)\right)+K_{xx}\left(c_2\left(\phi_x+\frac{\partial w}{\partial x}\right)\right)$$
$$+P_{xx}\left(c_1\left(\frac{\partial \phi_x}{\partial x}+\frac{\partial^2 w}{\partial x^2}\right)\right)+P_{yy}\left(c_1\left(\frac{\partial \phi_y}{\partial y}+\frac{\partial^2 w}{\partial y^2}\right)\right)$$
$$\left.+P_{xy}\left(\frac{\partial \phi_y}{\partial x}+\frac{\partial \phi_x}{\partial y}+2\frac{\partial^2 w}{\partial x\partial y}\right)\right)dA, \tag{9–7}$$

in which the stress resultant in term of deflection may be given as

$$\begin{Bmatrix} N_{xx} \\ N_{yy} \\ N_{xy} \end{Bmatrix}=\int_{-h/2}^{h/2}\begin{bmatrix} \sigma_{xx} \\ \sigma_{yy} \\ \sigma_{xy} \end{bmatrix}dz, \tag{9–8}$$

$$\begin{Bmatrix} M_{xx} \\ M_{yy} \\ M_{xy} \end{Bmatrix}=\int_{-h/2}^{h/2}\begin{bmatrix} \sigma_{xx} \\ \sigma_{yy} \\ \sigma_{xy} \end{bmatrix}zdz, \tag{9–9}$$

$$\begin{Bmatrix} P_{xx} \\ P_{yy} \\ P_{xy} \end{Bmatrix}=\int_{-h/2}^{h/2}\begin{bmatrix} \sigma_{xx} \\ \sigma_{yy} \\ \sigma_{xy} \end{bmatrix}z^3dz, \tag{9–10}$$

$$\begin{bmatrix} Q_{xx} \\ Q_{yy} \end{bmatrix}=\int_{-h/2}^{h/2}\begin{bmatrix} \sigma_{xz} \\ \sigma_{yz} \end{bmatrix}dz, \tag{9–11}$$

$$\begin{bmatrix} K_{xx} \\ K_{yy} \end{bmatrix}=\int_{-h/2}^{h/2}\begin{bmatrix} \sigma_{xz} \\ \sigma_{yz} \end{bmatrix}z^2dz. \tag{9–12}$$

Substituting Equations 9–1 through 9–5 into Equations 9–8 through 9–12, the stress resultant in terms of deflection may be derived as:

$$N_{xx} = A_{11}\left(\frac{\partial u}{\partial x}+\frac{1}{2}\left(\frac{\partial w}{\partial x}\right)^2\right)+A_{12}\left(\frac{\partial v}{\partial y}+\frac{1}{2}\left(\frac{\partial w}{\partial y}\right)^2\right)+A_{16}\left(\frac{\partial u}{\partial y}+\frac{\partial v}{\partial x}+\frac{\partial w}{\partial x}\frac{\partial w}{\partial y}\right)$$
$$+B_{11}\frac{\partial \varphi_x}{\partial x}+B_{12}\frac{\partial \varphi_y}{\partial y}+B_{16}\left(\frac{\partial \varphi_x}{\partial y}+\frac{\partial \varphi_y}{\partial x}\right)+E_{11}c_1\left(\frac{\partial \varphi_x}{\partial x}+\frac{\partial^2 w}{\partial x^2}\right)$$
$$+E_{12}c_1\left(\frac{\partial \varphi_y}{\partial y}+\frac{\partial^2 w}{\partial y^2}\right)+E_{16}c_1\left(\frac{\partial \varphi_y}{\partial x}+\frac{\partial \varphi_x}{\partial y}+2\frac{\partial^2 w}{\partial x\partial y}\right),$$
$$N_{yy} = A_{12}\left(\frac{\partial u}{\partial x}+\frac{1}{2}\left(\frac{\partial w}{\partial x}\right)^2\right)+A_{22}\left(\frac{\partial v}{\partial y}+\frac{1}{2}\left(\frac{\partial w}{\partial y}\right)^2\right)+A_{26}\left(\frac{\partial u}{\partial y}+\frac{\partial v}{\partial x}+\frac{\partial w}{\partial x}\frac{\partial w}{\partial y}\right)$$
$$+B_{12}\frac{\partial \varphi_x}{\partial x}+B_{22}\frac{\partial \varphi_y}{\partial y}+B_{26}\left(\frac{\partial \varphi_x}{\partial y}+\frac{\partial \varphi_y}{\partial x}\right)+E_{12}c_1\left(\frac{\partial \varphi_x}{\partial x}+\frac{\partial^2 w}{\partial x^2}\right)$$
$$+E_{22}c_1\left(\frac{\partial \varphi_y}{\partial y}+\frac{\partial^2 w}{\partial y^2}\right)+E_{26}c_1\left(\frac{\partial \varphi_y}{\partial x}+\frac{\partial \varphi_x}{\partial y}+2\frac{\partial^2 w}{\partial x\partial y}\right),$$
$$N_{xy} = A_{16}\left(\frac{\partial u}{\partial x}+\frac{1}{2}\left(\frac{\partial w}{\partial x}\right)^2\right)+A_{26}\left(\frac{\partial v}{\partial y}+\frac{1}{2}\left(\frac{\partial w}{\partial y}\right)^2\right)+A_{66}\left(\frac{\partial u}{\partial y}+\frac{\partial v}{\partial x}+\frac{\partial w}{\partial x}\frac{\partial w}{\partial y}\right)$$
$$+B_{16}\frac{\partial \varphi_x}{\partial x}+B_{26}\frac{\partial \varphi_y}{\partial y}+B_{66}\left(\frac{\partial \varphi_x}{\partial y}+\frac{\partial \varphi_y}{\partial x}\right)+E_{16}c_1\left(\frac{\partial \varphi_x}{\partial x}+\frac{\partial^2 w}{\partial x^2}\right)$$
$$+E_{26}c_1\left(\frac{\partial \varphi_y}{\partial y}+\frac{\partial^2 w}{\partial y^2}\right)+E_{66}c_1\left(\frac{\partial \varphi_y}{\partial x}+\frac{\partial \varphi_x}{\partial y}+2\frac{\partial^2 w}{\partial x\partial y}\right), \tag{9–13}$$

$$M_{xx} = B_{11}\left(\frac{\partial u}{\partial x}+\frac{1}{2}\left(\frac{\partial w}{\partial x}\right)^2\right)+B_{12}\left(\frac{\partial v}{\partial y}+\frac{1}{2}\left(\frac{\partial w}{\partial y}\right)^2\right)+B_{16}\left(\frac{\partial u}{\partial y}+\frac{\partial v}{\partial x}+\frac{\partial w}{\partial x}\frac{\partial w}{\partial y}\right)$$
$$+D_{11}\frac{\partial \varphi_x}{\partial x}+D_{12}\frac{\partial \varphi_y}{\partial y}+D_{16}\left(\frac{\partial \varphi_x}{\partial y}+\frac{\partial \varphi_y}{\partial x}\right)+F_{11}c_1\left(\frac{\partial \varphi_x}{\partial x}+\frac{\partial^2 w}{\partial x^2}\right)$$
$$+F_{12}c_1\left(\frac{\partial \varphi_y}{\partial y}+\frac{\partial^2 w}{\partial y^2}\right)+F_{16}c_1\left(\frac{\partial \varphi_y}{\partial x}+\frac{\partial \varphi_x}{\partial y}+2\frac{\partial^2 w}{\partial x\partial y}\right),$$
$$M_{yy} = B_{12}\left(\frac{\partial u}{\partial x}+\frac{1}{2}\left(\frac{\partial w}{\partial x}\right)^2\right)+B_{22}\left(\frac{\partial v}{\partial y}+\frac{1}{2}\left(\frac{\partial w}{\partial y}\right)^2\right)+B_{26}\left(\frac{\partial u}{\partial y}+\frac{\partial v}{\partial x}+\frac{\partial w}{\partial x}\frac{\partial w}{\partial y}\right)$$
$$+D_{12}\frac{\partial \varphi_x}{\partial x}+D_{22}\frac{\partial \varphi_y}{\partial y}+D_{26}\left(\frac{\partial \varphi_x}{\partial y}+\frac{\partial \varphi_y}{\partial x}\right)+F_{12}c_1\left(\frac{\partial \varphi_x}{\partial x}+\frac{\partial^2 w}{\partial x^2}\right)$$
$$+F_{22}c_1\left(\frac{\partial \varphi_y}{\partial y}+\frac{\partial^2 w}{\partial y^2}\right)+F_{26}c_1\left(\frac{\partial \varphi_y}{\partial x}+\frac{\partial \varphi_x}{\partial y}+2\frac{\partial^2 w}{\partial x\partial y}\right),$$

$$M_{xy}=B_{16}\left(\frac{\partial u}{\partial x}+\frac{1}{2}\left(\frac{\partial w}{\partial x}\right)^2\right)+B_{26}\left(\frac{\partial v}{\partial y}+\frac{1}{2}\left(\frac{\partial w}{\partial y}\right)^2\right)+B_{66}\left(\frac{\partial u}{\partial y}+\frac{\partial v}{\partial x}+\frac{\partial w}{\partial x}\frac{\partial w}{\partial y}\right)$$

$$+D_{16}\frac{\partial \varphi_x}{\partial x}+D_{26}\frac{\partial \varphi_y}{\partial y}+D_{66}\left(\frac{\partial \varphi_x}{\partial y}+\frac{\partial \varphi_y}{\partial x}\right)+F_{16}c_1\left(\frac{\partial \varphi_x}{\partial x}+\frac{\partial^2 w}{\partial x^2}\right)$$

$$+F_{26}c_1\left(\frac{\partial \varphi_y}{\partial y}+\frac{\partial^2 w}{\partial y^2}\right)+F_{66}c_1\left(\frac{\partial \varphi_y}{\partial x}+\frac{\partial \varphi_x}{\partial y}+2\frac{\partial^2 w}{\partial x\partial y}\right), \tag{9–14}$$

$$P_{xx}=E_{11}\left(\frac{\partial u}{\partial x}+\frac{1}{2}\left(\frac{\partial w}{\partial x}\right)^2\right)+E_{12}\left(\frac{\partial v}{\partial y}+\frac{1}{2}\left(\frac{\partial w}{\partial y}\right)^2\right)+E_{16}\left(\frac{\partial u}{\partial y}+\frac{\partial v}{\partial x}+\frac{\partial w}{\partial x}\frac{\partial w}{\partial y}\right)$$

$$+F_{11}\frac{\partial \varphi_x}{\partial x}+F_{12}\frac{\partial \varphi_y}{\partial y}+F_{16}\left(\frac{\partial \varphi_x}{\partial y}+\frac{\partial \varphi_y}{\partial x}\right)+H_{11}c_1\left(\frac{\partial \varphi_x}{\partial x}+\frac{\partial^2 w}{\partial x^2}\right)$$

$$+H_{12}c_1\left(\frac{\partial \varphi_y}{\partial y}+\frac{\partial^2 w}{\partial y^2}\right)+H_{16}c_1\left(\frac{\partial \varphi_y}{\partial x}+\frac{\partial \varphi_x}{\partial y}+2\frac{\partial^2 w}{\partial x\partial y}\right),$$

$$P_{yy}=E_{21}\left(\frac{\partial u}{\partial x}+\frac{1}{2}\left(\frac{\partial w}{\partial x}\right)^2\right)+E_{22}\left(\frac{\partial v}{\partial y}+\frac{1}{2}\left(\frac{\partial w}{\partial y}\right)^2\right)+E_{26}\left(\frac{\partial u}{\partial y}+\frac{\partial v}{\partial x}+\frac{\partial w}{\partial x}\frac{\partial w}{\partial y}\right)$$

$$+F_{12}\frac{\partial \varphi_x}{\partial x}+F_{22}\frac{\partial \varphi_y}{\partial y}+F_{26}\left(\frac{\partial \varphi_x}{\partial y}+\frac{\partial \varphi_y}{\partial x}\right)+H_{12}c_1\left(\frac{\partial \varphi_x}{\partial x}+\frac{\partial^2 w}{\partial x^2}\right)$$

$$+H_{22}c_1\left(\frac{\partial \varphi_y}{\partial y}+\frac{\partial^2 w}{\partial y^2}\right)+H_{26}c_1\left(\frac{\partial \varphi_y}{\partial x}+\frac{\partial \varphi_x}{\partial y}+2\frac{\partial^2 w}{\partial x\partial y}\right),$$

$$P_{xy}=E_{16}\left(\frac{\partial u}{\partial x}+\frac{1}{2}\left(\frac{\partial w}{\partial x}\right)^2\right)+E_{26}\left(\frac{\partial v}{\partial y}+\frac{1}{2}\left(\frac{\partial w}{\partial y}\right)^2\right)+E_{66}\left(\frac{\partial u}{\partial y}+\frac{\partial v}{\partial x}+\frac{\partial w}{\partial x}\frac{\partial w}{\partial y}\right)$$

$$+F_{16}\frac{\partial \varphi_x}{\partial x}+F_{26}\frac{\partial \varphi_y}{\partial y}+F_{66}\left(\frac{\partial \varphi_x}{\partial y}+\frac{\partial \varphi_y}{\partial x}\right)+H_{16}c_1\left(\frac{\partial \varphi_x}{\partial x}+\frac{\partial^2 w}{\partial x^2}\right)$$

$$+H_{26}c_1\left(\frac{\partial \varphi_y}{\partial y}+\frac{\partial^2 w}{\partial y^2}\right)+H_{66}c_1\left(\frac{\partial \varphi_y}{\partial x}+\frac{\partial \varphi_x}{\partial y}+2\frac{\partial^2 w}{\partial x\partial y}\right), \tag{9–15}$$

$$Q_{xx}=A_{55}\left(\frac{\partial w}{\partial x}+\varphi_x\right)+A_{45}\left(\frac{\partial w}{\partial y}+\varphi_y\right)+D_{55}c_2\left(\varphi_x+\frac{\partial w}{\partial x}\right)+D_{45}c_2\left(\frac{\partial w}{\partial y}+\varphi_y\right),$$

$$Q_{yy}=A_{45}\left(\frac{\partial w}{\partial y}+\varphi_y\right)+A_{44}\left(\frac{\partial w}{\partial y}+\varphi_y\right)+D_{45}c_2\left(\varphi_y+\frac{\partial w}{\partial y}\right)+D_{44}c_2\left(\frac{\partial w}{\partial y}+\varphi_y\right), \tag{9–16}$$

$$K_{xx} = D_{55}\left(\frac{\partial w}{\partial x}+\varphi_x\right)+D_{45}\left(\frac{\partial w}{\partial y}+\varphi_y\right)+F_{55}c_2\left(\varphi_x+\frac{\partial w}{\partial x}\right)+F_{45}c_2\left(\frac{\partial w}{\partial y}+\varphi_y\right),$$

$$K_{yy} = D_{45}\left(\frac{\partial w}{\partial y}+\varphi_y\right)+D_{44}\left(\frac{\partial w}{\partial y}+\varphi_y\right)+F_{45}c_2\left(\varphi_y+\frac{\partial w}{\partial y}\right)+F_{44}c_2\left(\frac{\partial w}{\partial y}+\varphi_y\right), \quad (9\text{–}17)$$

where

$$A_{ij} = \int_{-h/2}^{h/2} Q_{ij}\,dz, \qquad (i,j=1,2,6) \quad (9\text{–}18)$$

$$B_{ij} = \int_{-h/2}^{h/2} Q_{ij} z\,dz, \quad (9\text{–}19)$$

$$D_{ij} = \int_{-h/2}^{h/2} Q_{ij} z^2\,dz, \quad (9\text{–}20)$$

$$E_{ij} = \int_{-h/2}^{h/2} Q_{ij} z^3\,dz, \quad (9\text{–}21)$$

$$F_{ij} = \int_{-h/2}^{h/2} Q_{ij} z^4\,dz, \quad (9\text{–}22)$$

$$H_{ij} = \int_{-h/2}^{h/2} Q_{ij} z^6\,dz. \quad (9\text{–}23)$$

The total kinetic energy of the structure cam be given as

$$K = \frac{\rho}{2}\int\left(\dot{u}_1^{\,2}+\dot{u}_2^{\,2}+\dot{u}_3^{\,2}\right)dV. \quad (9\text{–}24)$$

The external force induced by harmonic load may be given as

$$W_e = -\int\int\left(F_0\cos(\omega\,\mathrm{t})\right)w\,dA, \quad (9\text{–}25)$$

where $\omega$ and $F_0$ are the excitation frequency and amplitude, respectively. The governing final equations may be calculated based on Hamilton's principle as:

$$\int_0^t (\delta U-\delta K-\delta W_e)dt = 0. \quad (9\text{–}26)$$

Substituting Equations 9–7, 9–24, and 9–25 into Equation 9–26 yields the below governing final equations:

$$\delta u: \frac{\partial N_{xx}}{\partial x}+\frac{\partial N_{xy}}{\partial y} = I_0\frac{\partial^2 u}{\partial t^2}+J_1\frac{\partial^2 \phi_x}{\partial t^2}-\frac{4I_3}{h^2}\frac{\partial^3 w}{\partial t^2\partial x}, \quad (9\text{–}27)$$

$$\delta v: \frac{\partial N_{xy}}{\partial x} + \frac{\partial N_{yy}}{\partial y} = I_0 \frac{\partial^2 v}{\partial t^2} + J_1 \frac{\partial^2 \phi_y}{\partial t^2} - \frac{4I_3}{h^2} \frac{\partial^3 w}{\partial t^2 \partial y}, \tag{9–28}$$

$$\delta w: \frac{\partial Q_{xx}}{\partial x} + \frac{\partial Q_{yy}}{\partial y} + c_2 \left( \frac{\partial K_{xx}}{\partial x} + \frac{\partial K_{yy}}{\partial y} \right) + N_{xx} \frac{\partial^2 w}{\partial x^2} + N_{yy} \frac{\partial^2 w}{\partial y^2}$$

$$-c_1 \left( \frac{\partial^2 P_{xx}}{\partial x^2} + 2\frac{\partial^2 P_{xy}}{\partial x \partial y} + \frac{\partial^2 P_{yy}}{\partial y^2} \right) - K_w w = I_0 \frac{\partial^2 w}{\partial t^2} - \left( \frac{4}{3h^2} \right)^2$$

$$I_6 \left( \frac{\partial^4 w}{\partial x^2 \partial t^2} + \frac{\partial^4 w}{\partial y^2 \partial t^2} \right) + \frac{4}{3h^2} \left( I_3 \frac{\partial^3 u}{\partial t^2 \partial x} + I_3 \frac{\partial^3 v}{\partial t^2 \partial y} + J_4 \left( \frac{\partial^3 \phi_x}{\partial t^2 \partial x} + \frac{\partial^3 \phi_\theta}{\partial t^2 \partial y} \right) \right), \tag{9–29}$$

$$\delta \phi_x: \frac{\partial M_{xx}}{\partial x} + \frac{\partial M_{xy}}{\partial y} + c_1 \left( \frac{\partial P_{xx}}{\partial x} + \frac{\partial P_{xy}}{\partial y} \right) - Q_{xx} - c_2 K_{xx} = J_1 \frac{\partial^2 u}{\partial t^2}$$

$$+K_2 \frac{\partial^2 \phi_x}{\partial t^2} - \frac{4}{3h^2} J_4 \frac{\partial^3 w}{\partial t^2 \partial x}, \tag{9–30}$$

$$\delta \phi_y: \frac{\partial M_{xy}}{\partial x} + \frac{\partial M_{yy}}{\partial y} + c_1 \left( \frac{\partial P_{xy}}{\partial x} + \frac{\partial P_{yy}}{\partial y} \right) - Q_{yy} - c_2 K_{yy} = J_1 \frac{\partial^2 v}{\partial t^2}$$

$$+K_2 \frac{\partial^2 \phi_\theta}{\partial t^2} - \frac{4}{3h^2} J_4 \frac{\partial^3 w}{\partial t^2 \partial y}, \tag{9–31}$$

where

$$I_i = \int_{-h/2}^{h/2} \rho z^i dz \qquad (i = 0,1,\ldots,6), \tag{9–32}$$

$$J_i = I_i - \frac{4}{3h^2} I_{i+2} \qquad (i = 1,4), \tag{9–33}$$

$$K_2 = I_2 - \frac{8}{3h^2} I_4 + \left( \frac{4}{3h^2} \right)^2 I_6. \tag{9–34}$$

Replacing Equations 9–13 to 9–17 into Equations 9–27 through 9–31, the governing final equations may be given as:

$$\frac{\partial}{\partial x} \left[ \begin{array}{l} A_{11} \left( \frac{\partial u}{\partial x} + \frac{1}{2} \left( \frac{\partial w}{\partial x} \right)^2 \right) + A_{12} \left( \frac{\partial v}{\partial y} + \frac{1}{2} \left( \frac{\partial w}{\partial y} \right)^2 \right) + A_{16} \left( \frac{\partial u}{\partial y} + \frac{\partial v}{\partial x} + \frac{\partial w}{\partial x} \frac{\partial w}{\partial y} \right) \\ + B_{11} \frac{\partial \varphi_x}{\partial x} + B_{12} \frac{\partial \varphi_y}{\partial y} + B_{16} \left( \frac{\partial \varphi_x}{\partial y} + \frac{\partial \varphi_y}{\partial x} \right) + E_{11} c_1 \left( \frac{\partial \varphi_x}{\partial x} + \frac{\partial^2 w}{\partial x^2} \right) \\ + E_{12} c_1 \left( \frac{\partial \varphi_y}{\partial y} + \frac{\partial^2 w}{\partial y^2} \right) + E_{16} c_1 \left( \frac{\partial \varphi_y}{\partial x} + \frac{\partial \varphi_x}{\partial y} + 2 \frac{\partial^2 w}{\partial x \partial y} \right) \end{array} \right]$$

$$+\frac{\partial}{\partial y}\left[\begin{array}{l} A_{16}\left(\frac{\partial u}{\partial x}+\frac{1}{2}\left(\frac{\partial w}{\partial x}\right)^2\right)+A_{26}\left(\frac{\partial v}{\partial y}+\frac{1}{2}\left(\frac{\partial w}{\partial y}\right)^2\right)+A_{66}\left(\frac{\partial u}{\partial y}+\frac{\partial v}{\partial x}+\frac{\partial w}{\partial x}\frac{\partial w}{\partial y}\right) \\ +B_{16}\frac{\partial \varphi_x}{\partial x}+B_{26}\frac{\partial \varphi_y}{\partial y}+B_{66}\left(\frac{\partial \varphi_x}{\partial y}+\frac{\partial \varphi_y}{\partial x}\right)+E_{16}c_1\left(\frac{\partial \varphi_x}{\partial x}+\frac{\partial^2 w}{\partial x^2}\right) \\ +E_{26}c_1\left(\frac{\partial \varphi_y}{\partial y}+\frac{\partial^2 w}{\partial y^2}\right)+E_{66}c_1\left(\frac{\partial \varphi_y}{\partial x}+\frac{\partial \varphi_x}{\partial y}+2\frac{\partial^2 w}{\partial x \partial y}\right) \end{array}\right]$$

$$=I_0\frac{\partial^2 u}{\partial t^2}+J_1\frac{\partial^2 \phi_x}{\partial t^2}-\frac{4I_3}{h^2}\frac{\partial^3 w}{\partial t^2 \partial x}, \tag{9–35}$$

$$\frac{\partial}{\partial x}\left[\begin{array}{l} A_{16}\left(\frac{\partial u}{\partial x}+\frac{1}{2}\left(\frac{\partial w}{\partial x}\right)^2\right)+A_{26}\left(\frac{\partial v}{\partial y}+\frac{1}{2}\left(\frac{\partial w}{\partial y}\right)^2\right)+A_{66}\left(\frac{\partial u}{\partial y}+\frac{\partial v}{\partial x}+\frac{\partial w}{\partial x}\frac{\partial w}{\partial y}\right) \\ +B_{16}\frac{\partial \varphi_x}{\partial x}+B_{26}\frac{\partial \varphi_y}{\partial y}+B_{66}\left(\frac{\partial \varphi_x}{\partial y}+\frac{\partial \varphi_y}{\partial x}\right)+E_{16}c_1\left(\frac{\partial \varphi_x}{\partial x}+\frac{\partial^2 w}{\partial x^2}\right) \\ +E_{26}c_1\left(\frac{\partial \varphi_y}{\partial y}+\frac{\partial^2 w}{\partial y^2}\right)+E_{66}c_1\left(\frac{\partial \varphi_y}{\partial x}+\frac{\partial \varphi_x}{\partial y}+2\frac{\partial^2 w}{\partial x \partial y}\right) \end{array}\right]$$

$$+\frac{\partial}{\partial y}\left[\begin{array}{l} A_{12}\left(\frac{\partial u}{\partial x}+\frac{1}{2}\left(\frac{\partial w}{\partial x}\right)^2\right)+A_{22}\left(\frac{\partial v}{\partial y}+\frac{1}{2}\left(\frac{\partial w}{\partial y}\right)^2\right)+A_{26}\left(\frac{\partial u}{\partial y}+\frac{\partial v}{\partial x}+\frac{\partial w}{\partial x}\frac{\partial w}{\partial y}\right) \\ +B_{12}\frac{\partial \varphi_x}{\partial x}+B_{22}\frac{\partial \varphi_y}{\partial y}+B_{26}\left(\frac{\partial \varphi_x}{\partial y}+\frac{\partial \varphi_y}{\partial x}\right)+E_{12}c_1\left(\frac{\partial \varphi_x}{\partial x}+\frac{\partial^2 w}{\partial x^2}\right) \\ +E_{22}c_1\left(\frac{\partial \varphi_y}{\partial y}+\frac{\partial^2 w}{\partial y^2}\right)+E_{26}c_1\left(\frac{\partial \varphi_y}{\partial x}+\frac{\partial \varphi_x}{\partial y}+2\frac{\partial^2 w}{\partial x \partial y}\right) \end{array}\right]$$

$$=I_0\frac{\partial^2 v}{\partial t^2}+J_1\frac{\partial^2 \phi_y}{\partial t^2}-\frac{4I_3}{h^2}\frac{\partial^3 w}{\partial t^2 \partial y}, \tag{9–36}$$

$$\frac{\partial}{\partial x}\left[A_{55}\left(\frac{\partial w}{\partial x}+\varphi_x\right)+A_{45}\left(\frac{\partial w}{\partial y}+\varphi_y\right)+D_{55}c_2\left(\varphi_x+\frac{\partial w}{\partial x}\right)+D_{45}c_2\left(\frac{\partial w}{\partial y}+\varphi_y\right)\right]$$

$$+\frac{\partial}{\partial y}\left[A_{45}\left(\frac{\partial w}{\partial y}+\varphi_y\right)+A_{44}\left(\frac{\partial w}{\partial y}+\varphi_y\right)+D_{45}c_2\left(\varphi_y+\frac{\partial w}{\partial y}\right)+D_{44}c_2\left(\frac{\partial w}{\partial y}+\varphi_y\right)\right]$$

$$+c_2\left(\begin{array}{l}\frac{\partial}{\partial x}\left[D_{55}\left(\frac{\partial w}{\partial x}+\varphi_x\right)+D_{45}\left(\frac{\partial w}{\partial y}+\varphi_y\right)+F_{55}c_2\left(\varphi_x+\frac{\partial w}{\partial x}\right)+F_{45}c_2\left(\frac{\partial w}{\partial y}+\varphi_y\right)\right]\\+\frac{\partial}{\partial y}\left[D_{45}\left(\frac{\partial w}{\partial y}+\varphi_y\right)+D_{44}\left(\frac{\partial w}{\partial y}+\varphi_y\right)+F_{45}c_2\left(\varphi_y+\frac{\partial w}{\partial y}\right)+F_{44}c_2\left(\frac{\partial w}{\partial y}+\varphi_y\right)\right]\end{array}\right)$$

$$+N_{xx}\frac{\partial^2 w}{\partial x^2}+N_{yy}\frac{\partial^2 w}{\partial y^2}-F_o\cos(\omega\,\mathrm{t})-c_1\left(\frac{\partial^2}{\partial x^2}\left[E_{11}\left(\frac{\partial u}{\partial x}+\frac{1}{2}\left(\frac{\partial w}{\partial x}\right)^2\right)\right.\right.$$

$$+E_{12}\left(\frac{\partial v}{\partial y}+\frac{1}{2}\left(\frac{\partial w}{\partial y}\right)^2\right)+E_{16}\left(\frac{\partial u}{\partial y}+\frac{\partial v}{\partial x}+\frac{\partial w}{\partial x}\frac{\partial w}{\partial y}\right)+F_{11}\frac{\partial\varphi_x}{\partial x}$$

$$+F_{12}\frac{\partial\varphi_y}{\partial y}+F_{16}\left(\frac{\partial\varphi_x}{\partial y}+\frac{\partial\varphi_y}{\partial x}\right)+H_{11}c_1\left(\frac{\partial\varphi_x}{\partial x}+\frac{\partial^2 w}{\partial x^2}\right)+H_{12}c_1\left(\frac{\partial\varphi_y}{\partial y}+\frac{\partial^2 w}{\partial y^2}\right)$$

$$+H_{16}c_1\left(\frac{\partial\varphi_y}{\partial x}+\frac{\partial\varphi_x}{\partial y}+2\frac{\partial^2 w}{\partial x\partial y}\right)\Bigg]+2\frac{\partial^2}{\partial x\partial y}\left[E_{16}\left(\frac{\partial u}{\partial x}+\frac{1}{2}\left(\frac{\partial w}{\partial x}\right)^2\right)\right.$$

$$+E_{26}\left(\frac{\partial v}{\partial y}+\frac{1}{2}\left(\frac{\partial w}{\partial y}\right)^2\right)+E_{66}\left(\frac{\partial u}{\partial y}+\frac{\partial v}{\partial x}+\frac{\partial w}{\partial x}\frac{\partial w}{\partial y}\right)+F_{16}\frac{\partial\varphi_x}{\partial x}+F_{26}\frac{\partial\varphi_y}{\partial y}$$

$$+F_{66}\left(\frac{\partial\varphi_x}{\partial y}+\frac{\partial\varphi_y}{\partial x}\right)+H_{16}c_1\left(\frac{\partial\varphi_x}{\partial x}+\frac{\partial^2 w}{\partial x^2}\right)+H_{26}c_1\left(\frac{\partial\varphi_y}{\partial y}+\frac{\partial^2 w}{\partial y^2}\right)$$

$$+H_{66}c_1\left(\frac{\partial\varphi_y}{\partial x}+\frac{\partial\varphi_x}{\partial y}+2\frac{\partial^2 w}{\partial x\partial y}\right)\Bigg]+\frac{\partial^2}{\partial y^2}\left[E_{21}\left(\frac{\partial u}{\partial x}+\frac{1}{2}\left(\frac{\partial w}{\partial x}\right)^2\right)\right.$$

$$+E_{22}\left(\frac{\partial v}{\partial y}+\frac{1}{2}\left(\frac{\partial w}{\partial y}\right)^2\right)+E_{26}\left(\frac{\partial u}{\partial y}+\frac{\partial v}{\partial x}+\frac{\partial w}{\partial x}\frac{\partial w}{\partial y}\right)+F_{12}\frac{\partial\varphi_x}{\partial x}$$

$$+F_{22}\frac{\partial\varphi_y}{\partial y}+F_{26}\left(\frac{\partial\varphi_x}{\partial y}+\frac{\partial\varphi_y}{\partial x}\right)+H_{12}c_1\left(\frac{\partial\varphi_x}{\partial x}+\frac{\partial^2 w}{\partial x^2}\right)+H_{22}c_1\left(\frac{\partial\varphi_y}{\partial y}+\frac{\partial^2 w}{\partial y^2}\right)$$

$$+H_{26}c_1\left(\frac{\partial\varphi_y}{\partial x}+\frac{\partial\varphi_x}{\partial y}+2\frac{\partial^2 w}{\partial x\partial y}\right)\Bigg]=I_0\frac{\partial^2 w}{\partial t^2}-\left(\frac{4}{3h^2}\right)^2 I_6\left(\frac{\partial^4 w}{\partial x^2\partial t^2}+\frac{\partial^4 w}{\partial y^2\partial t^2}\right)$$

$$+\frac{4}{3h^2}\left(I_3\frac{\partial^3 u}{\partial t^2\partial x}+I_3\frac{\partial^3 v}{\partial t^2\partial y}+J_4\left(\frac{\partial^3\phi_x}{\partial t^2\partial x}+\frac{\partial^3\phi_\theta}{\partial t^2\partial y}\right)\right),\qquad(9\text{–}37)$$

$$\frac{\partial}{\partial x}\left[B_{11}\left(\frac{\partial u}{\partial x}+\frac{1}{2}\left(\frac{\partial w}{\partial x}\right)^2\right)+B_{12}\left(\frac{\partial v}{\partial y}+\frac{1}{2}\left(\frac{\partial w}{\partial y}\right)^2\right)+B_{16}\left(\frac{\partial u}{\partial y}+\frac{\partial v}{\partial x}+\frac{\partial w}{\partial x}\frac{\partial w}{\partial y}\right)\right.$$

$$+D_{11}\frac{\partial \varphi_x}{\partial x}+D_{12}\frac{\partial \varphi_y}{\partial y}+D_{16}\left(\frac{\partial \varphi_x}{\partial y}+\frac{\partial \varphi_y}{\partial x}\right)+F_{11}c_1\left(\frac{\partial \varphi_x}{\partial x}+\frac{\partial^2 w}{\partial x^2}\right)$$

$$\left.+F_{12}c_1\left(\frac{\partial \varphi_y}{\partial y}+\frac{\partial^2 w}{\partial y^2}\right)+F_{16}c_1\left(\frac{\partial \varphi_y}{\partial x}+\frac{\partial \varphi_x}{\partial y}+2\frac{\partial^2 w}{\partial x\partial y}\right)\right]+\frac{\partial}{\partial y}$$

$$\left[B_{16}\left(\frac{\partial u}{\partial x}+\frac{1}{2}\left(\frac{\partial w}{\partial x}\right)^2\right)+B_{26}\left(\frac{\partial v}{\partial y}+\frac{1}{2}\left(\frac{\partial w}{\partial y}\right)^2\right)+B_{66}\left(\frac{\partial u}{\partial y}+\frac{\partial v}{\partial x}+\frac{\partial w}{\partial x}\frac{\partial w}{\partial y}\right)\right.$$

$$+D_{16}\frac{\partial \varphi_x}{\partial x}+D_{26}\frac{\partial \varphi_y}{\partial y}+D_{66}\left(\frac{\partial \varphi_x}{\partial y}+\frac{\partial \varphi_y}{\partial x}\right)+F_{16}c_1\left(\frac{\partial \varphi_x}{\partial x}+\frac{\partial^2 w}{\partial x^2}\right)$$

$$\left.+F_{26}c_1\left(\frac{\partial \varphi_y}{\partial y}+\frac{\partial^2 w}{\partial y^2}\right)+F_{66}c_1\left(\frac{\partial \varphi_y}{\partial x}+\frac{\partial \varphi_x}{\partial y}+2\frac{\partial^2 w}{\partial x\partial y}\right)\right]$$

$$+c_1\left(\frac{\partial}{\partial x}\left[E_{11}\left(\frac{\partial u}{\partial x}+\frac{1}{2}\left(\frac{\partial w}{\partial x}\right)^2\right)+E_{12}\left(\frac{\partial v}{\partial y}+\frac{1}{2}\left(\frac{\partial w}{\partial y}\right)^2\right)+E_{16}\left(\frac{\partial u}{\partial y}+\frac{\partial v}{\partial x}+\frac{\partial w}{\partial x}\frac{\partial w}{\partial y}\right)\right.\right.$$

$$+F_{11}\frac{\partial \varphi_x}{\partial x}+F_{12}\frac{\partial \varphi_y}{\partial y}+F_{16}\left(\frac{\partial \varphi_x}{\partial y}+\frac{\partial \varphi_y}{\partial x}\right)+H_{11}c_1\left(\frac{\partial \varphi_x}{\partial x}+\frac{\partial^2 w}{\partial x^2}\right)+H_{12}c_1\left(\frac{\partial \varphi_y}{\partial y}+\frac{\partial^2 w}{\partial y^2}\right)$$

$$\left.+H_{16}c_1\left(\frac{\partial \varphi_y}{\partial x}+\frac{\partial \varphi_x}{\partial y}+2\frac{\partial^2 w}{\partial x\partial y}\right)\right]$$

$$+\frac{\partial}{\partial y}\left[E_{16}\left(\frac{\partial u}{\partial x}+\frac{1}{2}\left(\frac{\partial w}{\partial x}\right)^2\right)+E_{26}\left(\frac{\partial v}{\partial y}+\frac{1}{2}\left(\frac{\partial w}{\partial y}\right)^2\right)+E_{66}\left(\frac{\partial u}{\partial y}+\frac{\partial v}{\partial x}+\frac{\partial w}{\partial x}\frac{\partial w}{\partial y}\right)+F_{16}\frac{\partial \varphi_x}{\partial x}\right.$$

$$+F_{26}\frac{\partial \varphi_y}{\partial y}+F_{66}\left(\frac{\partial \varphi_x}{\partial y}+\frac{\partial \varphi_y}{\partial x}\right)+H_{16}c_1\left(\frac{\partial \varphi_x}{\partial x}+\frac{\partial^2 w}{\partial x^2}\right)+H_{26}c_1\left(\frac{\partial \varphi_y}{\partial y}+\frac{\partial^2 w}{\partial y^2}\right)$$

$$\left.\left.+H_{66}c_1\left(\frac{\partial \varphi_y}{\partial x}+\frac{\partial \varphi_x}{\partial y}+2\frac{\partial^2 w}{\partial x\partial y}\right)\right]\right)-A_{55}\left(\frac{\partial w}{\partial x}+\varphi_x\right)+A_{45}\left(\frac{\partial w}{\partial y}+\varphi_y\right)+D_{55}c_2\left(\varphi_x+\frac{\partial w}{\partial x}\right)$$

$$+D_{45}c_2\left(\frac{\partial w}{\partial y}+\varphi_y\right)-c_2\left[{}_{55}\left(\frac{\partial w}{\partial x}+\varphi_x\right)+D_{45}\left(\frac{\partial w}{\partial y}+\varphi_y\right)+F_{55}c_2\left(\varphi_x+\frac{\partial w}{\partial x}\right)\right.$$

$$\left.+F_{45}c_2\left(\frac{\partial w}{\partial y}+\varphi_y\right)\right]=J_1\frac{\partial^2 u}{\partial t^2}+K_2\frac{\partial^2 \phi_x}{\partial t^2}-\frac{4}{3h^2}J_4\frac{\partial^3 w}{\partial t^2\partial x},\tag{9–38}$$

$$
\begin{aligned}
&\frac{\partial}{\partial x}\Bigg[B_{16}\left(\frac{\partial u}{\partial x}+\frac{1}{2}\left(\frac{\partial w}{\partial x}\right)^2\right)+B_{26}\left(\frac{\partial v}{\partial y}+\frac{1}{2}\left(\frac{\partial w}{\partial y}\right)^2\right)+B_{66}\left(\frac{\partial u}{\partial y}+\frac{\partial v}{\partial x}+\frac{\partial w}{\partial x}\frac{\partial w}{\partial y}\right)\\
&+D_{16}\frac{\partial \varphi_x}{\partial x}+D_{26}\frac{\partial \varphi_y}{\partial y}+D_{66}\left(\frac{\partial \varphi_x}{\partial y}+\frac{\partial \varphi_y}{\partial x}\right)+F_{16}c_1\left(\frac{\partial \varphi_x}{\partial x}+\frac{\partial^2 w}{\partial x^2}\right)\\
&+F_{26}c_1\left(\frac{\partial \varphi_y}{\partial y}+\frac{\partial^2 w}{\partial y^2}\right)+F_{66}c_1\left(\frac{\partial \varphi_y}{\partial x}+\frac{\partial \varphi_x}{\partial y}+2\frac{\partial^2 w}{\partial x\partial y}\right)\Bigg]\\
&+\frac{\partial}{\partial y}\Bigg[B_{12}\left(\frac{\partial u}{\partial x}+\frac{1}{2}\left(\frac{\partial w}{\partial x}\right)^2\right)+B_{22}\left(\frac{\partial v}{\partial y}+\frac{1}{2}\left(\frac{\partial w}{\partial y}\right)^2\right)\\
&+B_{26}\left(\frac{\partial u}{\partial y}+\frac{\partial v}{\partial x}+\frac{\partial w}{\partial x}\frac{\partial w}{\partial y}\right)+D_{12}\frac{\partial \varphi_x}{\partial x}+D_{22}\frac{\partial \varphi_y}{\partial y}\\
&+D_{26}\left(\frac{\partial \varphi_x}{\partial y}+\frac{\partial \varphi_y}{\partial x}\right)+F_{12}c_1\left(\frac{\partial \varphi_x}{\partial x}+\frac{\partial^2 w}{\partial x^2}\right)+F_{22}c_1\left(\frac{\partial \varphi_y}{\partial y}+\frac{\partial^2 w}{\partial y^2}\right)\\
&+F_{26}c_1\left(\frac{\partial \varphi_y}{\partial x}+\frac{\partial \varphi_x}{\partial y}+2\frac{\partial^2 w}{\partial x\partial y}\right)\Bigg]+c_1\Bigg(\frac{\partial}{\partial x}\Bigg[E_{16}\left(\frac{\partial u}{\partial x}+\frac{1}{2}\left(\frac{\partial w}{\partial x}\right)^2\right)\\
&+E_{26}\left(\frac{\partial v}{\partial y}+\frac{1}{2}\left(\frac{\partial w}{\partial y}\right)^2\right)+E_{66}\left(\frac{\partial u}{\partial y}+\frac{\partial v}{\partial x}+\frac{\partial w}{\partial x}\frac{\partial w}{\partial y}\right)+F_{16}\frac{\partial \varphi_x}{\partial x}+F_{26}\frac{\partial \varphi_y}{\partial y}\\
&+F_{66}\left(\frac{\partial \varphi_x}{\partial y}+\frac{\partial \varphi_y}{\partial x}\right)+H_{16}c_1\left(\frac{\partial \varphi_x}{\partial x}+\frac{\partial^2 w}{\partial x^2}\right)+H_{26}c_1\left(\frac{\partial \varphi_y}{\partial y}+\frac{\partial^2 w}{\partial y^2}\right)\\
&+H_{66}c_1\left(\frac{\partial \varphi_y}{\partial x}+\frac{\partial \varphi_x}{\partial y}+2\frac{\partial^2 w}{\partial x\partial y}\right)\Bigg]+\frac{\partial}{\partial y}\Bigg[E_{21}\left(\frac{\partial u}{\partial x}+\frac{1}{2}\left(\frac{\partial w}{\partial x}\right)^2\right)\\
&+E_{22}\left(\frac{\partial v}{\partial y}+\frac{1}{2}\left(\frac{\partial w}{\partial y}\right)^2\right)+E_{26}\left(\frac{\partial u}{\partial y}+\frac{\partial v}{\partial x}+\frac{\partial w}{\partial x}\frac{\partial w}{\partial y}\right)+F_{12}\frac{\partial \varphi_x}{\partial x}+F_{22}\frac{\partial \varphi_y}{\partial y}\\
&+F_{26}\left(\frac{\partial \varphi_x}{\partial y}+\frac{\partial \varphi_y}{\partial x}\right)+H_{12}c_1\left(\frac{\partial \varphi_x}{\partial x}+\frac{\partial^2 w}{\partial x^2}\right)+H_{22}c_1\left(\frac{\partial \varphi_y}{\partial y}+\frac{\partial^2 w}{\partial y^2}\right)\\
&+H_{26}c_1\left(\frac{\partial \varphi_y}{\partial x}+\frac{\partial \varphi_x}{\partial y}+2\frac{\partial^2 w}{\partial x\partial y}\right)\Bigg]\Bigg)-\Bigg[A_{45}\left(\frac{\partial w}{\partial y}+\varphi_y\right)+A_{44}\left(\frac{\partial w}{\partial y}+\varphi_y\right)\\
&+D_{45}c_2\left(\varphi_y+\frac{\partial w}{\partial y}\right)+D_{44}c_2\left(\frac{\partial w}{\partial y}+\varphi_y\right)\Bigg]-c_2\Bigg[D_{45}\left(\frac{\partial w}{\partial y}+\varphi_y\right)+D_{44}\left(\frac{\partial w}{\partial y}+\varphi_y\right)\\
&+F_{45}c_2\left(\varphi_y+\frac{\partial w}{\partial y}\right)+F_{44}c_2\left(\frac{\partial w}{\partial y}+\varphi_y\right)\Bigg]=J_1\frac{\partial^2 v}{\partial t^2}+K_2\frac{\partial^2 \phi_\theta}{\partial t^2}-\frac{4}{3h^2}J_4\frac{\partial^3 w}{\partial t^2\partial y}.
\end{aligned}
\tag{9–39}
$$

## 9.3 NUMERICAL RESULTS AND DISCUSSION

Based on HDQM and the Newmark method presented in Sections 2.2.2 and 2.3.1, respectively, forced vibration response of concrete plates armed by $SiO_2$ nanoparticles is investigated. The $SiO_2$ nanoparticles have Poisson's ratio of $v_r = 0.2$ and Young's modulus of $E_r = 70\ GPa$. The acceleration and deflection of the concrete structure subjected to harmonic transverse load are calculated as exposed in Figure 9.2.

### 9.3.1 Convergence of Numerical Solution

The convergence of numerical solution in estimating the maximum displacement of the concrete structure as a function of the grid points number is studied in Figure 9.3. As may be found, by enhancing the grid points number, the maximum displacement of the concrete structure reduces so far as, at $N = 15$, the displacement converges. Consequently, the outcomes studied in the below are based on 15 grid points for the numerical solution.

### 9.3.2 Validation

In this chapter, the outcomes of the FEM and HDQM are matched for validation. The displacement and acceleration of the concrete structure considered by the FEM and HDQM are studied in Figures 9.4a and 9.4b.

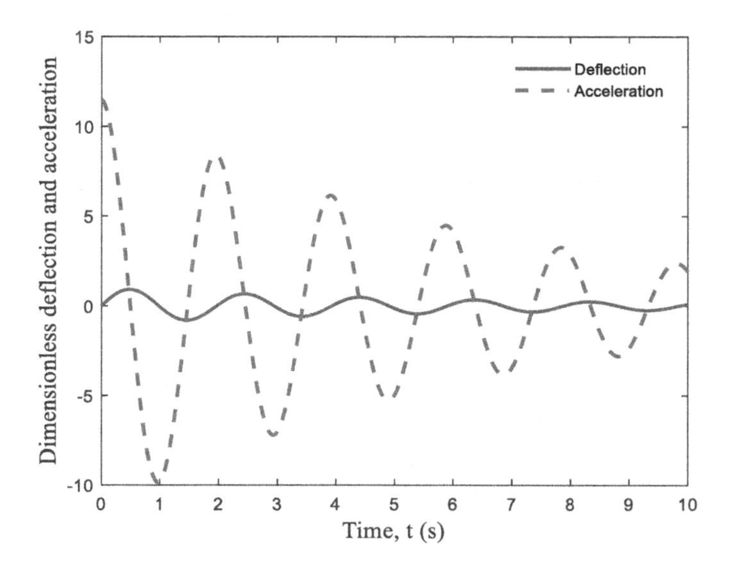


**FIGURE 9.2** Acceleration and deflection of the concrete structure subjected to harmonic transverse load.

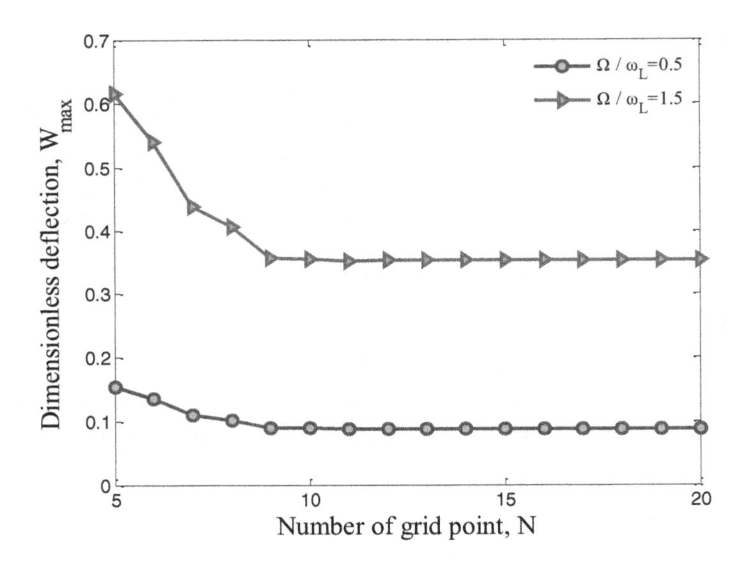


**FIGURE 9.3** Accuracy of HDQM.

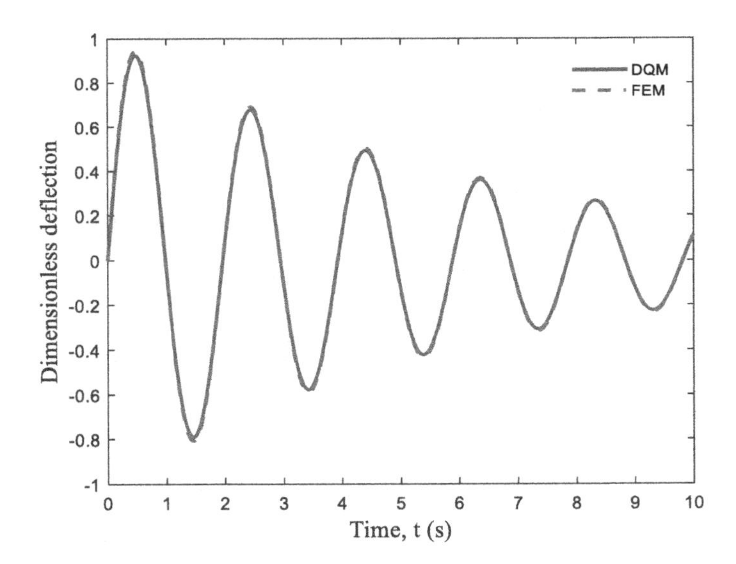


**FIGURE 9.4A** Deflection of the concrete structure achieved by the FEM and HDQM.

Furthermore, the frequency analysis of the concrete plate is presented in Figure 9.5 for the FEM and HDQM. As may be found, the outcomes of the FEM and HDQM are very close to, which displays the accuracy of the achieved outcomes.

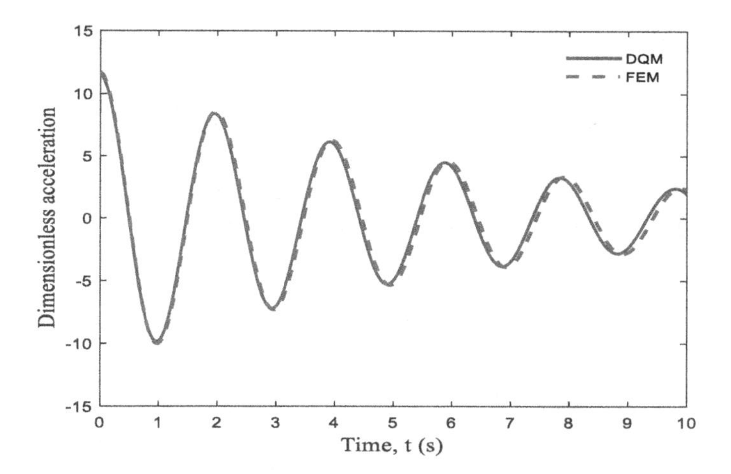


**FIGURE 9.4B** Acceleration of the concrete structure achieved by the FEM and HDQM.

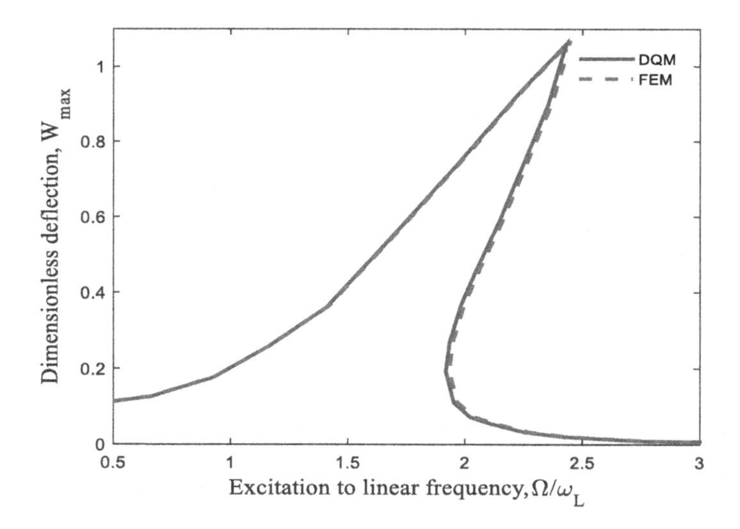


**FIGURE 9.5** Frequency curve of the concrete structure achieved by the FEM and HDQM.

### 9.3.3 Effects of Different Parameters

Figure 9.6 shows the influences of the $SiO_2$ nanoparticle volume percent on the non-dimensional frequency curve of the concrete structure ($\Omega = \omega L\sqrt{\rho_m / E_m}$). It may be found that by enhancing the values of $SiO_2$ nanoparticle volume percent to 0.3%, the linear frequency of the concrete structure is enhanced, and the dynamic maximum displacement is reduced. Furthermore, by enhancing the values of $SiO_2$ nanoparticles volume percent to 0.3%, the hardening effect of the analysis will be lesser.

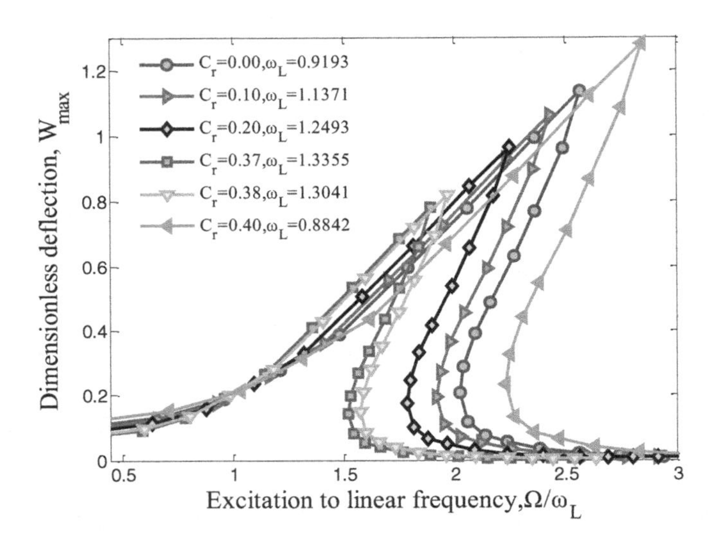


**FIGURE 9.6** Influence of $SiO_2$ nanoparticle volume fraction on the frequency curve.

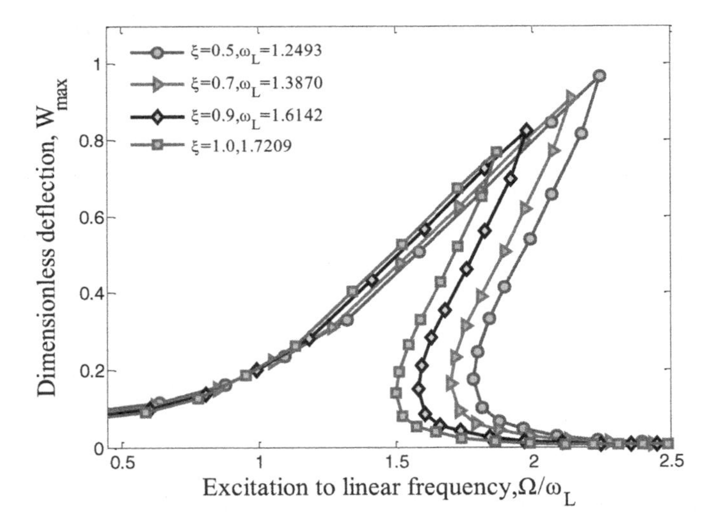


**FIGURE 9.7** Influence of $SiO_2$ nanoparticle agglomeration on the frequency curve.

It is since that the enhance of $SiO_2$ nanoparticles leads to a harder structure. But, with a further increase in the value of $SiO_2$ nanoparticles volume percent to 0.%, the outcomes are inverted; namely, the linear frequency of the concrete structure is reduced, and the dynamic maximum displacement is enhanced. Hence, it may be found that an enhance in the fraction of nanoparticles up to a special value, which is 0.3%, advances the dynamic response.

The influence of $SiO_2$ nanoparticle agglomeration on the non-dimensional frequency analysis of the concrete plate is established in Figure 9.7. As may be found, assuming agglomeration of $SiO_2$ nanoparticles reduces the linear frequency and enhances the dynamic maximum displacement of the concrete structure. It is since assuming the agglomeration of $SiO_2$ nanoparticle leads to a softer structure. In other words, considering $SiO_2$ nanoparticle agglomeration strengthens the hardening effect for the vibration response. These outcomes illustrates that the agglomeration of nanoparticles leads to enhances of 26.04% in the maximum deflection in the concrete structure.

The influence of the plate length on the non-dimensional frequency curve of the structure is showed in Figure 9.8. As may be found, the dynamic maximum displacement and linear frequency of the concrete structure are, respectively, reduced and improved when the plates length is enhance. This is since enhancing the length of plate makes the structure softer.

Figure 9.9 illustrates the non-dimensional frequency curve of the concrete structure for various plates widths. It may be seen that the dynamic maximum displacement and linear frequency the of the concrete structure reduces and enhances, respectively, when the width of plate enhances, since the structure has greater stiffness as the plate width is reduced. It is also found that the hardening effect of the analysis is enhanced by increasing the width plate.

The influence of the plates thickness on the non-dimensional frequency analysis is studied in Figure 9.10. It may be seen that by enhancing the thickness of plate, the frequency of the concrete structure is enhanced. Furthermore, by enhancing the thickness of plate, the hardening effect of the analysis is reduced, and the dynamic maximum displacement is decreased. This is since by enhancing the thickness of the concrete plate, the stiffness is improved.

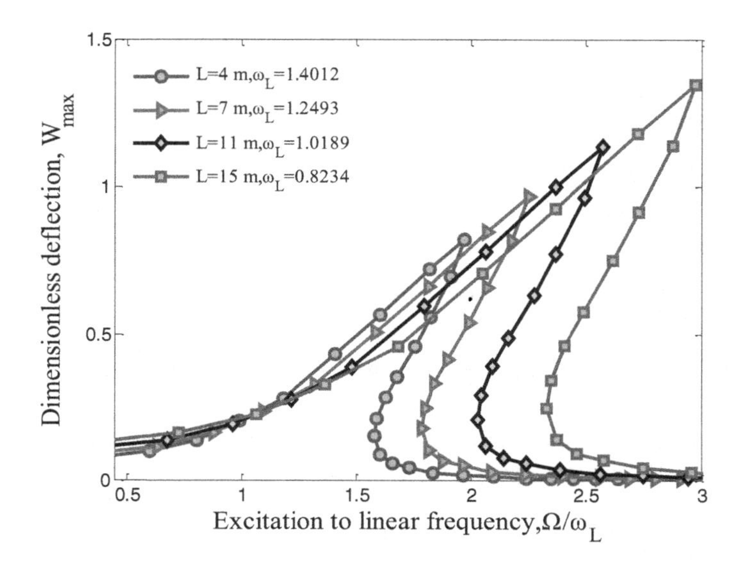


**FIGURE 9.8** Influence of the length of concrete plates on the frequency curve.

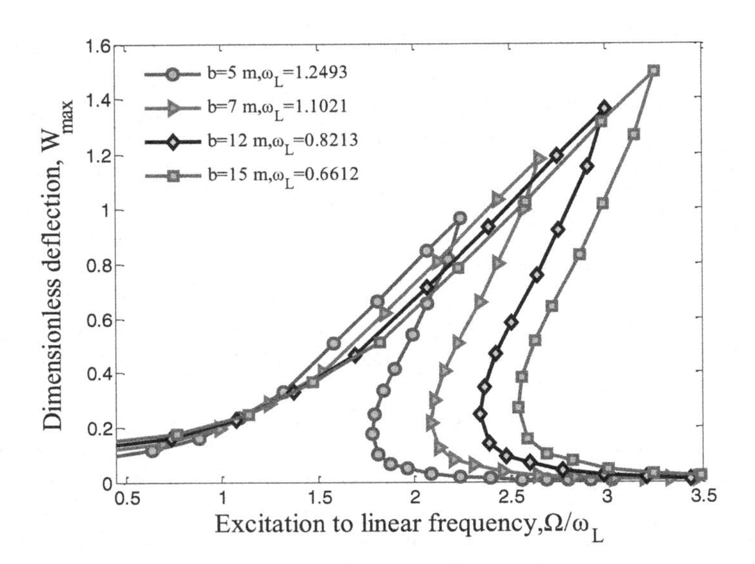


**FIGURE 9.9** Influence of plate width on the frequency curve.

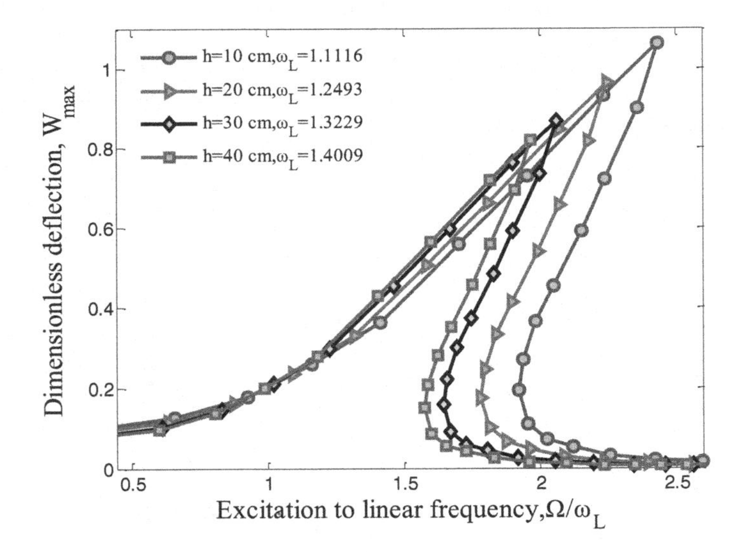


**FIGURE 9.10** Influence of plate thickness on the frequency curve.

The influence of the plate length-to-width ratio on the non-dimensional frequency curve is shown in Figure 9.11. As may be found, by enhancing the length-to-width ratio, the linear frequency reduces, and the hardening effect of the analysis, as well as maximum dynamic deflection, will enhances.

The effect of different boundary supports on the non-dimensional frequency response is illustrated in Figure 9.12.

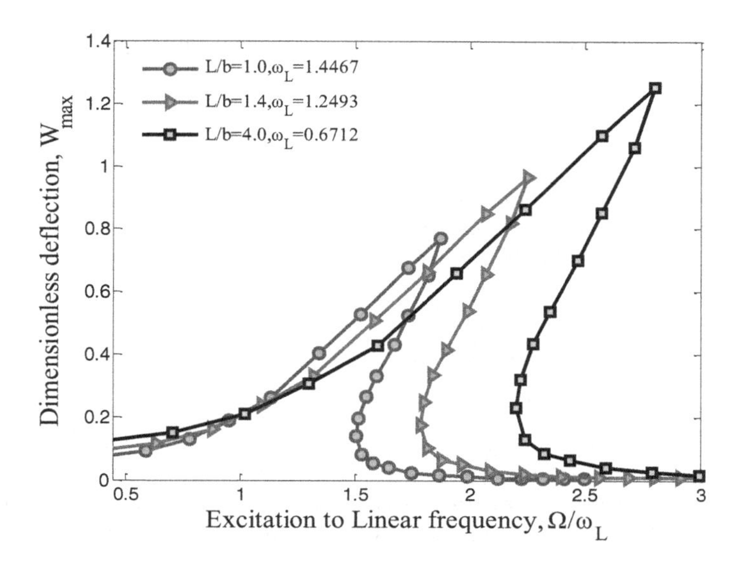


**FIGURE 9.11** Influence of the plates length-to-width ratio on the frequency curve.

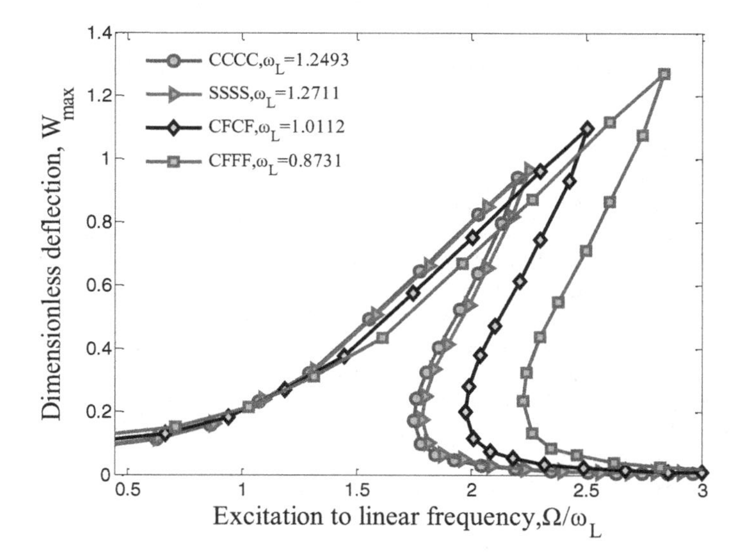


**FIGURE 9.12** Influence of boundary supports on the frequency curve.

Now, four types of boundary support with the following equations are assumed:

- Clamped-supported on four edges

$$\begin{aligned} &w = u = v = \phi_x = \phi_y = 0, && @ \quad x = 0, L \\ &w = u = v = \phi_x = \phi_y = 0. && @ \quad y = 0, b \end{aligned} \tag{9–40}$$

- Clamped-supported on two edges, free on two opposite edges

$$\begin{aligned} &w = u = v = \phi_x = \phi_y = 0, && @ \quad x = 0, L \\ &P_x = Q_x = K_x = N_x = M_x = 0. && @ \quad y = 0, b \end{aligned} \tag{9–41}$$

- Simply supported on four edges

$$\begin{aligned} &w = v = \phi_y = M_y = M_{xy} = 0, && @ \quad x = 0, L \\ &w = u = \phi_x = M_x = M_{xy} = 0. && @ \quad y = 0, b \end{aligned} \tag{9–42}$$

- Clamped-supported on one edge, free on three edges

$$\begin{aligned} &w = u = v = \phi_x = \phi_y = 0, && @ \quad x = 0 \\ &F_x = Q_x = N_x = M_x = 0. && @ \quad x = L, y = 0, b \end{aligned} \tag{9–43}$$

It may be found that the hardening effect of the analysis, as well as the dynamic maximum displacement, are maximum for the clamped boundary on one edge and free boundary on other three edges in comparison with other assumed boundary supports. Furthermore, the frequency of the concrete plates with clamped supports is higher because of high stiffness and bending rigidity for the concrete structure.

**Acknowledgments:** This chapter is a slightly modified version of Ref. [28] and has been reproduced here with the permission of the copyright holder.

## REFERENCES

[1] Akbarov SD, Zamanov AD, Suleimanov TR. Forced vibration of a prestretched two-layer plate on a rigid foundation. *Mech Compos Mater* 2005;41:229–240.

[2] Hussein MFM, Hunt HEM. A numerical model for calculating vibration due to a harmonic moving load on a floating-plate track with discontinuous plates in an underground railway tunnel. *J Sound Vib* 2009;321:363–374.

[3] Rajasekaran S. Structural dynamics of earthquake engineering: Forced vibration (harmonic force) of single-degree-of-freedom systems in relation to structural dynamics during earthquakes. *Elsevier* 2009:68–104.

[4] Batista M. Refined Mindlin–Reissner theory of forced vibrations of shear deformable plates. *Eng Struct* 2011;33:265–272.

[5] Ferreira AGM, Roque CMC, Neves AMA, Jorge RMN, Soares CMM. Buckling and vibration analysis of isotropic and laminated plates by radial basis functions. *Compos Part B- Eng* 2011;42:592–606.

[6] Eftekhari SA, Jafari AA. A mixed method for free and forced vibration of rectangular plates. *Appl Math Model* 2012;36:2814–2831.

[7] Lee SH, Lee KK, Woo SS, Cho SH. Global vertical mode vibrations due to human group rhythmic movement in a 39 story building structure. *Eng Struct* 2013;57:296–305.

[8] Khan AH, Patel BP. Nonlinear forced vibration response of bimodular laminated composite plates. *Compos Struct* 2014;108:524–537.

[9] Ansari R, Hasrati E, Faghih Shojaei M, Gholami R, Shahabodini A. Forced vibration analysis of functionally graded carbon nanotube-reinforced composite plates using a numerical strategy. *Physica E* 2015;69:294–305.

[10] Song ZG, Zhang LW, Liew KM. Active vibration control of CNT reinforced functionally graded plates based on a higher-order shear deformation theory. *Int J Mech Sci* 2016;105:90–101.
[11] Gromysz K. Nonlinear analytical model of composite plate free and forced vibrations. *Procedia Eng* 2017;193:281–288.
[12] Kulikov GM, Plotnikova SV, Kulikov MG. Strong SaS formulation for free and forced vibrations of laminated composite plates. *Compos Struct* 2017;180:286–297.
[13] Soufeiani L, Ghadyani G, Kueh ABH, Nguyen KTQ. The effect of laminate stacking sequence and fiber orientation on the dynamic response of FRP composite plates. *J Build Eng* 2017;13:41–52.
[14] Dey P, Haldar S, Sengupta D, Sheikh AH. An efficient plate element for the vibration of composite plates. *Appl Math Model* 2016;40:5589–5604.
[15] Darabi M, Ganesan R. Non-linear vibration and dynamic instability of internally-thickness-tapered composite plates under parametric excitation. *Compos Struct* 2017;176:82–104.
[16] Kumar S, Mitra A, Roy H. Forced vibration response of axially functionally graded non-uniform plates considering geometric nonlinearity. *Int J Mech Sci* 2017;128–129:194–205.
[17] Amin M, El-hassan KA. Effect of using different types of nano materials on mechanical properties of high strength. *Constr Build Mater* 2015;80:116–124.
[18] Mohamed AM. Influence of nano materials on flexural behavior and compressive strength of concrete. *HBRC J* 2016;12:212–225.
[19] Alrekabi S, Cundy AB, Lampropoulos A, Whitby RLD, Savina I. Mechanical performance of novel cement-based composites prepared with nano-fibres, and hybrid nano- and micro-fibres. *Compos Struct* 2017;178:145–156.
[20] Jafarian Arani A, Kolahchi R. Buckling analysis of embedded beams armed with carbon nanotubes. *Comput* 2016;17:567–578.
[21] Mahapatra TR, Panda SK. Nonlinear free vibration analysis of laminated composite spherical shell panel under elevated hygrothermal environment: A micromechanical approach. *Aerosp Sci Technol* 2016a;49:276–288.
[22] Mahapatra TR, Panda SK, Kar VR. Nonlinear flexural analysis of laminated composite panel under hygro-thermo-mechanical loading—A micromechanical approach. *Int J Computat Meth* 2016b;13:1650015.
[23] Mahapatra TR, Panda SK, Kar VR. Nonlinear hygro-thermo-elastic vibration analysis of doubly curved composite shell panel using finite element micromechanical model. *Mech Advan Mat Struct* 2016c;23:1343–1359.
[24] Mahapatra TR, Panda SK, Kar VR. Geometrically nonlinear flexural analysis of hygro-thermo-elastic laminated composite doubly curved shell panel. Int *J Mech Mat Des* 2016d;12:153–171.
[25] Zamanian M, Kolahchi R, Rabani Bidgoli M. Agglomeration effects on the buckling behavior of embedded beams reinforced with SiO2 nanoparticles. *Wind Struct* 2017;24:43–57.
[26] Arbabi A, Kolahchi R, Rabani Bidgoli M. Concrete columns reinforced with Zinc Oxide nanoparticles subjected to electric field: Buckling analysis. *Wind Struct* 2017;24:431–446.
[27] Dutta G, Panda SK, Mahapatra TR, Singh VK. Electro-magneto-elastic response of laminated composite plate: A finite element approach. *Int J Appl Computat Math* 2017;3:2573–2592.
[28] Jassas MR, Bidgoli MR. Forced vibration analysis of concrete slabs reinforced by agglomerated $SiO_2$ nanoparticles based on numerical methods. *Construct Build Mat* 2019;211:796–806.

# 10 Seismic Analysis of Plates Reinforced by Nanoparticles

## 10.1 INTRODUCTION

Seismic analysis is a subset of structural analysis in which the dynamic response of building structures (or nonbuilding structures such as bridges, etc.) against the earthquake is examined. This analysis is part of the structural engineering, earthquake engineering, and seismic retrofitting of the structures that should be constructed in earthquake-prone zones.

Mechanical analyses of nanostructures have been reported by many researchers. Liang and Parra-Montesinos [1] studied the seismic behavior of four reinforced concrete (RC) column–steel plates under various ground motions using experimental tests. Cheng and Chen [2] and Changwang et al. [3] studied the seismic behavior of steel-RC column–steel truss plates. They developed a design formula for the shear strength of the structure subjected to seismic activities using experimental tests. The effect of the cumulative damage on the seismic behavior of steel tube–reinforced concrete (ST-RC) columns through experimental testing was investigated by Ji et al. [4]. Six large-scale ST-RC column specimens were subjected to high axial forces and cyclic lateral loading. The effect of the plastic hinge relocation on the potential damage of a reinforced concrete frame subjected to different seismic levels was studied by Cao and Ronagh [5] based on current seismic designs. The optimal seismic retrofit method that uses fiber reinforced polymer (FRP) jackets for shear-critical RC frames was presented by Choi et al. [6]. This optimal method uses nondominated sorting genetic algorithm II to optimize the two conflicting objective functions of the retrofit cost, as well as the seismic performance, simultaneously. They examined various parameters such as failure mode, hysteresis curves, ductility, and reduction of stiffness. Liu et al. [7] focused on studying the seismic behavior of steel-reinforced concrete special-shaped column–plate joints. Six specimens, which are designed according to the principle of strong-member and weak-joint core, are tested under low cyclic reversed load.

In none of the mentioned articles is the nanocomposite structure considered. Wuite and Adali [8] performed stress analysis of carbon nanotube (CNT)–reinforced plates. They concluded that using CNTs during the reinforcing phase can increase the stiffness and stability of the system. Also, Matsuna [9] examined the stability of the composite cylindrical shell using third-order shear deformation theory (TSDT). Formica et al. [10] analyzed the vibration behavior of CNT-reinforced composites. They employed an equivalent continuum model based on the Eshelby–Mori–Tanaka

DOI: 10.1201/9781003349525-10

model to obtain the material properties of the composite. Liew et al. [11] studied the post-buckling of nanocomposite cylindrical panels. They used the extended rule of mixture to estimate the effective material properties of the nanocomposite structure. They also applied a meshless approach to examine the post-buckling response of the nanocomposite cylindrical panel. In another similar work, Lei et al. [12] studied the dynamic stability of a CNT-reinforced functionally graded (FG) cylindrical panel. They used the Eshelby–Mori–Tanaka model to estimate effective material properties of the resulting nanocomposite structure and employed the Ritz method to distinguish the instability regions of the structure. Static stress analysis of CNT-reinforced cylindrical shells is presented by Ghorbanpour Arani et al. [13]. In this work, the cylindrical shell was subjected to non-axisymmetric thermal-mechanical loads and uniform electromagnetic fields. Eventually, the stress distribution in the structure is determined analytically by the Fourier series. A buckling analysis of CNT-reinforced microplates was carried out by Kolahchi et al. [14]. They derived the governing equations of the structure based on the Mindlin plate theory and using Hamilton's principle. They obtained buckling load of the structure by applying the differential quadrature method (DQM). The dynamic response of FG circular cylindrical shells is examined by Davar et al. [15]. They developed the mathematical formulation of the structure according to first-order shear deformation theory (FSDT) and Love's first approximation theory. Also, Kolahchi et al. [16] investigated the dynamic stability of FG CNT-reinforced plates. The material properties of the plate are assumed to be a function of temperature and the structure is considered resting on orthotropic elastomeric medium. Jafarian Arani and Kolahchi [17] presented a mathematical model for a buckling analysis of a CNT-reinforced concrete column. They simulated the problem based on the Euler–Bernoulli and Timoshenko plate theories. The nonlinear vibration of laminated cylindrical shells was analyzed by Shen and Yang [18]. They examined the influences of temperature variation, shell geometric parameter, and applied voltage on the linear and nonlinear vibration of the structure. An investigation on the nonlinear dynamic response and vibration of an imperfect laminated three-phase polymer nanocomposite panel resting on elastic foundations was presented by Duc et al. [19]. Van Thu and Duc [20] presented an analytical approach to investigating the nonlinear dynamic response and vibration of an imperfect three-phase laminated nanocomposite cylindrical panel resting on elastic foundations in thermal environments. Alibeigloo [21] employed the theory of piezo-elasticity to study the bending behavior of FG CNT-reinforced composite cylindrical panels. They used an analytical method to study the effect of CNT volume fraction, temperature variation, and applied voltage on the bending behavior of the system. Feng et al. [22] studied the nonlinear bending behavior of a novel class of multilayer polymer nanocomposite plates reinforced with graphene platelets (GPLs) that are nonuniformly distributed along the thickness direction. Amoli et al. [23] presented the nonlinear dynamic response of a concrete plate retrofitted with Aluminum oxide ($AL_2O_3$) under seismic load and a magnetic field is examined. The plate is modeled using a higher-order shear deformation plate model. By employing nonlinear strains-displacements and stress-strain relationships, the energy equations for the column are derived.

In this chapter, the dynamic response of $AL_2O_3$ nanoparticle-reinforced concrete plates subjected to seismic excitation and a magnetic field is studied. So the results of this research are of great importance to civil engineering. The concrete plate is modeled by applying sinusoidal shear plate theory (SSPT), and the effective material properties of the concrete plate are obtained based on Mori–Tanaka model considering agglomeration of $AL_2O_3$ nanoparticles. The dynamic displacement of the structure is calculated by harmonic differential quadrature method (HDQM) in conjunction with the Newmark method. The effects of different parameters, such as volume fraction and agglomeration of $AL_2O_3$ nanoparticles, magnetic field, boundary conditions, and geometrical parameters of a concrete plate, are studied on the dynamic response of the structure.

## 10.2 STRESS–STRAIN RELATIONS

As shown in Figure 10.1, a concrete plate is reinforced by agglomerated $AL_2O_3$ nanoparticles subjected to the earthquake load and a magnetic field. The geometrical parameters of plate are length of $L$ and thickness of $h$.

Based on the sinusoidal shear deformation theory SSDT (see Section 1.4.4), the displacement field can be assumed. The constitutive equations of the orthotropic plate are considered as follows:

$$\begin{bmatrix} \sigma_{xx} \\ \sigma_{yy} \\ \sigma_{zz} \\ \sigma_{zy} \\ \sigma_{xz} \\ \sigma_{zy} \end{bmatrix} = \begin{bmatrix} C_{11} & C_{12} & C_{13} & 0 & 0 & 0 \\ C_{12} & C_{22} & C_{23} & 0 & 0 & 0 \\ C_{13} & C_{23} & C_{33} & 0 & 0 & 0 \\ 0 & 0 & 0 & C_{44} & 0 & 0 \\ 0 & 0 & 0 & 0 & C_{55} & 0 \\ 0 & 0 & 0 & 0 & 0 & C_{66} \end{bmatrix} \begin{bmatrix} \varepsilon_{xx} \\ \varepsilon_{yy} \\ \varepsilon_{zz} \\ \gamma_{zy} \\ \gamma_{xz} \\ \gamma_{xy} \end{bmatrix}, \tag{10–1}$$

where $C_{ij}$ are the elastic constants of the concrete plate. To obtain the effective material properties of the concrete plate and consider the agglomeration effect, the

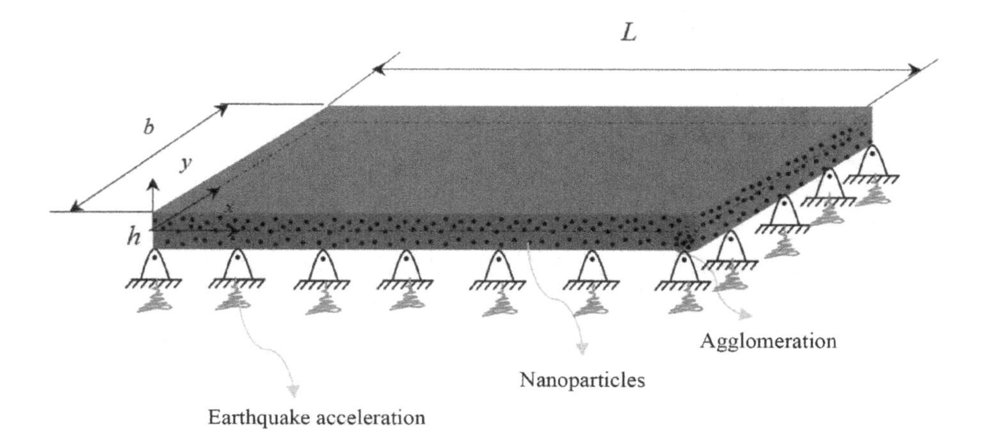


**FIGURE 10.1** Schematic of concrete plate reinforced by agglomerated $AL_2O_3$ nanoparticles under a magnetic field.

Mori–Tanaka model (see Section 4.3) is employed. The potential strain energy stored in the structure is given as follows:

$$U=\frac{1}{2}\int_A\int_{-\frac{h}{2}}^{\frac{h}{2}}\left(\sigma_{xx}\varepsilon_{xx}+\sigma_{yy}\varepsilon_{yy}+\sigma_{xy}\gamma_{xy}+\sigma_{xz}\gamma_{xz}+\sigma_{yz}\gamma_{yz}\right)dzdA. \tag{10–2}$$

Substituting Equations 1–36 and 1–37 into Equation 10–2, we have

$$\begin{aligned}U=\frac{1}{2}\int_A\Bigg(&N_{xx}\frac{\partial U}{\partial x}+N_{xy}\frac{\partial U}{\partial y}+N_{xy}\frac{\partial V}{\partial x}+N_{yy}\frac{\partial V}{\partial y}+Q_x\frac{\partial W_s}{\partial x}\\&+Q_y\frac{\partial W_s}{\partial y}-M_{xxS}\frac{\partial^2 W_s}{\partial x^2}-M_{yyS}\frac{\partial^2 W_s}{\partial y^2}-2M_{xyS}\frac{\partial^2 W_s}{\partial y\partial x}-M_{xxB}\frac{\partial^2 W_b}{\partial x^2}\\&-M_{yyB}\frac{\partial^2 W_b}{\partial y^2}-2M_{xyB}\frac{\partial^2 W_b}{\partial y\partial x}\Bigg)\mathrm{dA},\end{aligned} \tag{10–3}$$

where

$$\begin{bmatrix}N_{xx}\\N_{yy}\\N_{xy}\end{bmatrix}=\int_{-h}^{h}\begin{bmatrix}\sigma_{xx}\\\sigma_{yy}\\\sigma_{xy}\end{bmatrix}dz, \tag{10–4}$$

$$\begin{bmatrix}M_{xxB}\\M_{yyB}\\M_{xyB}\end{bmatrix}=\int_{-h}^{h}\begin{bmatrix}\sigma_{xx}\\\sigma_{yy}\\\sigma_{xy}\end{bmatrix}zdz, \tag{10–5}$$

$$\begin{bmatrix}M_{xxS}\\M_{yyS}\\M_{xyS}\end{bmatrix}=\int_{-h}^{h}\begin{bmatrix}\sigma_{xx}\\\sigma_{yy}\\\sigma_{xy}\end{bmatrix}fdz, \tag{10–6}$$

$$\begin{bmatrix}Q_x\\Q_y\end{bmatrix}=\int_{-h}^{h}\begin{bmatrix}\sigma_{xz}\\\sigma_{yz}\end{bmatrix}pdz. \tag{10–7}$$

By substituting Equations 10–1, 1–36, and 1–37 into Equations 10–4 through 10–7, the stress resultants of the plate take the following form:

$$\begin{aligned}N_{xx}=&A_{11}\frac{\partial}{\partial x}U-A_{11z}\frac{\partial^2}{\partial x^2}W_b-A_{11f}\frac{\partial^2}{\partial x^2}W_s\\&+A_{12}\frac{\partial}{\partial y}V-A_{12z}\frac{\partial^2}{\partial y^2}W_b-A_{12f}\frac{\partial^2}{\partial y^2}W_s,\end{aligned} \tag{10–8}$$

$$N_{yy} = A_{21}\frac{\partial}{\partial x}U - A_{21z}\frac{\partial^2}{\partial x^2}W_b - A_{21f}\frac{\partial^2}{\partial x^2}W_s + A_{22}\frac{\partial}{\partial y}V$$
$$-A_{22z}\frac{\partial^2}{\partial y^2}W_b - A_{22f}\frac{\partial^2}{\partial y^2}W_s, \tag{10–9}$$

$$N_{xy} = A_{44}\frac{\partial}{\partial y}U + A_{44}\frac{\partial}{\partial x}V - 2A_{44z}\frac{\partial^2}{\partial x\partial y}W_b - 2A_{44f}\frac{\partial^2}{\partial x\partial y}W, \tag{10–10}$$

$$Q_x = A_{55g}\frac{\partial}{\partial x}W_s + GA_{55g}\frac{\partial^2}{\partial x\partial t}W_s, \tag{10–11}$$

$$Q_y = A_{66g}\frac{\partial}{\partial y}W_s + GA_{66g}\frac{\partial^2}{\partial y\partial t}W_s, \tag{10–12}$$

$$M_{xxB} = A_{11z}\frac{\partial}{\partial x}U - B_{11}\frac{\partial^2}{\partial x^2}W_b - A_{11zf}\frac{\partial^2}{\partial x^2}W_s$$
$$+A_{12z}\frac{\partial}{\partial y}V - B_{12}\frac{\partial^2}{\partial y^2}W_b - A_{12zf}\frac{\partial^2}{\partial y^2}W_s, \tag{10–13}$$

$$M_{xxS} = A_{11f}\frac{\partial}{\partial x}U - A_{11zf}\frac{\partial^2}{\partial x^2}W_b - E_{11}\frac{\partial^2}{\partial x^2}W_s$$
$$+A_{12f}\frac{\partial}{\partial y}V - A_{12zf}\frac{\partial^2}{\partial y^2}W_b - E_{12}\frac{\partial^2}{\partial y^2}W_s, \tag{10–14}$$

$$M_{yyB} = A_{21z}\frac{\partial}{\partial x}U - B_{21}\frac{\partial^2}{\partial x^2}W_b - A_{21zf}\frac{\partial^2}{\partial x^2}W_s$$
$$+A_{22z}\frac{\partial}{\partial y}V - B_{22}\frac{\partial^2}{\partial y^2}W_b - A_{22zf}\frac{\partial^2}{\partial y^2}W_s, \tag{10–15}$$

$$M_{yyS} = A_{21f}\frac{\partial}{\partial x}U - A_{21zf}\frac{\partial^2}{\partial x^2}W_b - E_{21}\frac{\partial^2}{\partial x^2}W_s$$
$$+A_{22f}\frac{\partial}{\partial y}V - A_{22zf}\frac{\partial^2}{\partial y^2}W_b - E_{22}\frac{\partial^2}{\partial y^2}W_s, \tag{10–16}$$

$$M_{xyB} = 2A_{44z}\frac{\partial}{\partial y}U + 2A_{44z}\frac{\partial}{\partial x}V - 2B_{44}\frac{\partial^2}{\partial x\partial y}W_b - 2A_{44zf}\frac{\partial^2}{\partial x\partial y}W_s, \tag{10–17}$$

$$M_{xyS} = 2A_{44f}\frac{\partial}{\partial y}U + 2A_{44f}\frac{\partial}{\partial x}V - 2A_{44zf}\frac{\partial^2}{\partial x\partial y}W_b - 2E_{44}\frac{\partial^2}{\partial x\partial y}W_s, \tag{10–18}$$

in which

$$\left(A_{11},A_{12},A_{22},A_{44}\right)=\int_{-h}^{h}\left(C_{11},C_{12},C_{22},C_{44}\right)dz, \tag{10–19}$$

$$\left(A_{11z},A_{12z},A_{22z},A_{44z}\right)=\int_{-h}^{h}\left(C_{11},C_{12},C_{22},C_{44}\right)zdz, \tag{10–20}$$

$$\left(A_{11f},A_{12f},A_{22f},A_{44f}\right)=\int_{-h}^{h}\left(C_{11},C_{12},C_{22},C_{44}\right)fdz, \tag{10–21}$$

$$\left(A_{11zf},A_{12zf},A_{22zf},A_{44zf}\right)=\int_{-h}^{h}\left(C_{11},C_{12},C_{22},C_{44}\right)zfdz, \tag{10–22}$$

$$\left(A_{55g},A_{66g}\right)=\int_{-h}^{h}\left(C_{55},C_{66}\right)gdz, \tag{10–23}$$

$$\left(B_{11},B_{12},B_{22},B_{44}\right)=\int_{-h}^{h}\left(C_{11},C_{12},C_{22},C_{44}\right)z^2dz, \tag{10–24}$$

$$\left(E_{11},E_{12},E_{22},E_{44}\right)=\int_{-h}^{h}\left(C_{11},C_{12},C_{22},C_{44}\right)f^2dz. \tag{10–25}$$

The kinetic energy of the structure is defined as follows:

$$K=\frac{\rho}{2}\int\left(\dot{U}_1^{\,2}+\dot{U}_2^{\,2}+\dot{U}_3^{\,2}\right)dV. \tag{10–26}$$

By substituting Equation 1–35 into Equation 10–26, we have

$$K=0.5\int\left(\begin{array}{l}I_0\left(\left(\frac{\partial U}{\partial t}\right)^2+\left(\frac{\partial V}{\partial t}\right)^2+\left(\frac{\partial W_b}{\partial t}\right)^2+\left(\frac{\partial W_s}{\partial t}\right)^2\right)\\ -2I_1\left(\frac{\partial U}{\partial t}\frac{\partial^2 W_b}{\partial t\partial x}+\frac{\partial V}{\partial t}\frac{\partial^2 W_b}{\partial t\partial y}\right)+I_2\left(\left(\frac{\partial^2 W_b}{\partial t\partial x}\right)^2+\left(\frac{\partial^2 W_b}{\partial t\partial y}\right)^2\right)\\ +I_3\left(\left(\frac{\partial^2 W_s}{\partial t\partial x}\right)^2+\left(\frac{\partial^2 W_s}{\partial t\partial y}\right)^2\right)\begin{array}{l}+2I_4\left(\frac{\partial^2 w_b}{\partial t\partial x}\frac{\partial^2 W_s}{\partial t\partial x}+\frac{\partial^2 W_b}{\partial t\partial y}\frac{\partial^2 W_s}{\partial t\partial y}\right)\\ +2I_5\left(\frac{\partial U}{\partial t}\frac{\partial^2 W_s}{\partial t\partial x}+\frac{\partial V}{\partial t}\frac{\partial^2 W_s}{\partial t\partial y}\right)\end{array}\end{array}\right)dA. \tag{10–27}$$

where

$$\left(I_0,I_1,I_2,J_1,J_1,K_2\right)=\int_{-h}^{h}\rho\left(1,z,f,zf,z^2,f^2\right). \tag{10–28}$$

The external work for the earthquake and magnetic field can be calculated as follows:

$$W = \int \left( ma(t) + \mu \mathrm{H}_x^2 h \frac{\partial^2 w}{\partial x^2} \right) W dx, \tag{10–29}$$

where $m$ and $a(t)$ are the mass and the acceleration of the earth, respectively; $\mu$ and $H_x$ are magnetic permeability and magnetic field, respectively. To extract the governing equations of motion, Hamilton's principle is expressed as follows:

$$\int_0^t (\delta U - \delta K - \delta W) dt = 0, \tag{10–30}$$

where $\delta$ denotes the variational operator. Substituting Equations 10–3, 10–27, and 10–29 into Equation 10–30, the motion equations of the structure are obtained as follows:

$$\frac{\partial}{\partial x} N_{xx} + \frac{\partial}{\partial y} N_{xy} - I_0 \frac{\partial^2 U}{\partial t^2} + I_1 \frac{\partial^3 W_b}{\partial x \partial t^2} + J_1 \frac{\partial^3 W_s}{\partial x \partial t^2} = 0, \tag{10–31}$$

$$\frac{\partial}{\partial x} N_{xy} + \frac{\partial}{\partial y} N_{yy} - I_0 \frac{\partial^2 V}{\partial t^2} + I_1 \frac{\partial^3 W_b}{\partial y \partial t^2} + J_1 \frac{\partial^3 W_s}{\partial y \partial t^2} = 0, \tag{10–32}$$

$$\begin{aligned} &\frac{\partial^2}{\partial x^2} M_{xxB} + 2\frac{\partial^2}{\partial x \partial y} M_{xyB} + \frac{\partial^2}{\partial y^2} M_{yyB} - K_w w - I_0 \left( \frac{\partial^2 W_b}{\partial t^2} + \frac{\partial^2 W_s}{\partial t^2} \right) \\ &- I_1 \left( \frac{\partial^3 U}{\partial x \partial t^2} + \frac{\partial^3 V}{\partial y \partial t^2} \right) + I_2 \left( \frac{\partial^4 W_b}{\partial x^2 \partial t^2} + \frac{\partial^4 W_b}{\partial y^2 \partial t^2} \right) + J_2 \left( \frac{\partial^4 W_s}{\partial x^2 \partial t^2} + \frac{\partial^4 W_s}{\partial y^2 \partial t^2} \right) = 0, \end{aligned} \tag{10–33}$$

$$\begin{aligned} &\frac{\partial^2}{\partial x^2} M_{xxS} + 2\frac{\partial^2}{\partial x \partial y} M_{xyS} + \frac{\partial^2}{\partial y^2} M_{yyS} + \frac{\partial}{\partial x} Q_x + \frac{\partial}{\partial y} Q_y - K_w w - I_0 \\ &\left( \frac{\partial^2 W_b}{\partial t^2} + \frac{\partial^2 W_s}{\partial t^2} \right) - J_1 \left( \frac{\partial^3 U}{\partial x \partial t^2} + \frac{\partial^3 V}{\partial y \partial t^2} \right) + J_2 \left( \frac{\partial^4 W_b}{\partial x^2 \partial t^2} + \frac{\partial^4 W_b}{\partial y^2 \partial t^2} \right) \\ &+ K_2 \left( \frac{\partial^4 W_s}{\partial x^2 \partial t^2} + \frac{\partial^4 W_s}{\partial y^2 \partial t^2} \right) = 0. \end{aligned} \tag{10–34}$$

Also, the boundary conditions of the structure are considered as follows:

- Clamped–clamped supported

$$\begin{aligned} &W_s = W_b = U = V = \frac{\partial W_s}{\partial x} = \frac{\partial W_b}{\partial x} = 0, @ \quad x = 0, L \\ &W_s = W_b = U = V = \frac{\partial W_s}{\partial x} = \frac{\partial W_b}{\partial x} = 0. \ @ \quad y = 0, b \end{aligned} \tag{10–35}$$

- Clamped–simply supported

$$\begin{aligned} W_s = W_b = U = V = \frac{\partial W_s}{\partial x} = \frac{\partial W_b}{\partial x}, @ \quad x = 0, L \\ W_s = W_b = U = V = M_{xxB} = M_{xxs}. @ \quad y = 0, b \end{aligned} \tag{10–36}$$

- Simply–simply supported

$$\begin{aligned} W_s = W_b = U = V = M_{yyB} = M_{yys}, @ \quad x = 0, L \\ W_s = W_b = U = V = M_{xxB} = M_{xxs}. @ \quad y = 0, b \end{aligned} \tag{10–37}$$

## 10.3 NUMERICAL RESULTS AND DISCUSSION

In this section, the effect of various parameters on the dynamic response of the concrete plate reinforced by $AL_2O_3$ nanoparticles under seismic load and a magnetic field is examined. The length and the thickness of the concrete plate are $L = 3\ m$ and $h = 15\ cm$, respectively. The elastic moduli of concrete and $AL_2O_3$ nanoparticles are $E_c = 20\ GPa$ and $E_r = 165 GPa$, respectively. The earthquake acceleration is considered based on the Kobe earthquake that had a distribution of acceleration in 30 seconds as shown in Figure 10.2.

### 10.3.1 Convergence of DQM

Figure 10.3 shows the convergence of DQM in evaluating the maximum deflection of the structure versus time. As can be seen, with increasing the number of grid points $N$, the maximum deflection of the structure decreases until $N = 15$, at which point

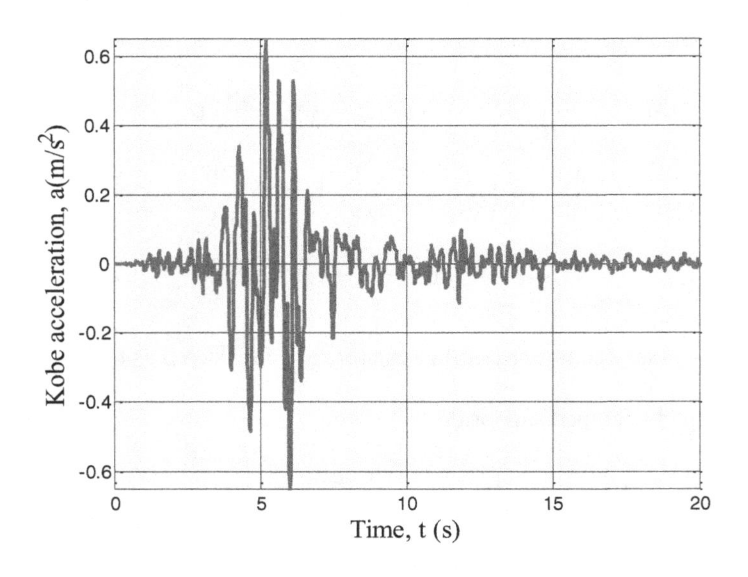


**FIGURE 10.2** Kobe earthquake acceleration.

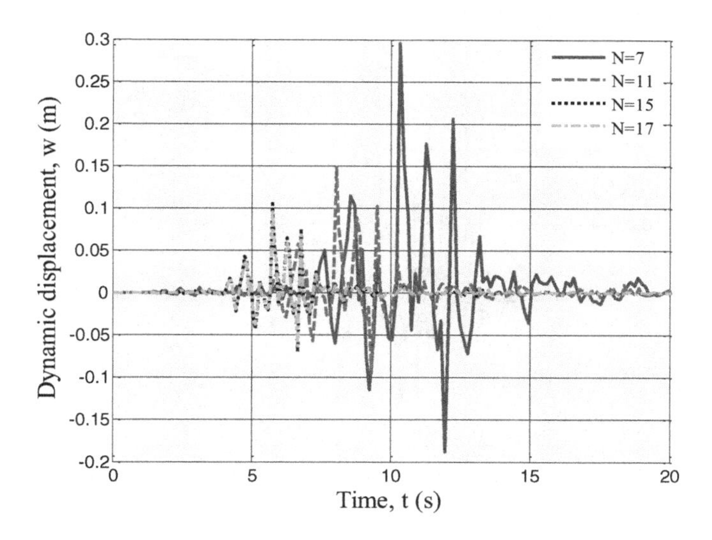


**FIGURE 10.3** Convergence and accuracy of DQM.

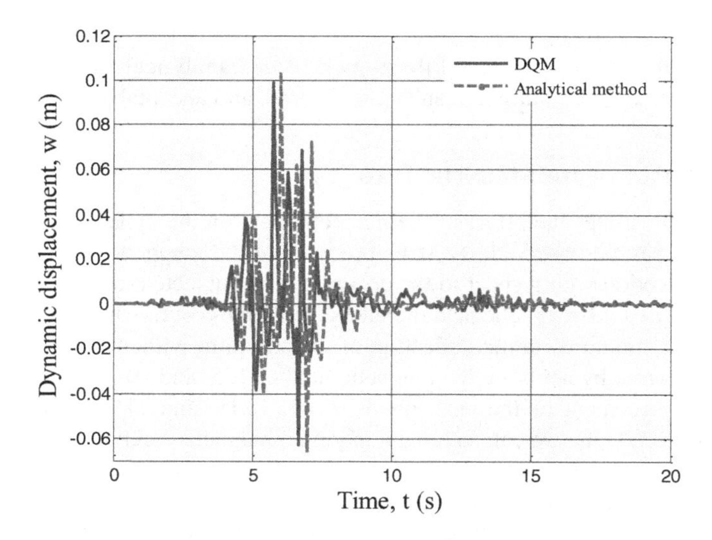


**FIGURE 10.4** Comparison of the numerical and analytical results.

the results converge to a constant value. So the following results are based on 15 grid points for the DQM solution.

### 10.3.2 Validation of Results

Since there is no similar work in the literature in the scope of this chapter, the validation of this chapter is done by comparing the numerical and analytical solutions. The results of the analytical and numerical (DQM) methods are depicted in Figure 10.4.

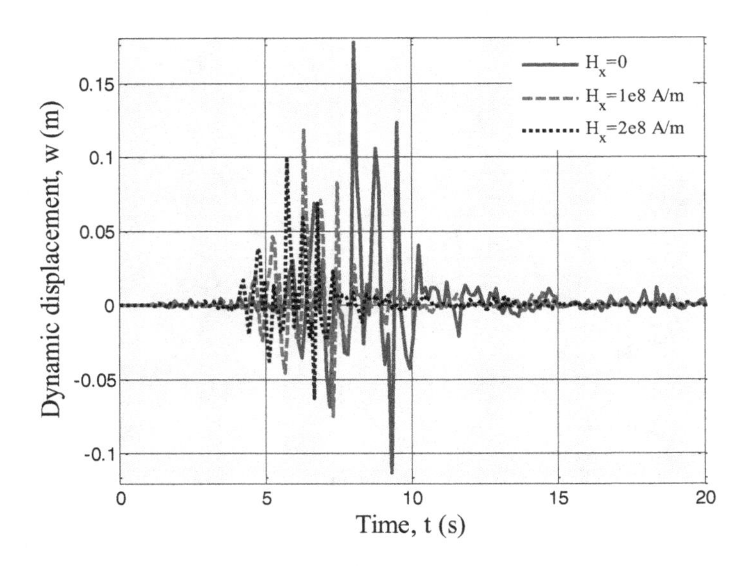


**FIGURE 10.5** Effect of a magnetic field on the dynamic response of the structure.

As can be observed, the results of the numerical and analytical methods are identical, and therefore, the obtained results are accurate and acceptable.

### 10.3.3 Effect of the Magnetic Field

Figure 10.5 illustrates the effect of the magnetic field on the dynamic deflection versus time. As can be observed, the structure without the magnetic field has a greater dynamic deflection with respect to the concrete plate subjected to the magnetic field. The reason is that the magnetic field increases the stiffness of the structure. Figure 10.5 shows the maximum dynamic deflection of the structure without the magnetic field equal to 39, while by applying the magnetic field of 1, 5, and 10 A/m, the maximum dynamic displacement of the structure is 27.05, 18.15, and 17.97, respectively. By comparing the results, we can say that applying a magnetic field of 1, 5, and 10 A/m decreases the maximum dynamic displacement of the structure up to 30.64, 53.46, and 53.92 percent, which is a remarkable result in the dynamic designing of the structures. Also, it should be noted that the excessive increase in the magnetic field increases costs while it does not have a noticeable effect on the dynamic response of the structure. Hence, the magnetic field of 5 A/m is the best choice for the present structure.

### 10.3.4 Effect of $AL_2O_3$ Nanoparticles

The effect of $AL_2O_3$ nanoparticle volume percent on the dynamic response of the structure is shown in Figure 10.6. It is apparent that the maximum dynamic displacement of the structure is equal to 32.3 for the case of $c_r = 0$ (without $AL_2O_3$ nanoparticles). By using $AL_2O_3$ nanoparticles in volume fractions of 0.05, 0.1, and 0.18, the amount of maximum dynamic displacement is 29.1, 27.05, and 30.83, respectively.

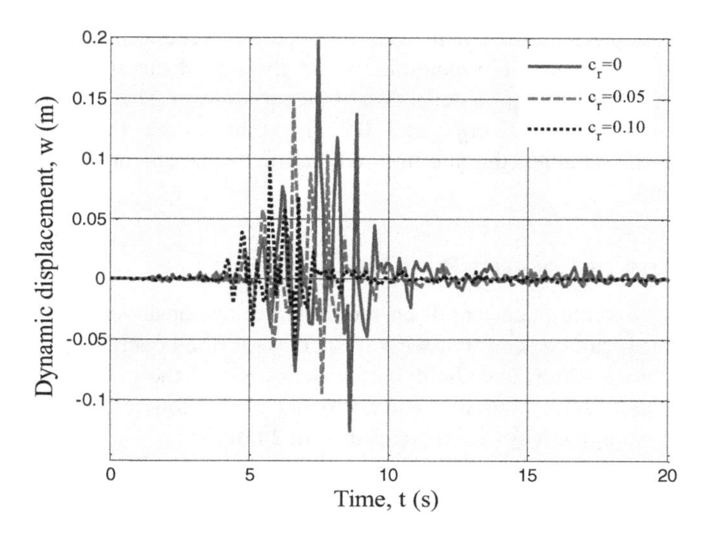


**FIGURE 10.6** Effect of $AL_2O_3$ nanoparticle volume percent on the dynamic response of the structure.

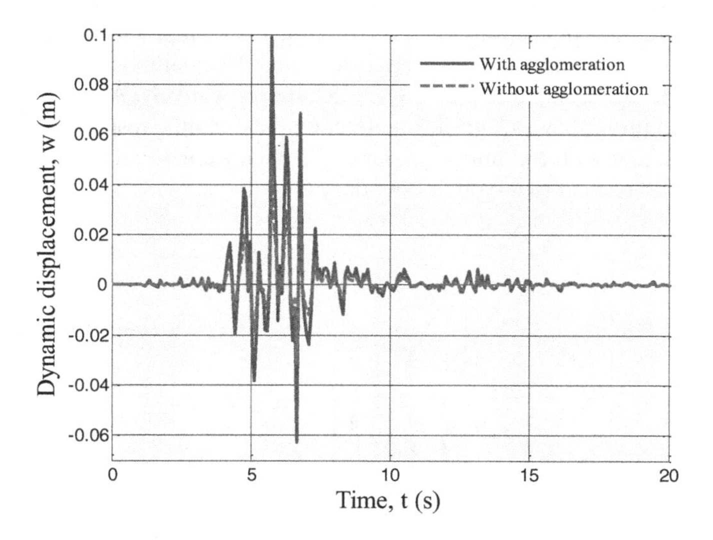


**FIGURE 10.7** Effect of $AL_2O_3$ nanoparticle agglomeration on the dynamic response of the structure.

Therefore, using $AL_2O_3$ nanoparticles in volume fractions of 0.05 and 0.1 increases the stiffness of the structure and reduces the maximum displacement of the structure by 9.91 and 16.25 percent, respectively, while the volume percent of 0.18 has a converse result and a 4.5 percent increase in the deflection.

The agglomeration effect of $AL_2O_3$ nanoparticles on the dynamic deflection of the structure versus time is illustrated in Figure 10.7. As can be observed, by considering the agglomeration effect, the stiffness of the structure is reduced while the dynamic displacement is increased. For example, in the absence of the agglomeration effect ($\xi = 1$), the maximum dynamic deflection of the structure is 22.22, while for $\xi = 0.5$, the maximum dynamic deflection is 27.05. The results reveal that the existence of the agglomeration changes the maximum dynamic displacement of the structure up to 21.74 percent.

### 10.3.5 Effect of Concrete Plate Length

The effect of concrete plate length on the dynamic response versus time is shown in Figure 10.8. It can be seen that with an increase in the concrete plate length, the structure becomes softer, and the dynamic deflection of the system increases. For example, the maximum dynamic displacements of the concrete plate increase by 72.75 percent when the length is increased from 2m to 3m.

### 10.3.6 Effect of Boundary Conditions on the Dynamic Response

Figure 10.9 illustrate the effect of various boundary conditions on the dynamic response versus time. Four boundary conditions including clamped–clamped, clamped–simply, simply–simply, and free–simply supported are considered. The maximum dynamic deflections of the structure for clamped–clamped, clamped–simply supported, simply–simply supported, and free–simply supported boundary conditions are 13.53, 18.94, 27.05, and 32.46, respectively. As can be observed, boundary conditions have a significant effect on the dynamic response of the system so that a structure with a clamped–clamped boundary condition has the lowest displacement with respect to the other boundary conditions.

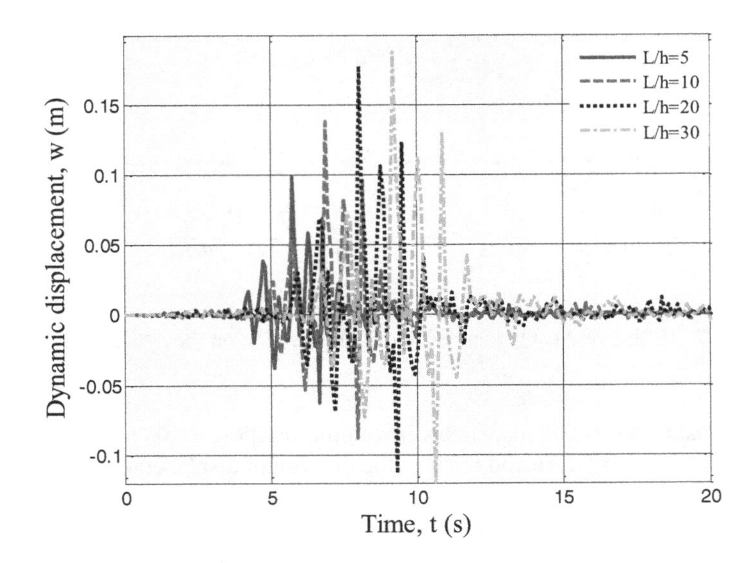


**FIGURE 10.8** Effect of concrete plate length on the dynamic response of the structure.

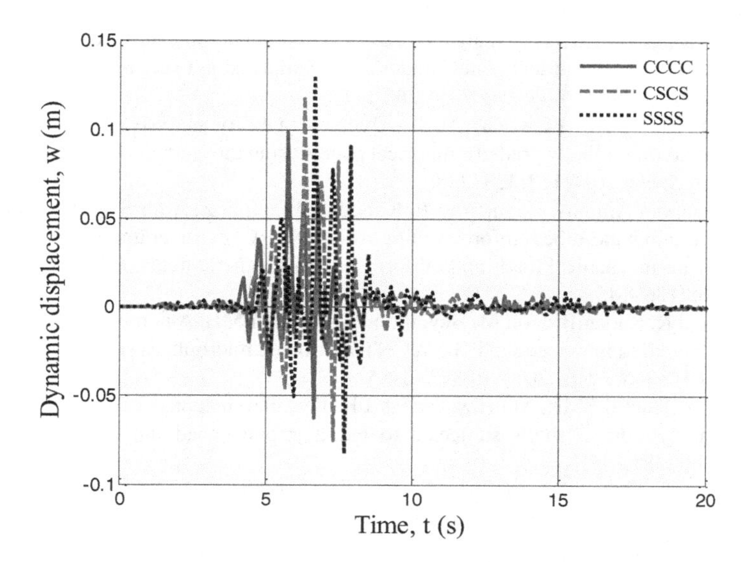


**FIGURE 10.9** Effect of different boundary conditions on the dynamic response of the structure.

**Acknowledgments:** This chapter is a slightly modified version of Ref. [23] and has been reproduced here with the permission of the copyright holder.

## REFERENCES

[1] Liang X, Parra-Montesinos GJ. Seismic behavior of reinforced concrete column-steel plate subassemblies and frame systems. *J Struct Eng* 2004;130:310:319.

[2] Cheng C, Chen C. Seismic behavior of steel plate and reinforced concrete column connections. *J Construct Steel Res* 2004;61:587–606.

[3] Changwang Y, Jinqing J, Ju Z. Seismic behavior of steel reinforced ultra high strength concrete column and reinforced concrete plate connection. *Trans Tianjin Univ* 2010;16:309–316.

[4] Ji X, Zhang M, Kang H, Qian J, Hu H. Effect of cumulative seismic damage to steel tube-reinforced concrete composite columns. *Earthq Struct* 2014;7:179–200.

[5] Cao VV, Ronagh HR. Reducing the potential seismic damage of reinforced concrete frames using plastic hinge relocation by FRP. *Compos Part B: Eng* 2014;60: 688–696.

[6] Choi SW, Yousok K, Park HS. Multi-objective seismic retrofit method for using FRP jackets in shear-critical reinforced concrete frames. *Compos Part B: Eng* 2014;56:207–216.

[7] Liu ZQ, Xue JY, Zhao, HT. Seismic behavior of steel reinforced concrete special-shaped column-plate joints. *Earthq Struct* 2016;11:665–680.

[8] Wuite J, Adali S. Deflection and stress behaviour of nanocomposite reinforced plates using a multiscale analysis. *Compos Struct* 2005;71:388–396.

[9] Matsuna H. Vibration and buckling of cross-ply laminated composite circular cylindrical shells according to a global higher-order theory. *Int J Mech Sci* 2007;49:1060–1075.

[10] Formica G, Lacarbonara W, Alessi R. Vibrations of carbon nanotube reinforced composites. *J Sound Vib* 2010;329:1875–1889.

[11] Liew KM, Lei ZX, Yu JL, Zhang LW. Postbuckling of carbon nanotube-reinforced functionally graded cylindrical panels under axial compression using a meshless approach. *Comput Meth Appl Mech Eng* 2014;268:1–17.

[12] Lei ZX, Zhang LW, Liew KM, Yu JL. Dynamic stability analysis of carbon nanotube-reinforced functionally graded cylindrical panels using the element-free kp-Ritz method. *Compos Struct* 2014;113:328–338.

[13] Ghorbanpour Arani A, Haghparast E, KhoddamiMaraghi Z, Amir S. Static stress analysis of carbon nano-tube reinforced composite (CNTRC) cylinder under non-axisymmetric thermo-mechanical loads and uniform electro-magnetic fields. *Compos Part B: Eng* 2015;68:136–145.

[14] Kolahchi R, RabaniBidgoli M, Beygipoor GH, Fakhar MH. A nonlocal nonlinear analysis for buckling in embedded FG-SWCNT-reinforced microplates subjected to magnetic field. *J Mech Sci Tech* 2013;5:2342–2355.

[15] Davar A, Khalili SMR, Malekzadeh Fard K. Dynamic response of functionally graded circular cylindrical shells subjected to radial impulse load. *Int J Mech Mater Des* 2013;9:65–81.

[16] Kolahchi R, Safari M, Esmailpour M. Dynamic stability analysis of temperature-dependent functionally graded CNT-reinforced visco-plates resting on orthotropic elastomeric medium. *Compos Struct* 2016;150:255–265.

[17] Jafarian Arani A, Kolahchi R. Buckling analysis of embedded concrete columns armed with carbon nanotubes. *Comput Concr* 2016;17:567–578.

[18] Shen HS, Yang DQ. Nonlinear vibration of anisotropic laminated cylindrical shells with piezoelectric fiber reinforced composite actuators. *Ocean Eng* 2014;80:36–49.

[19] Duc ND, Hadavinia H, Van Thu P, Quan TQ. Vibration and nonlinear dynamic response of imperfect three-phase polymer nanocomposite panel resting on elastic foundations under hydrodynamic loads. *Compos Struct* 2015;131:229–237.

[20] Van Thu P, Duc ND. Non-linear dynamic response and vibration of an imperfect three-phase laminated nanocomposite cylindrical panel resting on elastic foundations in thermal environment. *Sci Eng Compos Mat* 2016;24:951–962.

[21] Alibeigloo A. Thermoelastic analysis of functionally graded carbon nanotube reinforced composite cylindrical panel embedded in piezoelectric sensor and actuator layers. *Compos Part B: Eng* 2016;98:225–243.

[22] Feng CH, Kitipornchai S, Yang J. Nonlinear bending of polymer nanocomposite plates reinforced with non-uniformly distributed graphene platelets (GPLs). *Compos Part B: Eng* 2017;110:132–140.

[23] Amoli A, Kolahchi R, Bidgoli MR. Seismic analysis of $AL_2O_3$ nanoparticles-reinforced concrete plates based on sinusoidal shear deformation theory. *Earthquak Struct* 2018;(15)3:285–294.

# 11 Stress Analysis of Shells Reinforced with Nanoparticles

## 11.1 INTRODUCTION

Nanotechnology has a significant impact on the construction sector. Several applications have been developed for this specific sector to improve the durability and enhanced performance of construction components, the energy efficiency and safety of the buildings, facilitate the ease of maintenance, and provide increased living comfort. Nanoparticles of titanium dioxide ($TiO_2$), aluminum oxide ($AL_2O_3$), or zinc oxide (ZnO) are applied as a final coating on construction ceramics to bring these characteristics to the surfaces.

Stress analyses of structures have been done by many researchers. Mechanical and thermal stresses in a functionally graded hollow cylinder were investigated by Jabbari et al. [1]. An analysis of the thermal stress behavior of functionally graded hollow circular cylinders was presented by Liew et al. [2]. You et al. [3] presented an elastic analysis of internally pressurized thick-walled spherical pressure vessels of functionally graded materials. Dai et al. [4] studied the exact solutions for functionally graded pressure vessels in a uniform magnetic field. The coupled thermo-elasticity of functionally graded cylindrical shells was developed by Bahtui and Eslami [5]. Ghorbanpour Arani et al. [6] investigated the effect of material inhomogeneity on the electro-thermo-mechanical behaviors of a functionally graded piezoelectric rotating shaft. Also, they [7] studied the electro-thermo-mechanical behaviors of functionally graded piezoelectric material (FGPM) spheres using the analytical method and ANSYS software.

With respect to developmental works on analyzing cylinders, it should be noted that none of the research mentioned earlier considered composite structures and their specific characteristics. Singha and Daripa [8] analyzed the nonlinear vibration of a composite skew plate using a four-node shear flexible quadrilateral high-precision plate bending element. A nonlinear free vibration analysis of laminated composite skew thin plates using the differential quadrature method (DQM) was reported by Malekzadeh [9]. Qu et al. [10] employed the Rayleigh–Ritz method for vibration to analyze a damaged piezoelectric composite plate with different boundary conditions. Zhang et al. [11] presented a three-dimensional layer-wise DQM to analyze linear laminated piezo-elastic composite plates with different materials and boundary conditions. Thinh and Ngoc [12] carried out static behavior and vibration control of piezoelectric cantilever composite plates and compared the results with experiments using the finite element method (FEM) based on the first-order

DOI: 10.1201/9781003349525-11

shear deformation theory. Using the FEM, the free vibration of laminated piezoelectric composite plates based on an accurate theory was studied by Shu [13]. Using the Galerkin method, Shooshtari and Razavi [14] introduced a closed-form solution for the linear and nonlinear free vibrations of simply supported composite and fiber metal–laminated rectangular plates based on the first-order shear deformation theory. The electro-thermo-mechanical nonlinear vibration and the instability of a fluid, conveying a smart composite microtube made of polyvinylidene fluoride, were investigated by Ghorbanpour Arani et al. [15], using the modified couple stress theory and Timoshenko beam model. Heidarzadeh et al. [16] presented a stress analysis of concrete pipes reinforced with $AL_2O_3$ nanoparticles is provided, taking into account agglomeration effects. The pipe is exposed to mechanical, magnetic, and thermal loadings.

To the best of authors' knowledge, no research work has been found on the theoretical analysis of pipes reinforced with nanoparticles. However, a magneto-thermo-mechanical stress analysis of pipes reinforced with $AL_2O_3$ nanoparticles is presented. The agglomeration effects are considered based on the Mori–Tanaka approach. The governing equation is solved analytically to obtain the stresses and radial displacement of the structure. The effects of the volume percent of $AL_2O_3$ nanoparticles and agglomeration on the stress of the pipe are shown.

## 11.2 GOVERNING EQUATIONS

A pipe reinforced with $AL_2O_3$ nanoparticles is shown in Figure 11.1 in which the geometrical parameters of length $L$, inner radius $a$, and outer radius $b$ are also indicated.

The displacement is a function of radius; hence, the radial and circumferential strains can be expressed as

$$\varepsilon_{rr} = \frac{\partial u}{\partial r}, \tag{11–1}$$

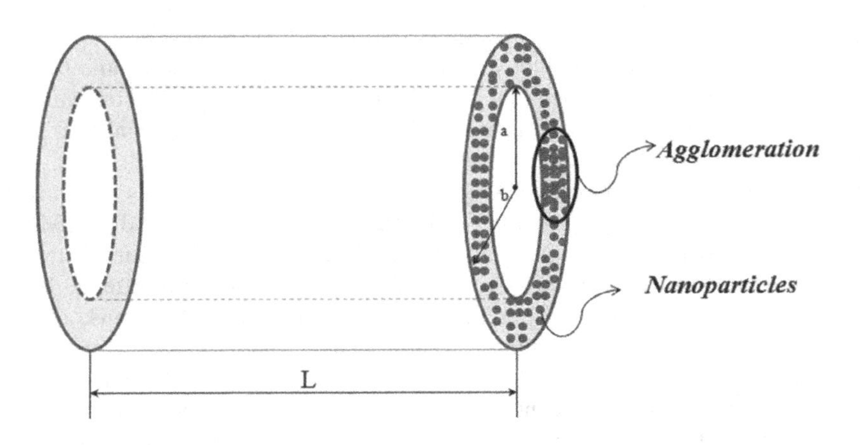


**FIGURE 11.1** A pipe reinforced with $AL_2O_3$ nanoparticles.

$$\varepsilon_{\theta\theta} = \frac{u}{r}. \tag{11–2}$$

The corresponding constitutive relations considering thermal strain may be written as

$$\sigma_{rr} = C_{11}\left(\varepsilon_{rr} - \alpha_{rr}T(r)\right) + C_{12}\left(\varepsilon_{\theta\theta} - \alpha_{\theta\theta}T(r)\right), \tag{11–3}$$

$$\sigma_{\theta\theta} = C_{12}\left(\varepsilon_{rr} - \alpha_{rr}T(r)\right) + C_{22}\left(\varepsilon_{\theta\theta} - \alpha_{\theta\theta}T(r)\right), \tag{11–4}$$

where $\sigma_{rr}$ and $\sigma_{\theta\theta}$ are the radial and circumferential stresses, respectively; $\alpha_{rr}$and $\beta_{\theta\theta}$ are the thermal expansion in the radial and circumferential directions, respectively; $C_{ij}$ are the elastic constants that can be obtained by the Mori–Tanaka model (see Section 4.3); and $T$ is the temperature distribution, which may be written as

$$T(r) = F_1 \ln(\mathrm{r}) + \mathrm{F}_2, \tag{11–5}$$

where $F_1$ and $F_2$ are integration constants that can be obtained by the thermal boundary conditions. The equation of equilibrium is

$$\frac{\partial \sigma_{rr}}{\partial r} + \frac{\sigma_{rr} - \sigma_{\theta\theta}}{r} + f_z = 0, \tag{11–6}$$

where $f_z$ is the Lorentz force, which can be obtained as [4]

$$f_z = \mu H_z^{\,2} \frac{\partial}{\partial r}\left(\frac{\partial u}{\partial r} + \frac{u}{r}\right), \tag{11–7}$$

where $H_z$ and $\mu$ are magnetic field and magnetic permeability constants, respectively. To develop the analytical solution, the following dimensionless quantities are introduced:

$$\sigma_r = \frac{\sigma_{rr}}{C_{22}}, \quad \sigma_\theta = \frac{\sigma_{\theta\theta}}{C_{22}}, \quad U = \frac{u}{a}, \quad \chi = \frac{r}{a}, \quad \beta = \frac{b}{a},$$
$$C_1 = \frac{C_{11}}{C_{22}}, C_2 = \frac{C_{12}}{C_{22}}, \quad \overline{\mathrm{H}}\mathrm{z} = \frac{\mu H_z^{\,2}}{C_{22}}. \tag{11–8}$$

Using these notations and the analytical method based on yields, the following inhomogeneity ordinary differential equations:

$$\chi^2 \frac{d^2U}{d\chi^2} + \chi L_1 \frac{dU}{d\chi} + L_2 U = -L_3\chi - L_4\chi \ln(\chi) - L_5, \tag{11–9}$$

where

$$L_1 = 1, \tag{11–10}$$

$$L_2 = \frac{-\overline{H}_z - 1}{C_1 + \overline{H}_z}, \tag{11–11}$$

$$L_3 = \frac{-F_2\left(C_2\alpha_\theta + C_2\alpha_r + C_1\alpha_{rr} - \alpha_\theta\right) - F_1\left(C_2\alpha_\theta + C_1\alpha_r\right)}{C_1 + \overline{H}_z}, \tag{11–12}$$

$$L_4 = \frac{F_1\left(\alpha_\theta - C_2\alpha_\theta - C_1\alpha_r\right)}{C_1 + \overline{H}_z}, \tag{11–13}$$

$$L_5 = \frac{F_3\Xi_2}{C_1 + \overline{H}_z}. \tag{11–14}$$

Equation 11–9, a nonhomogeneous second-order ordinary differential equation, is the governing (constitutive) equation for the displacement of the structure subjected to an axisymmetric thermo-mechanical. The corresponding general solution for the homogeneous differential equation can be written as

$$U_g = F_3 \underbrace{\chi^{\Gamma_1}}_{u_{g1}} + F_4 \underbrace{\chi^{\Gamma_2}}_{u_{g2}}, \tag{11–15}$$

in which $F_3$ and $F_4$ are integration constants that can be determined by the boundary conditions. $\Gamma_1$ and $\Gamma_2$ are the roots of the corresponding characteristic equation of Equation 11–15 and may be evaluated from

$$\Gamma_{1,2} = \frac{(1 - L_1) \pm \sqrt{(L_1 - 1)^2 - 4L_2}}{2}. \tag{11–16}$$

The particular solution for Equation 11–9 is assumed to have the form

$$U_p = \chi^{\Gamma_1} U_{p1} + \chi^{\Gamma_2} U_{p2}, \tag{11–17}$$

where

$$U_{p1} = -\int \frac{\chi^{\Gamma_2} R(\chi)}{W_p} d\chi, \tag{11–18}$$

$$U_{p2} = \int \frac{\chi^{\Gamma_1} R(\chi)}{W_p} d\chi, \tag{11–19}$$

where

$$W_p = \begin{bmatrix} U_{g1} & U_{g2} \\ \left(U_{g1}\right)' & \left(U_{g2}\right)' \end{bmatrix}, \tag{11–20}$$

$$R(\chi) = -L_3\chi - L_4\chi \ln(\chi) - L_5. \tag{11–21}$$

Finally, the redial displacement is

$$U = U_g + U_p. \tag{11–22}$$

Substituting $U$ from Equation 11–22 into Equations 11–3 and 11–4, the expressions for the radial and circumferential stresses can be obtained. The boundary conditions can be written as follows:

$$\sigma_r(1) = -1, \ \ \sigma_r(\chi) = 0, \ \ T(1) = 0, \ \ T(\chi) = 0. \tag{11–23}$$

## 11.3 NUMERICAL RESULTS AND DISCUSSION

The numerical results show the effects of the volume percent of $AL_2O_3$ and the agglomeration effects on the variation of the stress and radial displacement across the thickness of the pipe based on Matlab code presented in Appendix E. The presented results are for the boundary conditions with aspect ratios of $b/a = 2$. The plots in these figures correspond to $T(a) = 323$ K and $T(b) = 298$ K. The magnetic permittivity of the $AL_2O_3$ nanoparticles is selected as $\mu = 4\pi \times 10^{-7} N/A^2$, and the magnetic field intensity is $H_z = 1\times10^8 \ \ A/m$.

For different values of the volume percent of the $AL_2O_3$ nanoparticles ($C_r$), the graphs for the radial stress, circumferential stress, effective stress, and radial displacement, along the radial direction, are plotted in Figures 11.2 through 11.5, respectively.

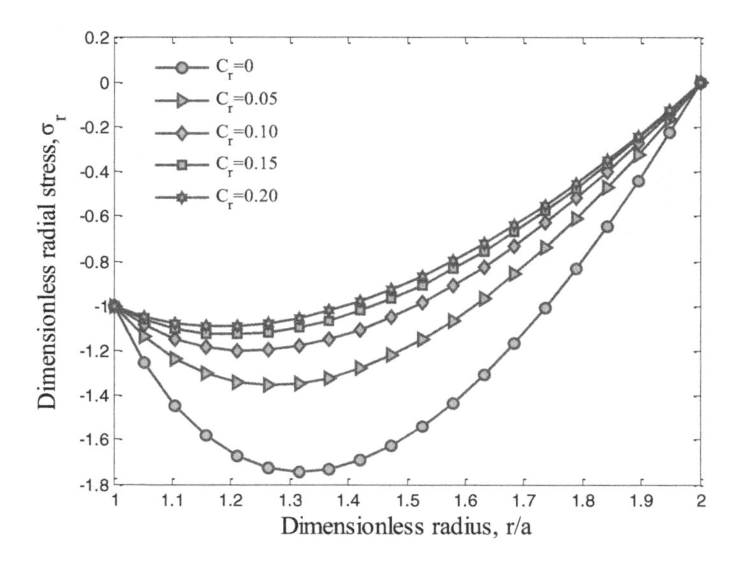


**FIGURE 11.2** Effect of volume percent of $Al_2O_3$ nanoparticles on the radial stress of structure.

These figures indicate clearly that the volume percent of $AL_2O_3$ nanoparticles has a major effect on the magneto-thermo-elastic stresses and the radial displacement.

Figure 11.2 depicts the distribution of the radial stress along the radius. As can be seen, the radial stresses at the internal and external surfaces of the pipe satisfy the given boundary conditions. Moreover, an increase in volume percent of $AL_2O_3$ nanoparticles results in a decrease in the radial stress.

The distributions of the hoop and effective stresses for different values of $C_r$ are displayed in Figures 11.3 and 11.4, where the tensile hoop and effective stresses

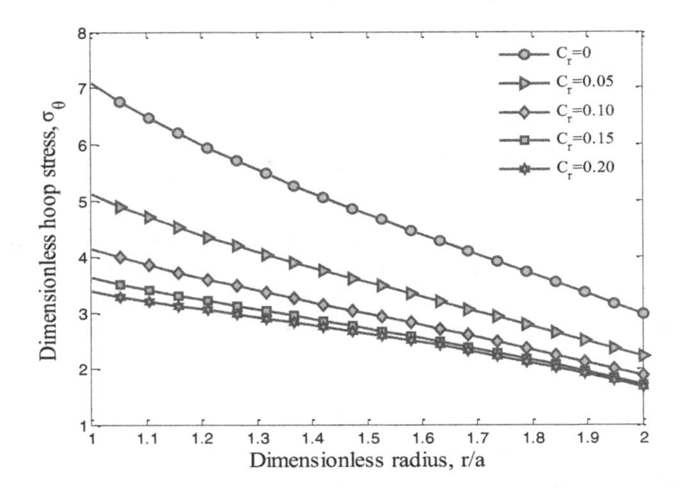


**FIGURE 11.3** Effect of volume percent of $Al_2O_3$ nanoparticles on the hoop stress of structure.

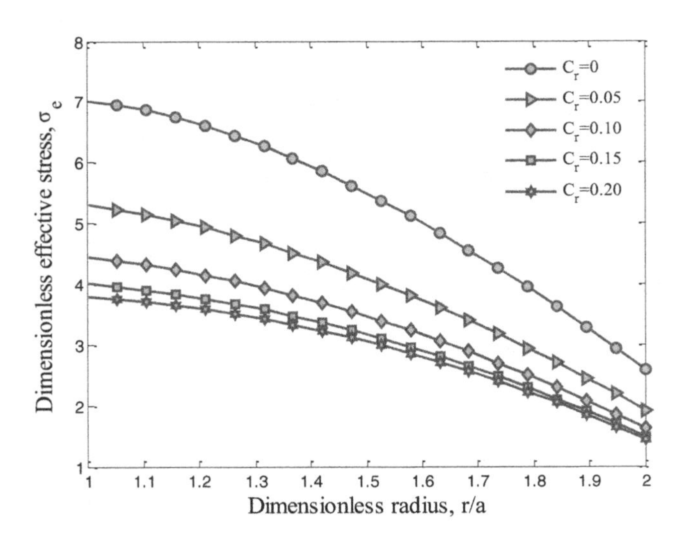


**FIGURE 11.4** Effect of volume percent of $Al_2O_3$ nanoparticles on the effective stress of the structure.

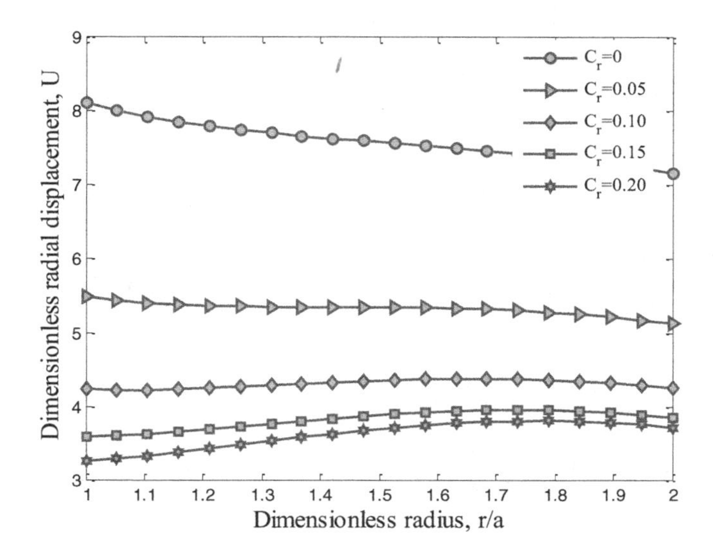


**FIGURE 11.5** Effect of volume percent of $Al_2O_3$ nanoparticles on the radial displacement of the structure.

monotonically decrease from the inner to the outer radius for different values of $C_r$. Furthermore, the hoop and effective stresses decrease with an increase in the volume percent of $AL_2O_3$ nanoparticles.

The variations of the radial displacement along the radius for different values of the volume percent of $Al_2O_3$ nanoparticles are demonstrated in Figure 11.5, which indicates that the radial displacement decreases with an increase in $C_r$. It is because with an increase in the volume percent of $Al_2O_3$ nanoparticles, the stability of the structure increases, and hence, the radial displacement decreases.

The effect of agglomeration ($\xi$), respectively, on the radial, circumferential, effective stresses and the radial displacement along the radial direction are demonstrated in Figures 11.6 through 11.9. It is worth noting that the compressive radial stresses plotted in Figure 11.6 are maximum at the inner surface. Also, the compressive radial stress increases with an increase in $\xi$. The figure demonstrates that the radial stresses at the internal and external surfaces of the nanocomposite pipe conform to the boundary conditions assumed here.

Figures 11.7 and 11.8 show the variations of hoop and effective stresses along the radius for different values of $\xi$. As can be seen, for the different values of the $\xi$, the hoop and effective stresses increase as $\xi$ is increased. It is also noted that for different values of $\xi$, the hoop and effective stresses always attain their maximum value at the inner surface. Figures 11.7 and 11.8 indicate that the agglomeration has a major effect on the stresses of structure.

The variations of the radial displacement along the radius for different values of $\xi$ are shown in Figure 11.9, indicating that the radial displacement increases as the index $\xi$ increases.

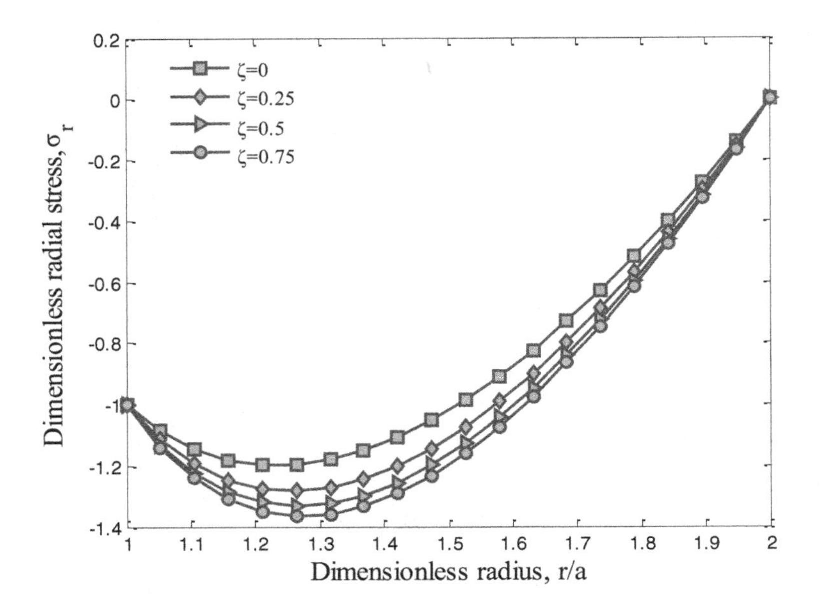


**FIGURE 11.6** Agglomeration effects on the radial stress of the structure.

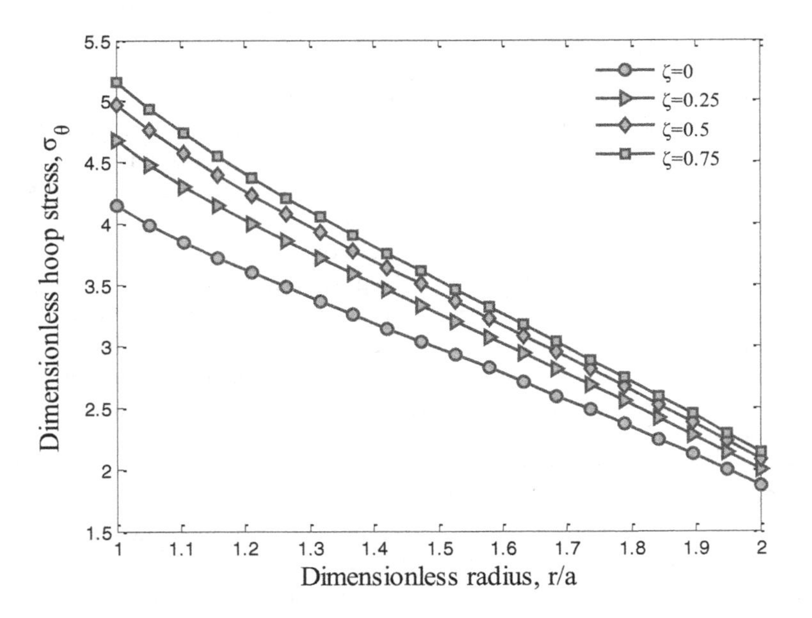


**FIGURE 11.7** Agglomeration effect on the hoop stress of the structure.

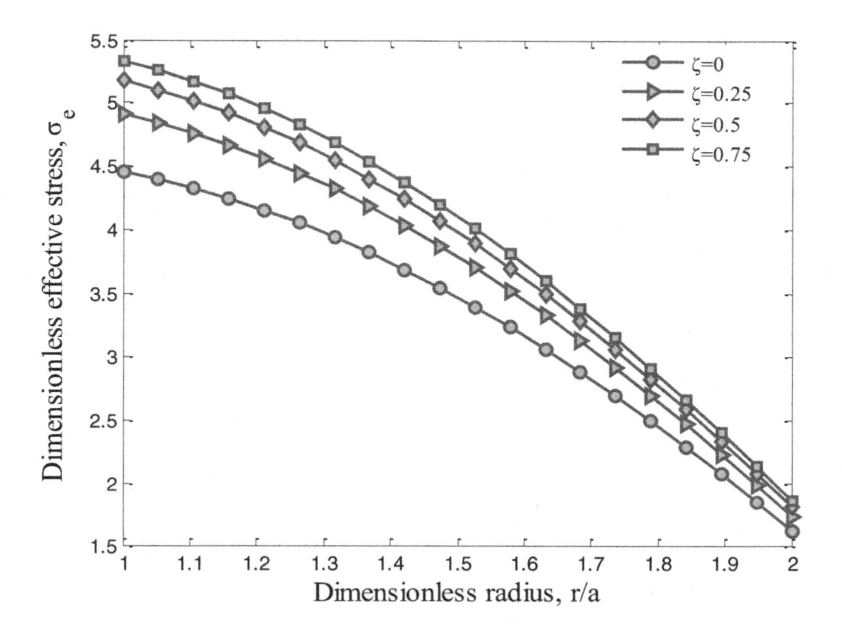


**FIGURE 11.8** Agglomeration effect on the effective stress of the structure.

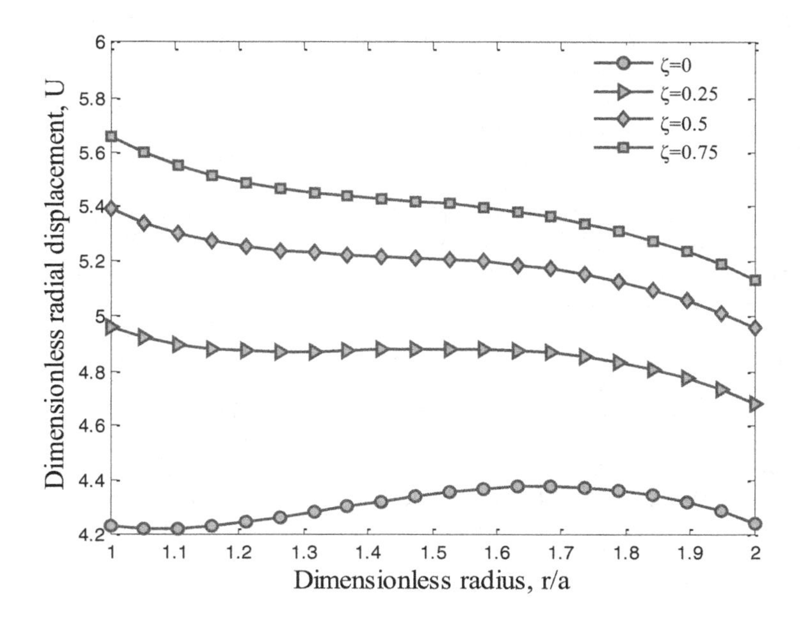


**FIGURE 11.9** Agglomeration effect on the radial displacement of the structure.

**Acknowledgments:** This chapter is a slightly modified version of Ref. [16] and has been reproduced here with the permission of the copyright holder.

## REFERENCES

[1] Jabbari M, Sohrabpour S, Eslami MR. Mechanical and thermal stresses in a functionally graded hollow cylinder due to radially symmetric loads. *Int J Press Ves Pip* 2002;79:493–497.
[2] Liew KM, Kitipornchai S, Zhang XZ, Lim CW. Analysis of the thermal stress behaviour of functionally graded hollow circular cylinders. *Int J Solids Struct* 2003;40:2355–2380.
[3] You LH, Zhang JJ, You XY. Elastic analysis of internally pressurized thick-walled spherical pressure vessels of functionally graded materials. *Int J Press Ves Pip* 2005;82:347–354.
[4] Dai HL, Fu YM, Dong ZM. Exact solutions for functionally graded pressure vessels in a uniform magnetic field. *Int J Solids Struct* 2006;43:5570–5580.
[5] Bahtui A, Eslami MR. Coupled thermoelasticity of functionally graded cylindrical shells. *Mech Res Commun* 2007;34:1–18.
[6] Ghorbanpour Arani A, Kolahchi R, Mosallaei Barzoki AA. Effect of material in-homogeneity on electro-thermo-mechanical behaviors of functionally graded piezoelectric rotating shaft. *Appl Math Model* 2011;35:2771–2789.
[7] Ghorbanpour Arani A, Kolahchi R, Mosallaie Barzoki AA, Loghman A. Electro-thermo-mechanical behaviors of FGPM spheres using analytical method and ANSYS software. *Appl Math Model* 2011;36:139–157.
[8] Singh MK, Darip R. Nonlinear vibration of symmetrically laminated composite skew-plates by finite element method. *Int J Non Linear Mech* 2007;42:1144–1152.
[9] Malekzadeh P. A differential quadrature nonlinear free vibration analysis of laminated composite skew thin plates. *Thin-Wall Struct* 2007;45:237–250.
[10] Qu GM, Li YY, Cheng L, Wang B. Vibration analysis of a piezoelectric composite plate with cracks. *Compos Struct* 2006;72:111–118.
[11] Zhang Z, Feng C, Liew KM. Three-dimensional vibration analysis of multilayered piezoelectric composite plates. *Int J Eng Sci* 2006;44:397–408.
[12] Thinh TI, Ngoc LK. Static behavior and vibration control of piezoelectric cantilever composite plates and comparison with experiments. *Comput Mater Sci* 2010;49:276–280.
[13] Shu X. Free vibration of laminated piezoelectric composite plates based on an accurate theory. *Compos Struct* 2005;67:375–382.
[14] Shooshtari A, Razavi S. A closed form solution for linear and nonlinear free vibrations of composite and fiber metal laminated rectangular plates. *Compos Struct* 2010;92:2663–2675.
[15] Ghorbanpour Arani A, Abdollahian M, Kolahchi R. Nonlinear vibration of embedded smart composite microtube conveying fluid based on modified couple stress theory. *Poly Compos* 2015;36:1314–1324.
[16] Heidarzadeh A, Kolahchi R, Bidgoli MR. Concrete Pipes Reinforced with $AL_2O_3$ Nanoparticles Considering Agglomeration: Magneto-Thermo-Mechanical Stress Analysis. *Int J Civil Eng* 2018;16:315–322.

# 12 Earthquake Response of Submerged Nanocomposite Shells Conveying Fluid

## 12.1 INTRODUCTION

Submerged fluid-conveying cylindrical pipes have been widely used in many civil and mechanical engineering applications, for example, in the submarine industry, the oil and gas industry, petrochemical systems, and so on. Such cylindrical structures have been analyzed for many of the failures and/or operating problems due to flow-induced vibrations and instabilities from previous decades (Housner [1]; Benjamin [2]; Païdoussisand and Issid [3]). In the last years, some studies have been done on the dynamic characteristics of pipelines (Amabiliet al. [4–8]; Païdoussis et al. [9–12]; Lopes et al. [13]; Semler et al. [14]; Wadham-Gagnon et al. [15]; Ibrahim [16,17]). Also, several studies have been performed on the dynamical response of submerged pipes and/or fluid-conveying pipes against underwater shock, the effect of the moving mass, fluid-induced vibrations, and ground-motion acceleration that are noted in the following discussion.

Mechanical analyses of nanostructures have been reported by many researchers (Zemri, [18]; Larbi Chaht [19]; Belkorissat [20]; Ahouel, [21]; Bounouara, [22]; Bouafia, [23]; Besseghier, [24]; Bellifa [25]; Mouffoki [26]; Khetir,[27]). Gong et al. [28] applied a computational method for a safety evaluation of submerged pipelines subjected to underwater shock. In this research, the fluid–structure interaction between the pipeline and seawater was considered based on the coupled boundary element and finite element programs by means of the doubly asymptotic approximation. Lee and Oh [29] developed a spectral element model for a pipe conveying fluid to study the flow-induced vibrations of the system by the exact constitutive dynamic stiffness matrix. Lam et al. [30] examined the dynamic response of a simply supported laminated underwater pipeline exposed to underwater explosion shock. They concluded that the strength of the radial direction for the pipe is weaker than the strengths in the longitudinal and circumferential directions. Consequently, the dynamic response of the radial direction is larger than those of other directions. Yoon and Son [31] studied the dynamic behavior of a simply supported fluid-conveying pipe due to the effect of the open crack and the moving mass. Lin and Qiao [32] explored the vibration and instability of an axially moving beam immersed in fluid with simply supported conditions along with torsional springs. Huang et al. [33] used Galerkin's method to obtain the eigenfrequencies of tubes conveying fluid

DOI: 10.1201/9781003349525-12

having different boundary conditions. Furthermore, they calculated the variation of system eigenfrequencies by the effect of the Coriolis forces and expressed a correlation between a pipe conveying fluid and an Euler–Bernoulli beam. Zhai et al. [34] used the Timoshenko beam model for obtaining the dynamic response of a fluid-conveying pipe under random excitation. They solved the governing equations by the pseudo-excitation method together with the complex mode superposition method. Also, they assumed that the parameters of the load are random. Liu et al. [35] analyzed fluid–solid interaction problem for an elastic cylinder by numerical simulations and acquired the vibration of a cylinder for both laminar and turbulent flows. Investigating the dynamic behavior of pipelines under earthquake acceleration is a research field with few works. The seismic response of natural gas and water pipelines in the Ji-Ji earthquake was considered by Chen et al. [36]. They conducted a statistical analysis to understand the relationship between seismic factors (spectrum intensity, peak ground acceleration, peak ground velocity) and repair rates. Also, Abdounet al. [37] studied influencing factors on the behavior of buried pipelines subjected to earthquake faulting. In none of the mentioned investigations is the structure a composite. The effect of using fiber-reinforced polymer composites for underwater steel pipeline repairs was studied by Shamsuddoha et al. [38]. They offered a widespread review of using fiber-reinforced polymer composites for in-air, underground, and underwater pipeline repairs. Ray and Reddy [39] made a study on the active damping of piezoelectric composite cylindrical shells conveying fluid. Alijani and Amabili [40] used an energy method with the Amabili–Reddy nonlinear higher order shear deformation theory for determining the nonlinear vibrations and multiple resonances of fluid filled arbitrary laminated cylindrical shells. They demonstrated that water-filled composite shells may exhibit complex nonlinear dynamic behavior. Thinh and Nguyen [41] investigated the free vibration of composite circular shells containing fluid. They used the dynamic stiffness method based on the Reissner–Mindlin theory and nonviscous incompressible fluid equations for modeling of the structure. The dynamic characteristics of steady fluid conveying in the periodical partially visco-elastic composite pipeline were studied by Zhou et al. [42]. It is shown that reducing the coverage fraction decreases the flutter velocity. The nonlinear vibration of laminated composite circular cylindrical shells using Donnell's shell theory and the incremental harmonic balance method was analyzed by Dey and Ramachandram [43]. Furthermore, the mechanical behavior of structures containing nanoparticles has been investigated experimentally and analytically by a number of researchers. The influences of nanoparticles on dynamic strength of ultra-high performance were tested by Su et al. [44]. JafarianArani and Kolahchi [45] analyzed the buckling of beams reinforced with carbon nanotubes (CNTs) by using the Euler–Bernoulli and Timoshenko beam models. The buckling of beams retrofitted with nanofiber-reinforced polymer was investigated by Safari Bilouei et al. [46]. Inozemtcev et al. [47] improved the properties of lightweight hollow microspheres with the nanoscale modifier. The mathematical modeling of pipes reinforced with CNTs conveying fluid for vibration and stability analysis was done by Zamani Nouri [48]. The vibration of silica nanoparticle–reinforced beams considering agglomeration effects was considered by Shokravi [49]. Also, Rabani Bidgoli and Saeidifar [50] studied the time-dependent buckling of silicon dioxide ($SiO_2$) nanoparticle–reinforced beams exposed to fire.

Maleki et al. [51] investigated the dynamic response of an underwater nanocomposite submerged pipeline conveying fluid is explored. The structure is exposed to dynamic loads induced by earthquakes, and the governing equations of the system are derived using a mathematical model based on classical shell theory and Hamilton's principle. In this research, the pipes were not submerged.

Hitherto, the dynamic behavior of the submerged nanocomposite pipes conveying fluid under earthquake load has not been investigated by any researcher. In this chapter, the seismic response of the submerged nanocomposite pipe conveying fluid under earthquake load is analytically considered as the importance of the subject. The Mori–Tanaka method is used to evaluate the material properties of the nanocomposite. The governing equations of the structure are derived using the energy method and according to classical theory. The dynamic displacement of the structure is derived using the differential quadrature method (DQM) and the Newmark method. In the present study, the effect of various parameters like the volume percent of $SiO_2$ nanoparticles, the boundary conditions, the geometrical parameters of the pipe, the internal and external fluid pressures, and earthquake intensity on the dynamic displacement of the structure is presented.

## 12.2 MATHEMATICAL MODELING

As shown in Figure 12.1, an underwater nanocomposite cylindrical pipe conveying fluid with length $a$, radius $R$, and thickness $h$ is considered. The geometrical properties of the nanocomposite pipe are set as length-to-radius ratio, $a/R = 10$, and thickness-to-radius ratio, $h/R = 0.03$. The boundary conditions are simply supported, and the $SiO_2$ nanoparticle volume percent is 0.05 unless otherwise specified.

To calculate the middle-surface strain and curvatures and using Kirchhoff–Law assumptions, the displacement components of a cylindrical shell in the axial $x$, circumferential $\theta$, and radial $z$ directions can be written based on Section 1.5.1.

The total potential energy, $V$, of the underwater cylindrical shell conveying fluid is the sum of strain energy $U$, kinetic energy $K$, and the work done by the fluid $W$. The strain energy can be written as

$$U = \int_V \left(\sigma_{xx}\varepsilon_{xx} + \sigma_{\theta\theta}\varepsilon_{\theta\theta} + \sigma_{x\theta}\gamma_{x\theta}\right) dV. \tag{12–1}$$

Substituting Equations 1–39a through 1–39c into Equation 12–1 yields

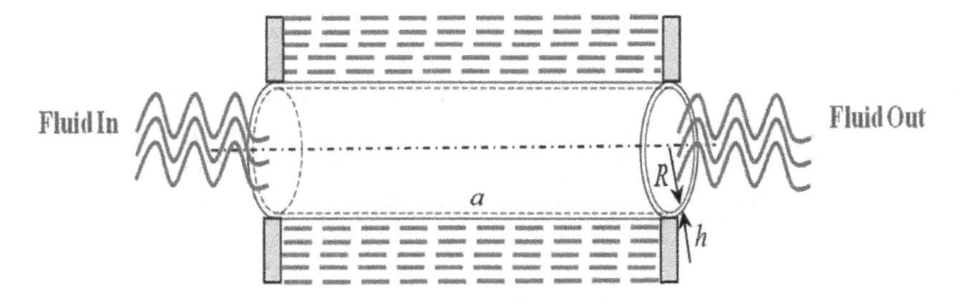


**FIGURE 12.1** Schematic of an underwater pipe conveying fluid.

$$U=\int_{-\frac{h}{2}}^{\frac{h}{2}}\int_{A}\left(\sigma_x\left(\frac{\partial u}{\partial x}+0.5\left(\frac{\partial w}{\partial x}\right)^2-z\frac{\partial^2 w}{\partial x^2}\right)+\sigma_\theta\left(\frac{\partial v}{R\partial\theta}+\frac{w}{R}+0.5\left(\frac{\partial w}{R\partial\theta}\right)^2\right.\right.$$
$$\left.\left.-z\frac{\partial^2 w}{R^2\,\partial\theta^2}\right)+\sigma_{x\theta}\left(\frac{\partial u}{R\partial\theta}+\frac{\partial v}{\partial x}+\frac{\partial w}{R\partial\theta}\frac{\partial w}{\partial x}-2z\frac{\partial^2 w}{R\,\partial\theta\,\partial x}\right)\right)dz\,dA. \quad (12–2)$$

By introducing force and moment resultants in Equation 12–2, we have:

$$\begin{Bmatrix} N_x \\ N_\theta \\ N_{x\theta} \end{Bmatrix}=\int_{-\frac{h}{2}}^{\frac{h}{2}}\begin{Bmatrix} \sigma_x \\ \sigma_\theta \\ \tau_{x\theta} \end{Bmatrix}dz, \quad (12–3)$$

$$\begin{Bmatrix} M_x \\ M_\theta \\ M_{x\theta} \end{Bmatrix}=\int_{-\frac{h}{2}}^{\frac{h}{2}}\begin{Bmatrix} \sigma_x \\ \sigma_\theta \\ \tau_{x\theta} \end{Bmatrix}zdz, \quad (12–4)$$

$$U=\int_{A}\left(N_x\left(\frac{\partial u}{\partial x}+0.5\left(\frac{\partial w}{\partial x}\right)^2\right)-M_x\frac{\partial^2 w}{\partial x^2}+N_\theta\left(\frac{\partial v}{R\partial\theta}+\frac{w}{R}+0.5\left(\frac{\partial w}{R\partial\theta}\right)^2\right)\right.$$
$$\left.-M_\theta\frac{\partial^2 w}{R^2\,\partial\theta^2}+N_{x\theta}\left(\frac{\partial u}{R\partial\theta}+\frac{\partial v}{\partial x}+\frac{\partial w}{R\partial\theta}\frac{\partial w}{\partial x}\right)-2M_{x\theta}\frac{\partial^2 w}{R\,\partial\theta\,\partial x}\right)dA. \quad (12–5)$$

The kinetic energy may be expressed as

$$K=\int\left(\frac{\rho}{2}\left(\frac{h^3}{12}\left((\frac{\partial^2 u}{\partial t\,\partial x})^2+(\frac{\partial^2 w}{\partial t\partial\theta})^2\right)\right)+h\left((\frac{\partial u}{\partial t})^2+(\frac{\partial v}{\partial t})^2+(\frac{\partial w}{\partial t})^2\right)\right)dA. \quad (12–6)$$

The external work due to the internal Newtonian fluid can be obtained using the well-known Navier–Stokes equation:

$$\rho_f\frac{d\mathrm{V}}{dt}=-\nabla\mathrm{P}+\mu\nabla^2\mathrm{V}+\mathrm{F}_{body}, \quad (12–7)$$

where $V=(v_x,v_\theta,v_z)$ is the flow velocity vector in the cylindrical coordinate system with components in longitudinal $x$, circumferential $\theta$, and radial $z$ directions. Also, P, $\theta$, and $\rho_f$ are the pressure, viscosity, and density of the fluid, respectively, and $F_{body}$ denotes the body forces. In the Navier–Stokes equation, the total derivative operator with respect to $t$ is

$$\frac{d}{dt}=\frac{\partial}{\partial t}+v_x\frac{\partial}{\partial x}+v_\theta\frac{\partial}{\partial\theta}+v_z\frac{\partial}{\partial z}. \quad (12–8)$$

At the point of contact between the fluid and the core, the relative velocity and the acceleration in the radial direction are equal. So

$$v_z = \frac{dw}{dt}. \tag{12–9}$$

By employing Equations 12–8 and 12–9 and substituting them into Equation 12–7, the pressure inside the pipe can be computed as

$$\frac{\partial p_z}{\partial z} = -\rho_f \left( \frac{\partial^2 w}{\partial t^2} + 2 v_x \frac{\partial^2 w}{\partial x \, \partial t} + v_x^{\ 2} \frac{\partial^2 w}{\partial x^2} \right)$$
$$+ \mu \left( \frac{\partial^3 w}{\partial x^2 \, \partial t} + \frac{\partial^3 w}{R^2 \, \partial \theta^2 \, \partial t} + v_x \left( \frac{\partial^3 w}{\partial x^3} + \frac{\partial^3 w}{R^2 \, \partial \theta^2 \, \partial x} \right) \right). \tag{12–10}$$

By multiplying two sides of Equation 12–10 in the inside area of the pipe ($A$), the radial force in the pipe is calculated as shown:

$$F_{fluid} = A \frac{\partial p_z}{\partial z} = -\rho_f \left( \frac{\partial^2 w}{\partial t^2} + 2 v_x \frac{\partial^2 w}{\partial x \, \partial t} + v_x^{\ 2} \frac{\partial^2 w}{\partial x^2} \right)$$
$$+ \mu \left( \frac{\partial^3 w}{\partial x^2 \, \partial t} + \frac{\partial^3 w}{R^2 \, \partial \theta^2 \, \partial t} + v_x \left( \frac{\partial^3 w}{\partial x^3} + \frac{\partial^3 w}{R^2 \, \partial \theta^2 \, \partial x} \right) \right). \tag{12–11}$$

Finally, the external work due to the pressure of the fluid may be obtained as follows:

$$W_f = \int (F_{fluid}) w dA = \int \left( -\rho_f \left( \frac{\partial^2 w}{\partial t^2} + 2 v_x \frac{\partial^2 w}{\partial x \, \partial t} + v_x^{\ 2} \frac{\partial^2 w}{\partial x^2} \right) \right.$$
$$\left. + \mu \left( \frac{\partial^3 w}{\partial x^2 \, \partial t} + \frac{\partial^3 w}{R^2 \, \partial \theta^2 \, \partial t} + v_x \left( \frac{\partial^3 w}{\partial x^3} + \frac{\partial^3 w}{R^2 \, \partial \theta^2 \, \partial x} \right) \right) \right) w dA. \tag{12–12}$$

Also, the external work due to the outside fluid can be obtained as follows:

$$F_v = -\alpha \frac{\partial w}{\partial t}, \tag{12–13}$$

where

$$\alpha = \frac{2 v_f \pi (\eta^2 - 1)}{(1 - \eta^2 + (\eta^2 + 1) \ln \eta)} \quad and \, \eta = \frac{R_0}{R_1}. \tag{12–14}$$

It should be noted that the parameter $\alpha$ is positive $(0 < \eta < 1)$. Here, $R_0$ is the shell's outer radius, and $R_1$ is the distance from the center line to the position where the induced viscous flow vanished. To couple the elastic deformation of the shell and the viscous flow of the external fluid, it is assumed that the surface traction of the external fluid along the interface is equal to the external force exerted on the shell:

$$q = F_v. \tag{12–15}$$

The external work due to the earthquake loads can be computed as follows:

$$W_s = \int \underbrace{(ma(t))}_{F_{Seismic}} w dA, \tag{12–16}$$

where $m$ and $a(t)$ are the mass and the acceleration of the ground. The governing equations of the structure are derived using Hamilton's principle, which is considered as follows:

$$\int_0^t (\delta U - \delta K - \delta W) dt = 0. \tag{12–17}$$

Now, by applying Hamilton's principle and after integration by part and some algebraic manipulation, three equations of motion can be derived as follows:

$$\frac{\partial N_x}{\partial x} + \frac{\partial N_{x\theta}}{R \partial \theta} = \rho h \frac{\partial^2 u}{\partial t^2}, \tag{12–18}$$

$$\frac{\partial N_\theta}{R \partial \theta} + \frac{\partial N_{x\theta}}{\partial x} = \rho h \frac{\partial^2 v}{\partial t^2}, \tag{12–19}$$

$$\begin{aligned} &\frac{\partial^2 M_x}{\partial x^2} + \frac{2\partial^2 M_{x\theta}}{R \partial x \partial \theta} + \frac{\partial^2 M_\theta}{R^2 \partial \theta^2} - \frac{N_\theta}{R} + N_x \frac{\partial^2 w}{\partial x^2} + N_\theta \frac{\partial^2 w}{R^2 \partial \theta^2} \\ &+ N_{x\theta} \frac{2\partial^2 w}{R \partial x \partial \theta} + F_V + F_{fluid} = \rho h \frac{\partial^2 w}{\partial t^2} + F_{Seismic}. \end{aligned} \tag{12–20}$$

By integrating the stress–strain relations of the structure, we have

$$N_x = h\left(C_{11}\left(\frac{\partial u}{\partial x} + 0.5\left(\frac{\partial w}{\partial x}\right)^2\right) + C_{12}\left(\frac{\partial v}{R\partial \theta} + \frac{w}{R} + 0.5\left(\frac{\partial w}{R\partial \theta}\right)^2\right)\right), \tag{12–21}$$

$$N_\theta = h\left(C_{12}\left(\frac{\partial u}{\partial x} + 0.5\left(\frac{\partial w}{\partial x}\right)^2\right) + C_{22}\left(\frac{\partial v}{R\partial \theta} + \frac{w}{R} + 0.5\left(\frac{\partial w}{R\partial \theta}\right)^2\right)\right), \tag{12–22}$$

$$N_{x\theta} = h\left(C_{66}\left(\frac{\partial u}{R\partial \theta} + \frac{\partial v}{\partial x} + \frac{\partial w}{R\partial \theta}\frac{\partial w}{\partial x}\right)\right), \tag{12–23}$$

$$M_x = \frac{h^3}{12}\left(C_{11}\left(-z\frac{\partial^2 w}{\partial x^2}\right) + C_{12}\left(-z\frac{\partial^2 w}{R^2 \partial \theta^2}\right)\right), \tag{12–24}$$

$$M_\theta = \frac{h^3}{12}\left(C_{12}\left(-z\frac{\partial^2 w}{\partial x^2}_x\right) + C_{22}\left(-z\frac{\partial^2 w}{R^2\,\partial\theta^2}\right)\right), \tag{12–25}$$

$$M_{x\theta} = \frac{h^3}{12}C_{66}\left(-2z\frac{\partial^2 w}{R\,\partial\theta\,\partial x}\right). \tag{12–26}$$

By substituting stress resultants, Equations 12–21 through 12–26, into the governing equations, Equations 12–18 through 12–20, the relations can be obtained in terms of only the displacement fields. Also, the boundary conditions are taken into account as follows:

- Clamped–clamped supported

$$\begin{aligned} &w = v = u = 0 && @ \quad x = 0, L \\ &\frac{\partial w}{\partial x} = 0 && @ \quad x = 0, L \end{aligned} \tag{12–27}$$

- Simply–simply supported

$$\begin{aligned} &w = v = \frac{\partial^2 w}{\partial x^2} = 0 && @ \quad x = 0 \\ &w = v = \frac{\partial^2 w}{\partial x^2} = 0 && @ \quad x = L \end{aligned} \tag{12–28}$$

- Clamped–simply supported

$$\begin{aligned} &w = v = u = \frac{\partial w}{\partial x} = 0 && @ \quad x = 0 \\ &w = v = \frac{\partial^2 w}{\partial x^2} = 0 && @ \quad x = L \end{aligned} \tag{12–29}$$

## 12.3 NUMERICAL RESULTS AND DISCUSSION

A computer program based on DQM (see Appendix F) to solve the nonlinear motion equations. For this purpose, a nanocomposite pipe of length $a$, radius $R$, thickness $h$, Young's modulus of 20 GPa, and Poisson's ratio of 0.3 as shown in Figure 12.1 is considered. Also, it is assumed the flowing liquid is water. The mass density and the viscosity of water are equal to $998.2\ kg/m^3$ and $10^{-3}\ Pa.s$, respectively. Also, for the strengthening of $SiO_2$ nanoparticles with the density of $\rho_{np} = 3970\ kg/m^3$ is used. For considering earthquake effects, the acceleration of the earthquake according to the Tabas earthquake is considered that the distribution of acceleration in 20 seconds is shown in Figure 12.2. Furthermore, the cylindrical shell is investigated with three kinds of boundary conditions: two edges simply supported (SS), clamped (CC), and simply supported and clamped (SC). In addition, in the following figures, the deflection at the middle point of the pipe is studied.

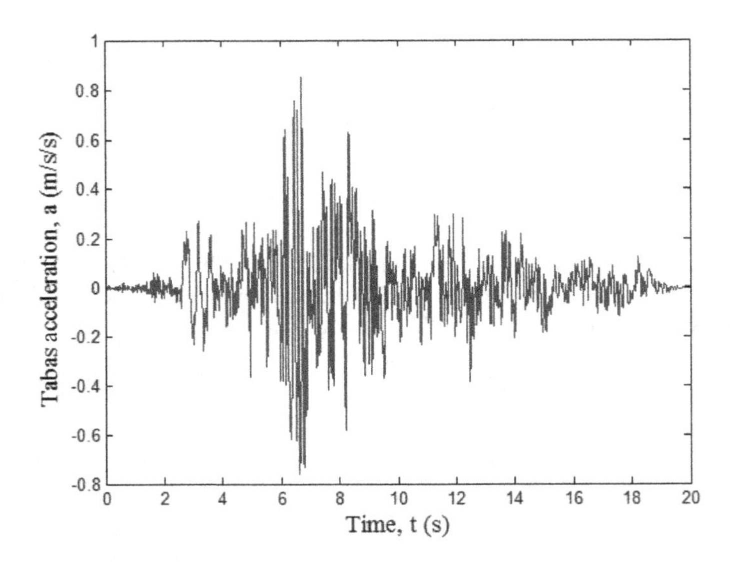


**FIGURE 12.2** The accelerogram of the 1978 Tabas earthquake.

### 12.3.1 Validation

In the absence of similar publications in the literature covering the same scope of the problem, one cannot directly validate the results found here. Therefore, the present work could be partially validated based on a simplified analysis without considering the nonlinear terms of the governing equations and by comparing the linear dynamic response of the pipe, which was obtained using DQM and the exact method. It can be concluded that DQM is accurate and acceptable for the present problem because the present results closely match the analytical method illustrated in Figure 12.3.

### 12.3.2 Convergence of the Present Method

The convergence and accuracy of the DQM in evaluating the maximum deflection of the underwater pipe conveying fluid with the CC boundary condition are illustrated in Figure 12.4. The results are offered for different values of the DQM grid points. it is found that 15 DQM grid points can yield accurate results. So, the following results are based on 15 grid points for the DQM solution.

### 12.3.3 Effects of Various Parameters

In this chapter, four types of CC pipelines were modeled and computed considering the existence of external and internal fluids. Figure 12.5 shows the dynamic displacement of the submerged pipeline under Tabas earthquake load for two cases of an empty and fluid-conveying pipeline with dashed and solid lines, respectively. It can be found that considering the interior fluid decreases the stiffness of the structure, and as a result, the displacement of the structure increases. Furthermore, the

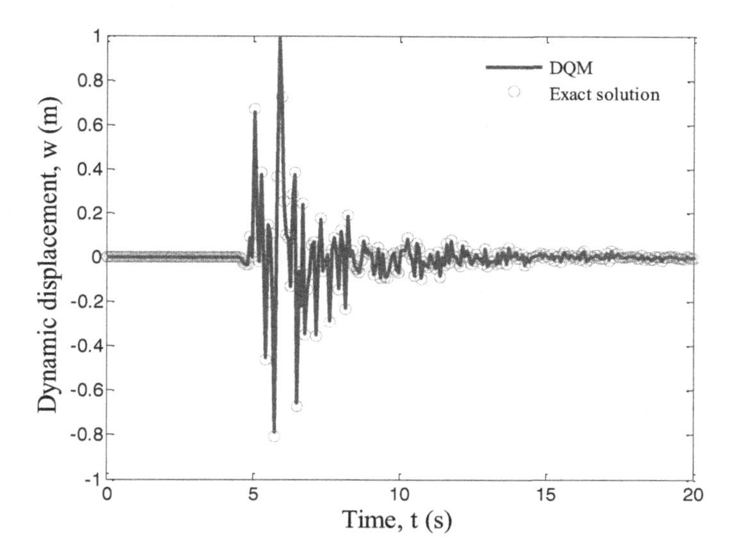


**FIGURE 12.3** Comparison of the present work with the exact solution.

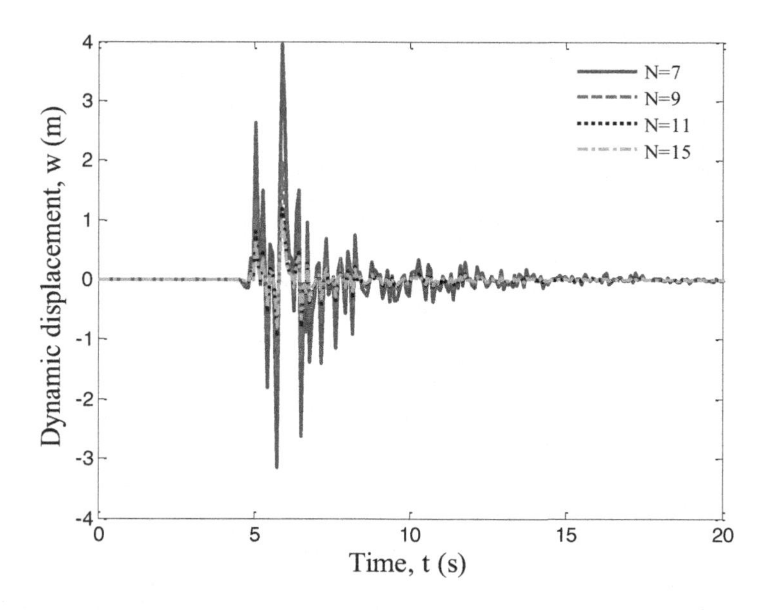


**FIGURE 12.4** Accuracy of the DQM for determining the dynamic displacement.

maximum displacement value for the submerged pipeline conveying fluid is almost five times more than the submerged pipeline without fluid.

The dynamic displacement of the CC pipeline conveying fluid under the Tabas earthquake conditions for two cases of the existence of external fluid and without external fluid is shown in Figure 12.6. It can be observed in the presence of external

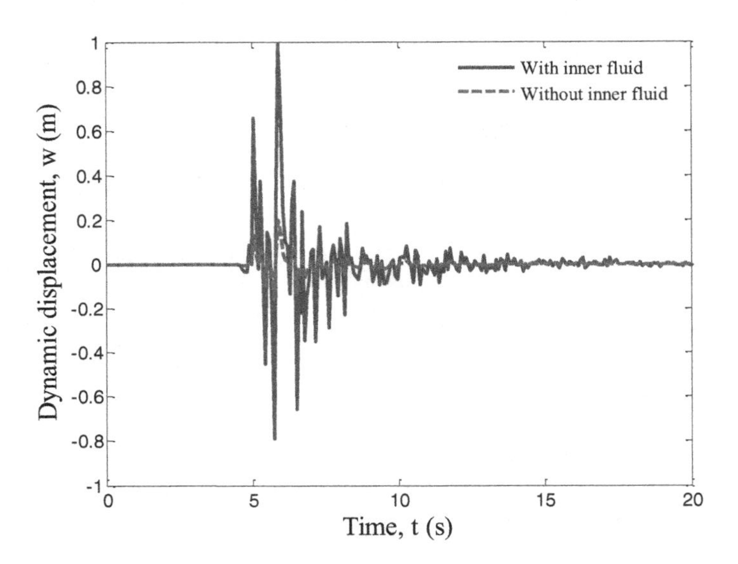


**FIGURE 12.5** Effect of internal fluid on the dynamic displacement of submerged pipe versus time.

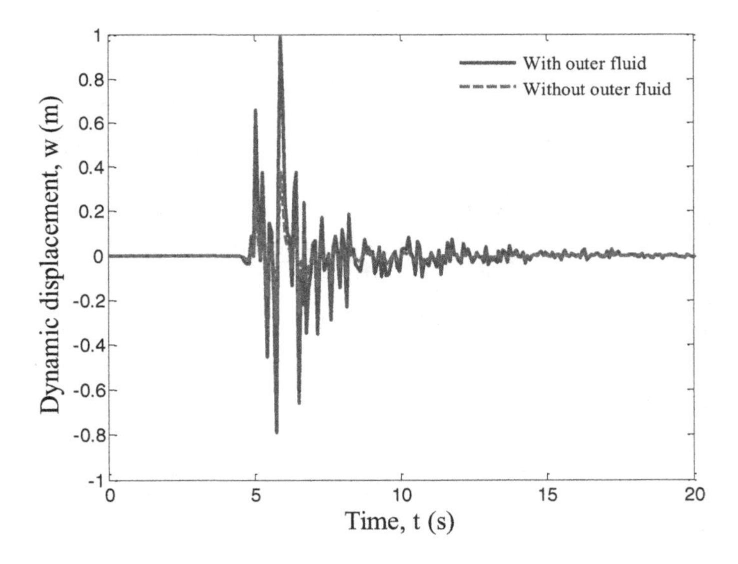


**FIGURE 12.6** Effect of external fluid on the dynamic displacement of fluid-conveying pipe versus time.

fluid, the dynamic deflection of the system increases because the stiffness of the structure decreases. Also, the maximum deflection value for the submerged pipeline conveying fluid is almost three times more than the pipeline conveying fluid without external fluid. By comparing Figures 12.5 and 12.6, it can be observed that the maximum deflection for the submerged pipeline without internal fluid is less than the pipeline conveying fluid without external fluid.

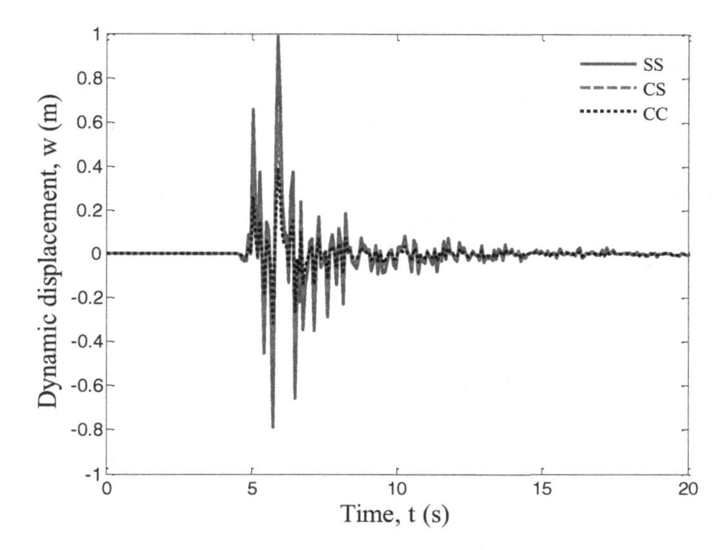


**FIGURE 12.7** Boundary condition effects on the dynamic displacement versus time.

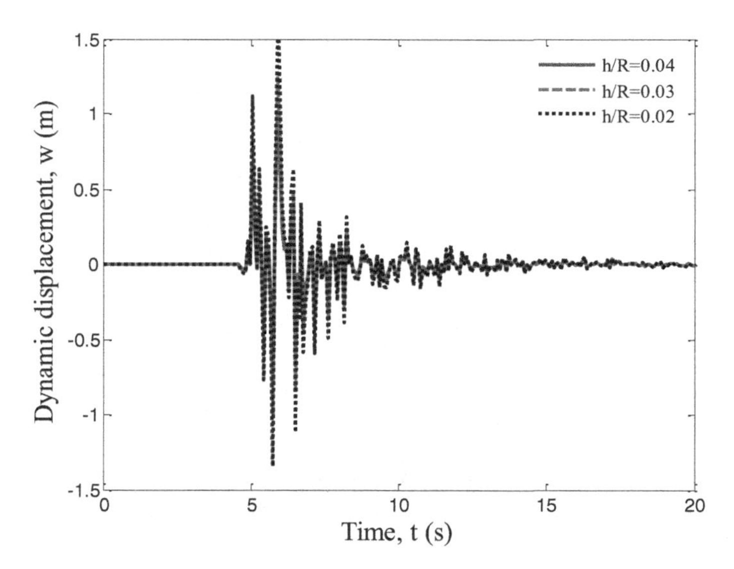


**FIGURE 12.8** Effect of thickness-to-radius ratio on the dynamic response of pipe versus time.

The changes in deflection versus time for various boundary conditions are shown in Figure 12.7. By investigating the boundary condition effects on the dynamic displacement of the structure, it is found that the pipe with simply–simply boundary condition has the most deflection in comparison to the other ones. Because the simple boundary condition has a lower constraint, consequently, the structure is softer.

Figure 12.8 indicates the effect of the thickness-to-radius ratio of the CC pipe on the dynamic deflection versus time. As can be seen, by increasing the thickness of the structure, dynamic deflection of system decreases. It is because the structure becomes stiffer.

The effect of aspect ratio (a/R) on the dynamic deflection of the CC pipe versus time is shown in Figure 12.9. It is obvious that by increasing the length-to-radius ratio, the dynamic deflection of the system increases.

The effect of $SiO_2$ nanoparticle volume percent on the dynamic displacement of the system versus time is shown in Figure 12.10. The results show that by increasing

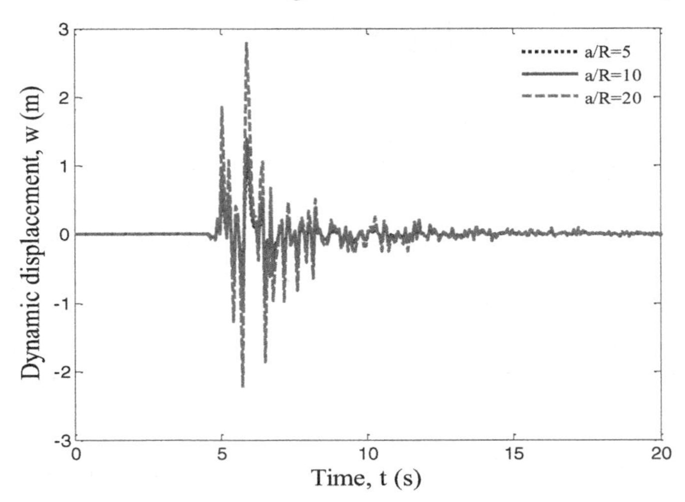


**FIGURE 12.9** Effect of length-to-radius ratio of the CC pipe on the dynamic response of pipe versus time.

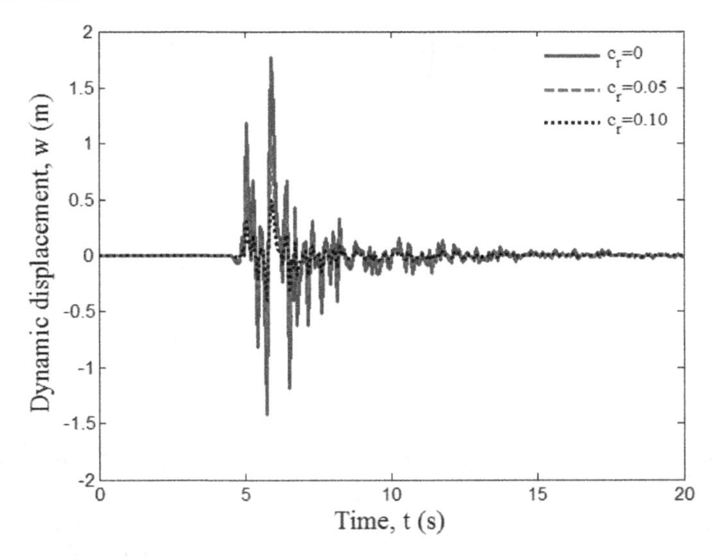


**FIGURE 12.10** The effect of $SiO_2$ nanoparticle volume percent on the dynamic response of pipe versus time.

of $SiO_2$ nanoparticle volume percent leads to a decrease in the dynamic deflection of the system for the reason that the stiffness of the structure increases.

**Acknowledgments:** This chapter is a slightly modified version of Ref. [51] and has been reproduced here with the permission of the copyright holder.

## REFERENCES

[1] Housner GW. Bending vibrations of a pipe line containing flowing fluid. *J Appl Mech* 1952;19:205–208.
[2] Benjamin TB. Dynamics of a system of articulated pipes conveying fluid. *Proc Royal Soc A* 1961;261:457–486.
[3] Paidoussis MP, Issid NT. Dynamic stability of pipes conveying fluid. *J Sound Vib* 1974;33:267–294.
[4] Amabili M, Pellicano F, Païdoussis MP. Non-linear dynamics and stability of circular cylindrical shells containing flowing fluid Part I: stability. *J Sound Vib* 1999a;225: 655–699.
[5] Amabili M, Pellicano F, Païdoussis MP. Non-linear dynamics and stability of circular cylindrical shells containing flowing fluid Part II: Large-amplitude vibrations without flow. *J Sound Vib* 1999b;228:1103–1124.
[6] Amabili M, Pellicano F, Païdoussis MP. Non-linear dynamics and stability of circular cylindrical shells containing flowing fluid. Part III: Truncation effect without flow and experiments. *J Sound Vib* 2000;237:617–640.
[7] Amabili M, Païdoussis MP. Review of studies on geometrically nonlinear vibrations and dynamics of circular cylindrical shells and panels, with and without fluid–structure interaction. *Appl Mech Rev* 2003;56:349–381.
[8] Amabili M. *Nonlinear vibrations and stability of shells and plates*. Cambridge University Press, Cambridge, 2008.
[9] Païdoussis MP. *Fluid–structure interactions, slender structures and axial flow*. Vol. 2. Elsevier Academic Press, London, 2004.
[10] Païdoussis MP. Some unresolved issues in fluid–structure interactions. *J Fluid Struct* 2005;20:871–890.
[11] Païdoussis MP, Grinevich E, Adamovic D, Semler C. Linear and nonlinear dynamics of cantilevered cylinders in axial flow part 1: Physical dynamics. *J Fluid Struct* 2007a;16:691–713.
[12] Païdoussis MP, Semler C, Wadham-Gagnon M, Saaid S. Dynamics of cantilevered pipes conveying fluid part 2: Dynamics of the system with intermediate spring support. *J Fluid Struct* 2007b;23:569–587.
[13] Lopes JL, Païdoussis MP, Semler C. Linear and nonlinear dynamics of cantilevered cylinders in axial flow part 2: The equations of motion. *J Fluid Struct* 2002;16:715–737.
[14] Semler C, Lopes JL, Augu N, Païdoussis MP. Linear and nonlinear dynamics of cantilevered cylinders in axial flow part 3: Nonlinear dynamics. *J Fluid Struct* 2002;16:739–759.
[15] Wadham-Gagnon M, Païdoussis MP, Semler C. Dynamics of cantilevered pipes conveying fluid part 1: Nonlinear equations of three-dimensional motion. *J Fluid Struct* 2007;23:545–567.
[16] Ibrahim RA. Overview of mechanics of pipes conveying fluids part I: Fundamental studies. *J Press Vessel Technol* 2010;132:034001.
[17] Ibrahim RA. Mechanics of pipes conveying fluids—Part II: Applications and fluidelastic problems. *J Press Vessel Technol* 2011;133:024001.

[18] Zemri A, Houari MSA, Bousahla AA, Tounsi A. A mechanical response of functionally graded nanoscale beam: An assessment of a refined nonlocal shear deformation theory beam theory. *Struct Eng Mech* 2015;54:693–710.
[19] Larbi Chaht F, Kaci A, Houari MSA, Hassan S. Bending and buckling analyses of functionally graded material (FGM) size-dependent nanoscale beams including the thickness stretching effect. *Steel Compos Struct* 2015;18:425–442.
[20] Belkorissat I, Houari MSA, Tounsi A, Hassan S. On vibration properties of functionally graded nano-plate using a new nonlocal refined four variable model. *Steel Compos Struct* 2015;18:1063–1081.
[21] Ahouel M, Houari MSA, Adda Bedia EA, Tounsi A. Size-dependent mechanical behavior of functionally graded trigonometric shear deformable nanobeams including neutral surface position concept. *Steel Compos Struct* 2016;20:963–981.
[22] Bounouara F, Benrahou KH, Belkorissat I, Tounsi A. A nonlocal zeroth-order shear deformation theory for free vibration of functionally graded nanoscale plates resting on elastic foundation. *Steel Compos Struct* 2016;20:227–249.
[23] Bouafia KH, Kaci A, Houari MSA, Tounsi A. A nonlocal quasi-3D theory for bending and free flexural vibration behaviors of functionally graded nanobeams. *Smart Struct Syst* 2017;19:115–126.
[24] Besseghier A, Houari MSA, Tounsi A, Hassan S. Free vibration analysis of embedded nanosize FG plates using a new nonlocal trigonometric shear deformation theory. *Smart Struct Syst* 2017;19:601–614.
[25] Bellifa H, Benrahou KH, Bousahla AA, Tounsi A, Mahmoud SR. A nonlocal zeroth-order shear deformation theory for nonlinear postbuckling of nanobeams. *Struct Eng Mech* 2017;62:695–702.
[26] Mouffoki A, Adda Bedia EA, Houari MSA, Hassan S. Vibration analysis of nonlocal advanced nanobeams in hygro-thermal environment using a new two-unknown trigonometric shear deformation beam theory. *Smart Struct Syst* 2017;20:369–383.
[27] Khetir H, Bouiadjra MB, Houari MSA, Tounsi A, Mahmoud SR. A new nonlocal trigonometric shear deformation theory for thermal buckling analysis of embedded nanosize FG plates. *Struct Eng Mech* 2017;64:391–402.
[28] Gong SW, Lam KY, Lu C. Structural analysis of a submarine pipeline subjected to underwater shock. *Int J Pres Ves Pip* 2000;77:417–423.
[29] Lee U, Oh H. The spectral element model for pipelines conveying internal steady flow. *Eng Struct* 2003;25:1045–1055.
[30] Lam KY, Zong Z, Wang QX. Dynamic response of a laminated pipeline on the seabed subjected to underwater shock. *Compos Part B-Eng* 2003;34:59–66.
[31] Yoon HI, Son I. Dynamic response of rotating flexible cantilever fluid with tip mass. *Int J Mech Sci* 2007;49:878–887.
[32] Lin W, Qiao N. Vibration and stability of an axially moving beam immersed in fluid. *Int J Solids Struct* 2008;45:1445–1457.
[33] Huang YM, Liu YS, Li BH, Li YJ, Yue ZF. Natural frequency analysis of fluid conveying pipeline with different boundary conditions. *Nucl Eng Des* 2010;240:461–467.
[34] Zhai H, Wu Z, Liu Y, Yue Z. Dynamic response of pipeline conveying fluid to random excitation. *Nucl Eng Des* 2011;241:2744–2749.
[35] Liu ZG, Liu Y, Lu J. Fluid–structure interaction of single flexible cylinder in axial flow. *Comput Fluids* 2012;56:143–151.
[36] Chen W, Shih BJ, Chen YC, Hung JH, Hwang HH. Seismic response of natural gas and water pipelines in the Ji-Ji earthquake. *Soil Dyn Earthq Eng* 2002;22:1209–1214.

[37] Abdoun TH, Ha D, O'Rourke M, Symans M, O'Rourke T, Palmer M, Harry E. Factors influencing the behavior of buried pipelines subjected to earthquake faulting. *Soil Dyn Earthq Eng* 2009;29:415–427.
[38] Shamsuddoha M, Islam MM, Aravinthan T, Manalo A, Lau KT. Effectiveness of using fibre-reinforced polymer composites for underwater steel pipeline repairs. *Compos Struct* 2013;100:40–54.
[39] Ray MC, Reddy JN. Active damping of laminated cylindrical shells conveying fluid using 1–3 piezoelectric composites. *Compos Struct* 2013;98:261–271.
[40] Alijani F, Amabili M. Nonlinear vibrations and multiple resonances of fluid filled arbitrary laminated circular cylindrical shells. *Compos Struct* 2014;108:951–962.
[41] Thinh TI, Nguyen MC. Dynamic Stiffness Method for free vibration of composite cylindrical shells containing fluid. *Appl Math Model* 2016;40:9286–9301.
[42] Zhou XQ, Yu DY, Shao XY, Zhang CY, Wang S. Dynamics characteristic of steady fluid conveying in the periodical partially viscoelastic composite pipeline. *Compos Part B-Eng* 2017;111:387–408.
[43] Dey T, Ramachandra LS. Non-linear vibration analysis of laminated composite circular cylindrical shells. *Compos Struct* 2017;163:89–100.
[44] Su Y, Li J, Wu C, Li ZX. Influences of nano-particles on dynamic strength of ultra-high performance. *Compos Part B-Eng* 2016;91:595–609.
[45] Jafarian Arani A, Kolahchi R. Buckling analysis of embedded beams armed with carbon nanotubes. *Comput Concr* 2016;17:567–578.
[46] Safari Bilouei B, Kolahchi R, Rabanibidgoli M. Buckling of beams retrofitted with Nano-Fiber Reinforced Polymer (NFRP). *Comput Concr* 2016;18:1053–1063.
[47] Inozemtcev AS, Korolev EV, Smirnov VA. Nanoscale modifier as an adhesive for hollow microspheres to increase the strength of high-strength lightweight. *Struct* 2017;18:67–74.
[48] Zamani Nouri A. Mathematical modeling of pipes reinforced with CNTs conveying fluid for vibration and stability analyses. *Comput Concr* 2017;19:325–331.
[49] Shokravi M. Vibration analysis of silica nanoparticles-reinforced beams considering agglomeration effects. *Comput Concr* 2017;19:333–338.
[50] Rabani Bidgoli M, Saeidifar M. Time-dependent buckling analysis of SiO2 nanoparticles reinforced beams exposed to fire. *Comput Concr* 2017;20:119–127.
[51] Maleki M, Bidgoli MR, Kolahchi R. Earthquake response of nanocomposite concrete pipes conveying and immersing in fluid using numerical methods. *Comput Concrete* 2019;(24)2:125–135.

# 13 Vibration and Instability Analysis of Shells Reinforced by Nanoparticles

## 13.1 INTRODUCTION

Pipes conveying fluid can be used in city wastewater management. The fluid velocity in the pipe can induce a dynamic load, and it is very important for the instability of the structure. However, improving the stiffness of the structure is important for the city wastewater pipes. In this chapter, $SiO_2$ nanoparticles are used to reinforce the pipes.

With respect to the developed works on pipes conveying fluid, Paidoussis and Li [1] studied the dynamic analysis of pipes conveying fluid. A dynamic analysis of anisotropic cylindrical shells containing flowing fluid was presented by Toorani and Lakis [2]. Based on Sanders's nonlinear theory, Zhang et al. [3] investigated the dynamics of initially tensioned orthotropic thin-walled cylindrical tubes conveying fluid. Kadoli and Ganesan [4] studied the vibration and buckling behavior of composite cylindrical shells conveying fluid. The stability of pipes conveying fluid was studied by Zhang et al. [5]. Meng et al. [6]. studied the three-dimensional nonlinear dynamics of a fluid-conveying pipe. Dai et al. [7] presented the vibration analysis of three-dimensional (3D) pipelines conveying fluid using the transfer matrix method. Based on the Timoshenko beam model, Gu et al. [8] studied the dynamic response of a fluid-conveying pipe.

Mechanical analyses of nanostructures have been reported by many researchers [9–18]. In the field of nanocomposite structures, Vodenitcharova and Zhang [19] studied the bending of a nanocomposite beam using the Airy stress function method. The large amplitude vibration behavior of nanocomposite cylindrical shells was studied by Shen and Xiang [20]. Rafiee and Moghadam [21] studied the impact analysis of nanocomposite plates using a 3D finite element model. A nonlinear buckling analysis of embedded polymeric temperature-dependent functionally graded (FG) carbon nanotube reinforced composite (CNTRC) microplates resting on an elastic matrix on orthotropic temperature-dependent elastomeric medium was investigated by Kolahchi et al. [22]. Thomas and Roy [23] studied vibration analysis of FG carbon nanotube (CNT)–reinforced composite shell structures. Mehri et al. [24] studied the bifurcation and vibration responses of a composite truncated conical shell with embedded single-walled CNTs (SWCNTs) subjected to external pressure and axial compression simultaneously. Bayat et al. [25] presented the nonlinear analysis of

DOI: 10.1201/9781003349525-13

the impact response of nanocomposites cylindrical shells reinforced by SWCNTs as FG in thermal environments. Safari Bilouei et al. [26] investigated the nonlinear buckling of straight beams armed with SWCNTs resting on a foundation. Kolahchi et al. [27] investigated the nonlinear dynamic stability of embedded temperature-dependent viscoelastic plates reinforced with CNTs. Golabchi et al. [28] presented an analysis of fluid velocity effects on the instability of pipes reinforced by silica nanoparticles ($SiO_2SiO_2$). The effective material properties of the nanocomposite structure are determined using the Mori-Tanaka model, taking into account agglomeration effects.

None of the mentioned works studied the instability of pipes. In the present chapter, fluid velocity analysis on the nonlinear instability of pipes reinforced by $SiO_2$ nanoparticles is presented. The Mori–Tanaka model is used in order to obtain the equivalent material properties of the pipe to consider the agglomeration effects. Based on the Reddy shell theory, the nonlinear motion equations are obtained based on Hamilton's principle. The differential quadrature method (DQM) is applied for obtaining the frequency and critical fluid velocity of the structure. The effects of the volume percent and agglomeration of $SiO_2$ nanoparticles, boundary conditions, and geometrical parameters of the pipes on the frequency and critical fluid velocity of the pipe are shown.

## 13.2 FORMULATION

In Figure 13.1, a pipe reinforced by $SiO_2$ nanoparticles conveying fluid is shown. The agglomeration effects of $SiO_2$ nanoparticles are considered.

There are many new theories for modeling different structures. Some of the new theories have been used by Tounsi and his coauthors [29–48]. Based on the Reddy shell theory, the displacement field can be expressed as presented in Section 1.5.3.

The stress–strain relations based on the Mori–Tanaka model can be written as follows:

$$\begin{Bmatrix} \sigma_{11} \\ \sigma_{22} \\ \sigma_{33} \\ \sigma_{23} \\ \sigma_{13} \\ \sigma_{12} \end{Bmatrix} = \begin{bmatrix} k+m & l & k-m & 0 & 0 & 0 \\ l & n & l & 0 & 0 & 0 \\ k-m & l & k+m & 0 & 0 & 0 \\ 0 & 0 & 0 & p & 0 & 0 \\ 0 & 0 & 0 & 0 & m & 0 \\ 0 & 0 & 0 & 0 & 0 & p \end{bmatrix} \begin{Bmatrix} \varepsilon_{11} \\ \varepsilon_{22} \\ \varepsilon_{33} \\ \gamma_{23} \\ \gamma_{13} \\ \gamma_{12} \end{Bmatrix}, \tag{13–1}$$

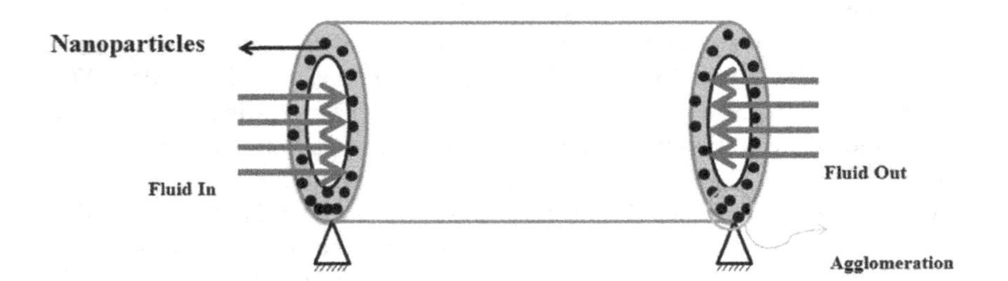


**FIGURE 13.1** Schematic of the pipe reinforced by $SiO_2$ nanoparticles conveying fluid.

where $k, m, n, l, p$ are defined in Section 4.3. The strain energy can be written as

$$U=\frac{1}{2}\int_{\Omega_0}\int_{-h/2}^{h/2}\left(\sigma_{xx}\varepsilon_{xx}+\sigma_{\theta\theta}\varepsilon_{\theta\theta}+\sigma_{x\theta}\gamma_{x\theta}+\sigma_{xz}\gamma_{xz}+\sigma_{\theta z}\gamma_{\theta z}\right)dV. \tag{13–2}$$

Combining of Equations 1–43 and 1–44 into Equation 13–2 yields

$$\begin{aligned}U=\frac{1}{2}\int_{\Omega_0}\Bigg(&N_{xx}\left(\frac{\partial u}{\partial x}+\frac{1}{2}\left(\frac{\partial w}{\partial x}\right)^2\right)+N_{\theta\theta}\left(\frac{\partial v}{\partial\theta}+\frac{w}{R}+\frac{1}{2}\left(\frac{\partial w}{R\partial\theta}\right)^2\right)\\
&+Q_\theta\left(\frac{\partial w}{R\partial\theta}+\psi_\theta\right)+Q_x\left(\frac{\partial w}{\partial x}+\psi_x\right)+N_{x\theta}\left(\frac{\partial v}{\partial x}+\frac{\partial u}{R\partial\theta}+\frac{\partial w}{\partial x}\frac{\partial w}{R\partial\theta}\right)\\
&+M_{xx}\frac{\partial\psi_x}{\partial x}+M_{\theta\theta}\frac{\partial\psi_\theta}{R\partial\theta}+M_{x\theta}\left(\frac{\partial\psi_x}{R\partial\theta}+\frac{\partial\psi_\theta}{\partial x}\right)\\
&+K_\theta\left(\frac{-4}{h^2}\left(\psi_\theta+\frac{\partial w}{R\partial\theta}\right)\right)+K_x\left(\frac{-4}{h^2}\left(\psi_x+\frac{\partial w}{\partial x}\right)\right)\\
&+P_{xx}\left(\frac{-4}{3h^2}\left(\frac{\partial\psi_x}{\partial x}+\frac{\partial^2 w}{\partial x^2}\right)\right)+P_{\theta\theta}\left(\frac{-4}{3h^2}\left(\frac{\partial\psi_\theta}{R\partial\theta}+\frac{\partial^2 w}{R^2\partial\theta^2}\right)\right)\\
&+P_{x\theta}\left(\frac{\partial\psi_\theta}{\partial x}+\frac{\partial\psi_x}{R\partial\theta}+2\frac{\partial^2 w}{R\partial x\partial\theta}\right)\Bigg)dxd\theta,\end{aligned} \tag{13–3}$$

where the stress resultant–displacement relations can be written as

$$\begin{Bmatrix}N_{xx}\\N_{\theta\theta}\\N_{x\theta}\end{Bmatrix}=\int_{-h/2}^{h/2}\begin{bmatrix}\sigma_{xx}\\\sigma_{\theta\theta}\\\sigma_{x\theta}\end{bmatrix}dz, \tag{13–4}$$

$$\begin{Bmatrix}M_{xx}\\M_{\theta\theta}\\M_{x\theta}\end{Bmatrix}=\int_{-h/2}^{h/2}\begin{bmatrix}\sigma_{xx}\\\sigma_{\theta\theta}\\\sigma_{x\theta}\end{bmatrix}zdz, \tag{13–5}$$

$$\begin{Bmatrix}P_{xx}\\P_{\theta\theta}\\P_{x\theta}\end{Bmatrix}=\int_{-h/2}^{h/2}\begin{bmatrix}\sigma_{xx}\\\sigma_{\theta\theta}\\\sigma_{x\theta}\end{bmatrix}z^3dz, \tag{13–6}$$

$$\begin{bmatrix}Q_x\\Q_\theta\end{bmatrix}=\int_{-h/2}^{h/2}\begin{bmatrix}\sigma_{xz}\\\sigma_{\theta z}\end{bmatrix}dz, \tag{13–7}$$

$$\begin{bmatrix} K_x \\ K_\theta \end{bmatrix} = \int_{-h/2}^{h/2} \begin{bmatrix} \sigma_{xz} \\ \sigma_{\theta z} \end{bmatrix} z^2 dz. \tag{13–8}$$

The kinetic energy of the system may be written as

$$K = \frac{\rho}{2} \int_{\Omega_0} \int_{-h/2}^{h/2} \left( (\dot{u}_x)^2 + (\dot{u}_\theta)^2 + (\dot{u}_z)^2 \right) dV. \tag{13–9}$$

The external work due to fluid can be written as

$$W = \int_0^L (P_{Fluid}) w dx, \tag{13–10}$$

where

$$P_{fluid} = h_f \frac{\partial p_z}{\partial z} = -\rho_f h_f \left( \frac{\partial^2 w}{\partial t^2} + 2 v_x \frac{\partial^2 w}{\partial x \partial t} + v_x{}^2 \frac{\partial^2 w}{\partial x^2} \right)$$
$$+ \mu h_f \left( \frac{\partial^3 w}{\partial x^2 \partial t} + \frac{\partial^3 w}{R^2 \partial \theta^2 \partial t} + v_x \left( \frac{\partial^3 w}{\partial x^3} + \frac{\partial^3 w}{R^2 \partial \theta^2 \partial x} \right) \right), \tag{13–11}$$

where $\rho_b$ and $P$ are fluid mass and flow fluid pressure, respectively; $v_x$ is the mean flow velocity. The governing equations can be derived using Hamilton's principle as follows:

$$\int_0^t (\delta U - \delta W - \delta K) dt = 0. \tag{13–12}$$

Substituting Equations 13–3, 13–9, and 13–10 into Equation 13–12 yields the following governing equations:

$$\delta u : \frac{\partial N_{xx}}{\partial x} + \frac{\partial N_{x\theta}}{R \partial \theta} = I_0 \frac{\partial^2 u}{\partial t^2} + J_1 \frac{\partial^2 \psi_x}{\partial t^2} - \frac{4 I_3}{h^2} \frac{\partial^3 w}{\partial t^2 \partial x}, \tag{13–13}$$

$$\delta v : \frac{\partial N_{x\theta}}{\partial x} + \frac{\partial N_{\theta\theta}}{R \partial \theta} = I_0 \frac{\partial^2 v}{\partial t^2} + J_1 \frac{\partial^2 \psi_\theta}{\partial t^2} - \frac{4 I_3}{h^2} \frac{\partial^3 w}{R \partial t^2 \partial \theta}, \tag{13–14}$$

$$\delta w : \frac{\partial Q_x}{\partial x} + \frac{\partial Q_\theta}{R \partial \theta} - \frac{4}{h^2} \left( \frac{\partial K_x}{\partial x} + \frac{\partial K_\theta}{R \partial \theta} \right) + \frac{\partial}{\partial x} \left( N_{xx} \frac{\partial w}{\partial x} + N_{x\theta} \frac{\partial w}{R \partial \theta} \right)$$
$$+ \frac{\partial}{R \partial \theta} \left( N_{x\theta} \frac{\partial w}{\partial x} + N_{\theta\theta} \frac{\partial w}{R \partial \theta} \right) + \frac{4}{3h^2} \left( \frac{\partial^2 P_{xx}}{\partial x^2} + 2 \frac{\partial^2 P_{x\theta}}{R \partial x \partial \theta} + \frac{\partial^2 P_{\theta\theta}}{R^2 \partial \theta^2} \right)$$
$$- \frac{N_{\theta\theta}}{R} + q = I_0 \frac{\partial^2 w}{\partial t^2} - \left( \frac{4}{3h^2} \right)^2 I_6 \left( \frac{\partial^4 w}{\partial x^2 \partial t^2} + \frac{\partial^4 w}{R^2 \partial \theta^2 \partial t^2} \right)$$
$$+ \frac{4}{3h^2} \left( I_3 \frac{\partial^3 u}{\partial t^2 \partial x} + I_3 \frac{\partial^3 v}{R \partial t^2 \partial \theta} + J_4 \left( \frac{\partial^3 \psi_x}{\partial t^2 \partial x} + \frac{\partial^3 \psi_\theta}{R \partial t^2 \partial \theta} \right) \right), \tag{13–15}$$

$$\delta\psi_x : \frac{\partial M_{xx}}{\partial x} + \frac{\partial M_{x\theta}}{R\partial\theta} - \frac{4}{3h^2}\left(\frac{\partial P_{xx}}{\partial x} + \frac{\partial P_{x\theta}}{R\partial\theta}\right)$$

$$-Q_x + \frac{4}{h^2}K_x = J_1\frac{\partial^2 u}{\partial t^2} + K_2\frac{\partial^2 \psi_x}{\partial t^2} - \frac{4}{3h^2}J_4\frac{\partial^3 w}{\partial t^2 \partial x}, \tag{13–16}$$

$$\delta\psi_\theta : \frac{\partial M_{x\theta}}{\partial x} + \frac{\partial M_{\theta\theta}}{R\partial\theta} - \frac{4}{3h^2}\left(\frac{\partial P_{x\theta}}{\partial x} + \frac{\partial P_{\theta\theta}}{R\partial\theta}\right)$$

$$-Q_\theta + \frac{4}{h^2}K_\theta = J_1\frac{\partial^2 v}{\partial t^2} + K_2\frac{\partial^2 \psi_\theta}{\partial t^2} - \frac{4}{3h^2}J_4\frac{\partial^3 w}{R\partial t^2 \partial\theta}, \tag{13–17}$$

where

$$I_i = \int_{-h/2}^{h/2} \rho z^i dz \qquad (i = 0,1,\dots,6), \tag{13–18}$$

$$J_i = I_i - \frac{4}{3h^2}I_{i+2} \qquad (i = 1,4), \tag{13–19}$$

$$K_2 = I_2 - \frac{8}{3h^2}I_4 + \left(\frac{4}{3h^2}\right)^2 I_6. \tag{13–20}$$

Substituting stress relations into Equations 13–3 through 13–8, the stress resultant–displacement relations can be obtained as follow:

$$N_{xx} = A_{11}\left(\frac{\partial u}{\partial x} + \frac{1}{2}\left(\frac{\partial w}{\partial x}\right)^2\right) + A_{12}\left(\frac{\partial v}{R\partial\theta} + \frac{w}{R} + \frac{1}{2}\left(\frac{\partial w}{R\partial\theta}\right)^2\right)$$
$$+B_{11}\left(\frac{\partial\psi_x}{\partial x}\right) + B_{12}\left(\frac{\partial\psi_\theta}{R\partial\theta}\right) + E_{11}\left(\frac{-4}{3h^2}\left(\frac{\partial\psi_x}{\partial x} + \frac{\partial^2 w}{\partial x^2}\right)\right)$$
$$+E_{12}\left(\frac{-4}{3h^2}\left(\frac{\partial\psi_\theta}{R\partial\theta} + \frac{\partial^2 w}{R^2\partial\theta^2}\right)\right) - N_{xx}^T,$$

$$N_{\theta\theta} = A_{12}\left(\frac{\partial u}{\partial x} + \frac{1}{2}\left(\frac{\partial w}{\partial x}\right)^2\right) + A_{22}\left(\frac{\partial v}{R\partial\theta} + \frac{w}{R} + \frac{1}{2}\left(\frac{\partial w}{R\partial\theta}\right)^2\right)$$
$$+B_{12}\left(\frac{\partial\psi_x}{\partial x}\right) + B_{22}\left(\frac{\partial\psi_\theta}{R\partial\theta}\right) + E_{12}\left(\frac{-4}{3h^2}\left(\frac{\partial\psi_x}{\partial x} + \frac{\partial^2 w}{\partial x^2}\right)\right)$$
$$+E_{22}\left(\frac{-4}{3h^2}\left(\frac{\partial\psi_\theta}{R\partial\theta} + \frac{\partial^2 w}{R^2\partial\theta^2}\right)\right) - N_{\theta\theta}^T,$$

$$N_{x\theta} = A_{66}\left(\frac{\partial u}{R\partial\theta} + \frac{\partial v}{\partial x} + \frac{\partial w}{\partial x}\frac{\partial w}{R\partial\theta}\right) + B_{66}\left(\frac{\partial\psi_x}{R\partial\theta} + \frac{\partial\psi_\theta}{\partial x}\right)$$
$$+E_{66}\left(\frac{-4}{3h^2}\left(\frac{\partial\psi_\theta}{\partial x} + \frac{\partial\psi_x}{R\partial\theta} + 2\frac{\partial^2 w}{R\partial x\partial\theta}\right)\right), \tag{13–21}$$

$$
\begin{aligned}
M_{xx} &= B_{11}\left(\frac{\partial u}{\partial x}+\frac{1}{2}\left(\frac{\partial w}{\partial x}\right)^2\right)+B_{12}\left(\frac{\partial v}{R\partial\theta}+\frac{w}{R}+\frac{1}{2}\left(\frac{\partial w}{R\partial\theta}\right)^2\right) \\
&+D_{11}\left(\frac{\partial\psi_x}{\partial x}\right)+D_{12}\left(\frac{\partial\psi_\theta}{R\partial\theta}\right)+F_{11}\left(\frac{-4}{3h^2}\left(\frac{\partial\psi_x}{\partial x}+\frac{\partial^2 w}{\partial x^2}\right)\right) \\
&+F_{12}\left(\frac{-4}{3h^2}\left(\frac{\partial\psi_\theta}{R\partial\theta}+\frac{\partial^2 w}{R^2\partial\theta^2}\right)\right)-M_{xx}^T, \\
M_{\theta\theta} &= B_{12}\left(\frac{\partial u}{\partial x}+\frac{1}{2}\left(\frac{\partial w}{\partial x}\right)^2\right)+B_{22}\left(\frac{\partial v}{R\partial\theta}+\frac{w}{R}+\frac{1}{2}\left(\frac{\partial w}{R\partial\theta}\right)^2\right) \\
&+D_{12}\left(\frac{\partial\psi_x}{\partial x}\right)+D_{22}\left(\frac{\partial\psi_\theta}{R\partial\theta}\right)+F_{12}\left(\frac{-4}{3h^2}\left(\frac{\partial\psi_x}{\partial x}+\frac{\partial^2 w}{\partial x^2}\right)\right) \\
&+F_{22}\left(\frac{-4}{3h^2}\left(\frac{\partial\psi_\theta}{R\partial\theta}+\frac{\partial^2 w}{R^2\partial\theta^2}\right)\right)-M_{\theta\theta}^T, \\
M_{x\theta} &= B_{66}\left(\frac{\partial u}{R\partial\theta}+\frac{\partial v}{\partial x}+\frac{\partial w}{\partial x}\frac{\partial w}{R\partial\theta}\right)+D_{66}\left(\frac{\partial\psi_x}{R\partial\theta}+\frac{\partial\psi_\theta}{\partial x}\right) \\
&+F_{66}\left(\frac{-4}{3h^2}\left(\frac{\partial\psi_\theta}{\partial x}+\frac{\partial\psi_x}{R\partial\theta}+2\frac{\partial^2 w}{R\partial x\partial\theta}\right)\right),
\end{aligned}
\tag{13–22}
$$

$$
\begin{aligned}
P_{xx} &= E_{11}\left(\frac{\partial u}{\partial x}+\frac{1}{2}\left(\frac{\partial w}{\partial x}\right)^2\right)+E_{12}\left(\frac{\partial v}{R\partial\theta}+\frac{w}{R}+\frac{1}{2}\left(\frac{\partial w}{R\partial\theta}\right)^2\right) \\
&+F_{11}\left(\frac{\partial\psi_x}{\partial x}\right)+F_{12}\left(\frac{\partial\psi_\theta}{R\partial\theta}\right)+H_{11}\left(\frac{-4}{3h^2}\left(\frac{\partial\psi_x}{\partial x}+\frac{\partial^2 w}{\partial x^2}\right)\right) \\
&+H_{12}\left(\frac{-4}{3h^2}\left(\frac{\partial\psi_\theta}{R\partial\theta}+\frac{\partial^2 w}{R^2\partial\theta^2}\right)\right)-P_{xx}^T, \\
P_{\theta\theta} &= E_{12}\left(\frac{\partial u}{\partial x}+\frac{1}{2}\left(\frac{\partial w}{\partial x}\right)^2\right)+E_{22}\left(\frac{\partial v}{R\partial\theta}+\frac{w}{R}+\frac{1}{2}\left(\frac{\partial w}{R\partial\theta}\right)^2\right) \\
&+F_{12}\left(\frac{\partial\psi_x}{\partial x}\right)+F_{22}\left(\frac{\partial\psi_\theta}{R\partial\theta}\right)+H_{12}\left(\frac{-4}{3h^2}\left(\frac{\partial\psi_x}{\partial x}+\frac{\partial^2 w}{\partial x^2}\right)\right) \\
&+H_{22}\left(\frac{-4}{3h^2}\left(\frac{\partial\psi_\theta}{R\partial\theta}+\frac{\partial^2 w}{R^2\partial\theta^2}\right)\right)-P_{\theta\theta}^T, \\
P_{x\theta} &= E_{66}\left(\frac{\partial u}{R\partial\theta}+\frac{\partial v}{\partial x}+\frac{\partial w}{\partial x}\frac{\partial w}{R\partial\theta}\right)+F_{66}\left(\frac{\partial\psi_x}{R\partial\theta}+\frac{\partial\psi_\theta}{\partial x}\right) \\
&+H_{66}\left(\frac{-4}{3h^2}\left(\frac{\partial\psi_\theta}{\partial x}+\frac{\partial\psi_x}{R\partial\theta}+2\frac{\partial^2 w}{R\partial x\partial\theta}\right)\right),
\end{aligned}
\tag{13–23}
$$

$$Q_x = A_{44}\left(\frac{\partial w}{\partial x}+\psi_x\right)+D_{44}\left(\frac{-4}{h^2}\left(\psi_x+\frac{\partial w}{\partial x}\right)\right),$$
$$Q_\theta = A_{55}\left(\frac{\partial w}{R\partial\theta}+\psi_\theta\right)+D_{55}\left(\frac{-4}{h^2}\left(\psi_\theta+\frac{\partial w}{R\partial\theta}\right)\right), \tag{13–24}$$

$$K_x = D_{44}\left(\frac{\partial w}{\partial x}+\psi_x\right)+F_{44}\left(\frac{-4}{h^2}\left(\psi_x+\frac{\partial w}{\partial x}\right)\right),$$
$$K_\theta = D_{55}\left(\frac{\partial w}{R\partial\theta}+\psi_\theta\right)+F_{55}\left(\frac{-4}{h^2}\left(\psi_\theta+\frac{\partial w}{R\partial\theta}\right)\right), \tag{13–25}$$

where

$$A_{ij} = \int_{-h/2}^{h/2} C_{ij}dz, \qquad (i,j=1,2,6) \tag{13–26}$$

$$B_{ij} = \int_{-h/2}^{h/2} C_{ij}zdz, \tag{13–27}$$

$$D_{ij} = \int_{-h/2}^{h/2} C_{ij}z^2dz, \tag{13–28}$$

$$E_{ij} = \int_{-h/2}^{h/2} C_{ij}z^3dz, \tag{13–29}$$

$$F_{ij} = \int_{-h/2}^{h/2} C_{ij}z^4dz, \tag{13–30}$$

$$H_{ij} = \int_{-h/2}^{h/2} C_{ij}z^6dz. \tag{13–31}$$

## 13.3 NUMERICAL RESULTS AND DISCUSSION

In this section, the numerical results of nonlinear vibration and instability of nanocomposite pipe conveying fluid are presented. Here, polyethylene (PE) is selected for the matrix, which had a constant Poisson's ratio of $v_m = 0.3$ and a Young's modulus of $E_m = 2\ GPa$. In addition, $SiO_2$ nanoparticles are selected as reinforcements with Poisson's ratio of $v_r = 0.2$ and Young's modulus of $E_r = 66\,GPa$. The results of this chapter are based on DQM, which is presented in Section 2.2.1.

### 13.3.1 DQM Convergence

The effect of the grid point number in DQM on the imaginary and real parts of dimensionless frequency ($\Omega = \omega L^2\sqrt{\rho / C_{11}}$) is demonstrated in Figures 13.2 and 13.3 versus the dimensionless fluid velocity ($V = \sqrt{\rho_f / C_{11}}\, v_x$). As can be seen, the fast rate of convergence of the method is quite evident, and it is found that 15 DQM grid points can yield accurate results.

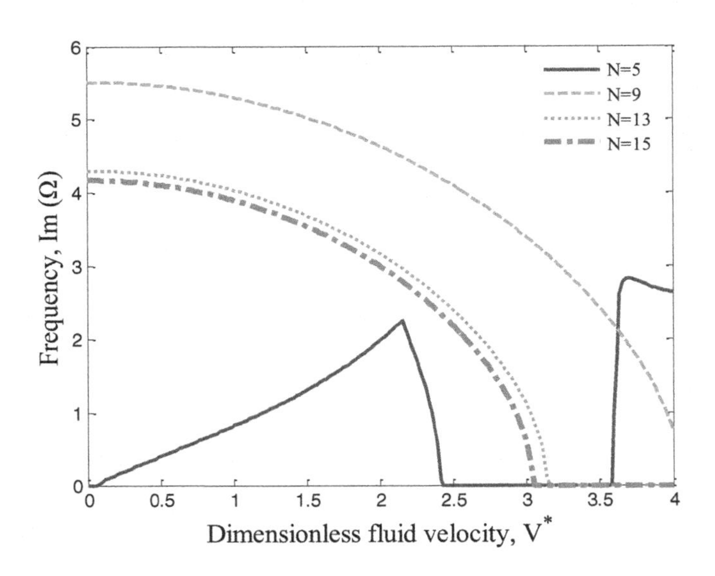


**FIGURE 13.2** Convergence of DQM for the imaginary part of the frequency.

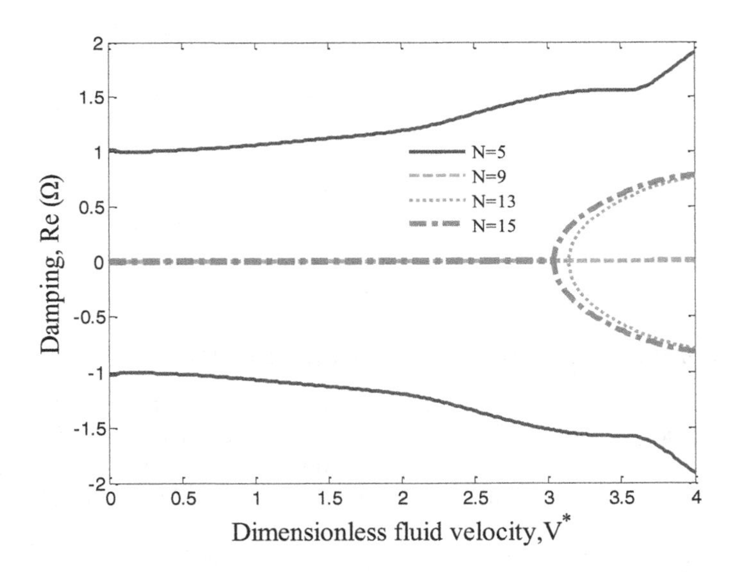


**FIGURE 13.3** Convergence of DQM for the real part of the frequency.

### 13.3.2 Effects of Different Parameters

Figures 13.4 and 13.5 show the effect of $SiO_2$ nanoparticle volume percent on the imaginary and real parts of a dimensionless eigenvalue. Note that the imaginary and real parts of the dimensionless eigenvalue are dimensionless natural frequency ($Im(\Omega)$) and damping ($Re(\Omega)$), respectively. As can be seen, with an increase in

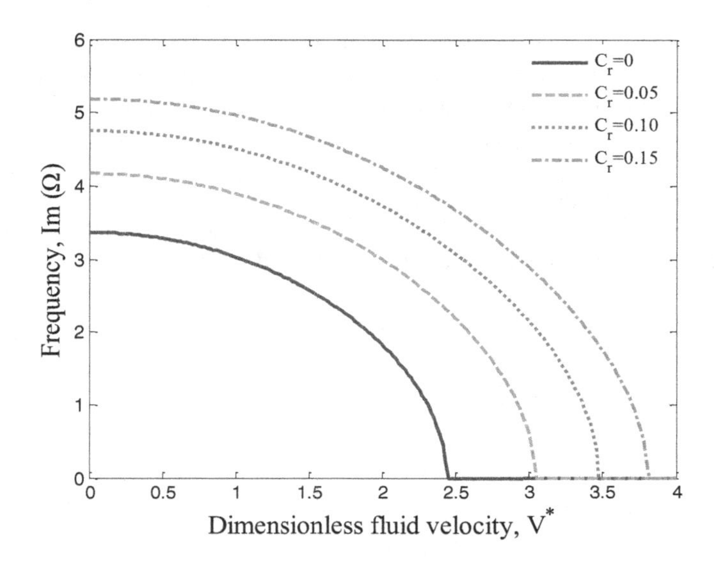


**FIGURE 13.4** Effect of $SiO_2$ nanoparticle volume percent on the imaginary part of the frequency.

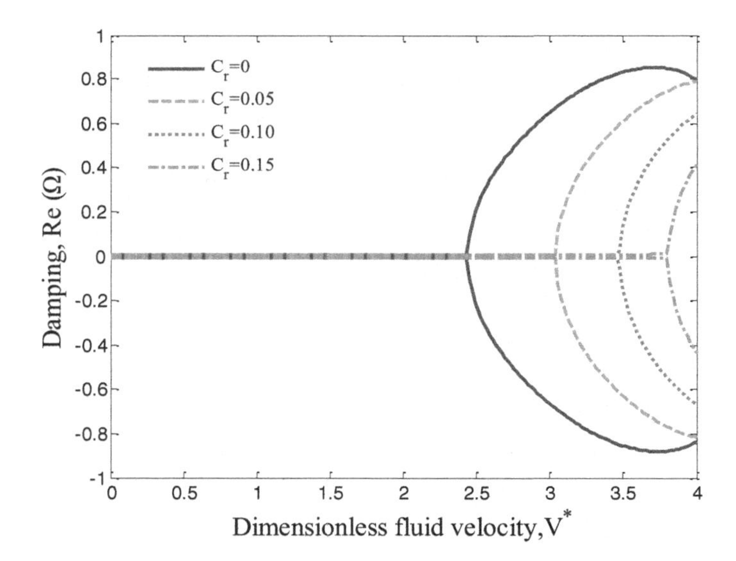


**FIGURE 13.5** Effect of $SiO_2$ nanoparticle volume percent on the real part of the frequency.

the $SiO_2$ nanoparticle volume percent, the frequency and critical fluid velocity will increase. It is due to the fact that the stiffness of pipes increases with an increase in the $SiO_2$ nanoparticle volume percent. In addition, $\mathrm{Im}(\Omega)$ decreases with an increase in $V$, while the $\mathrm{Re}(\Omega)$ remains zero. These imply that the system is stable. When the natural frequency becomes zero, critical velocity is reached, the system loses its stability due to the divergence via a pitchfork bifurcation. Hence, the Re(Ω) has the positive real parts, which the system becomes unstable. In this state, both real and imaginary parts of frequency become zero at the same point. Therefore, with an increased flow velocity, the system's stability decreases and becomes susceptible to buckling.

The effect of the $SiO_2$ nanoparticle agglomeration on the dimensionless frequency and damping of the pipe with respect to dimensionless flow velocity is shown in Figures 13.6 and 13.7. It is observed that considering the $SiO_2$ nanoparticle agglomeration decreases the dimensionless frequency and the critical fluid velocity of the

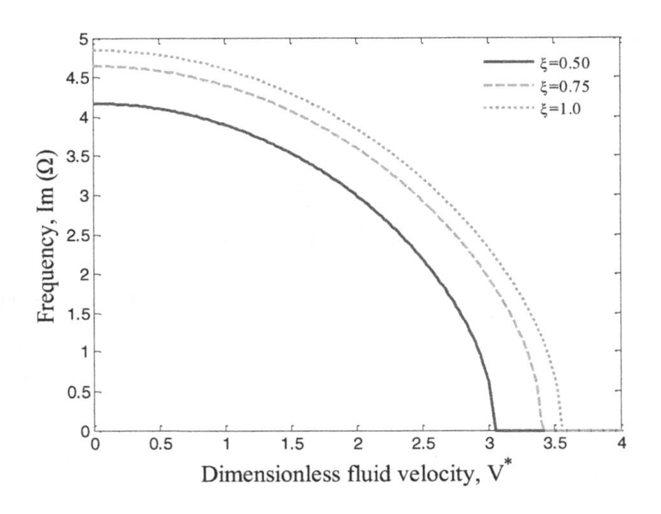


**FIGURE 13.6** Effect of $SiO_2$ nanoparticle agglomeration on the imaginary part of the frequency.

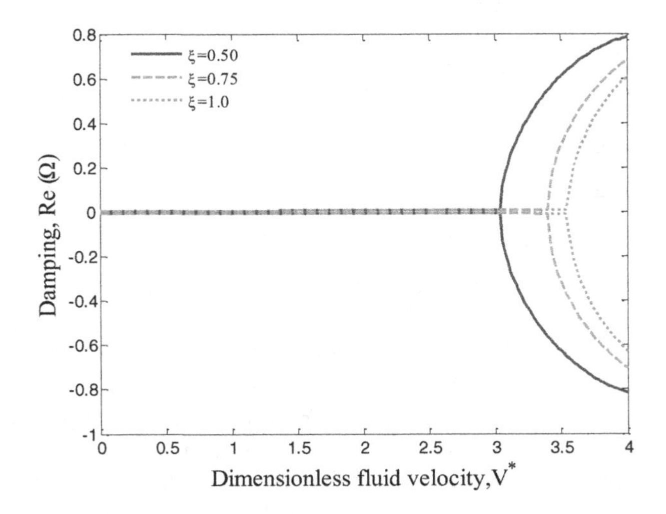


**FIGURE 13.7** Effect of $SiO_2$ nanoparticle agglomeration on the real part of the frequency.

pipe. This is due to the fact that considering $SiO_2$ nanoparticles agglomeration leads to a nonhomogeneous structure.

The dimensionless frequency and the damping of the structure versus the dimensionless flow velocity are shown in Figures 13.8 and 13.9 for the different boundary conditions of clamped (CC), simply supported and clamped (SC), and two edges simply supported (SS). The dimensionless frequency and critical fluid velocity of the pipe

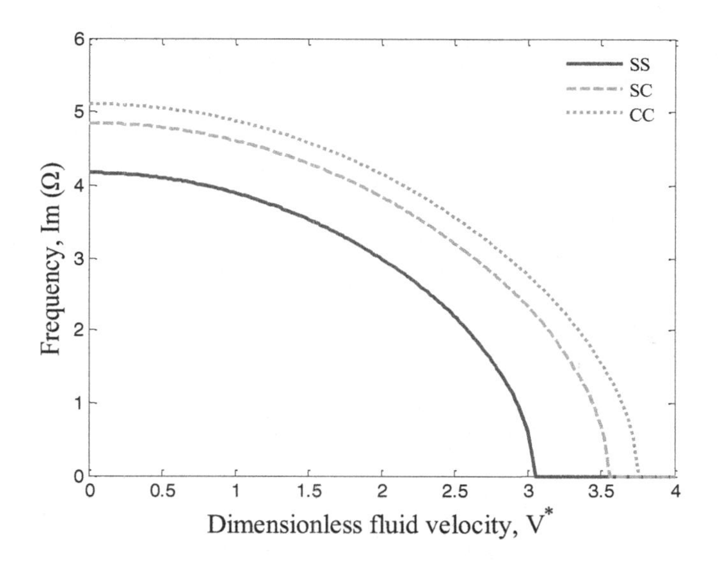


**FIGURE 13.8** Boundary condition effect on the imaginary part of the frequency.

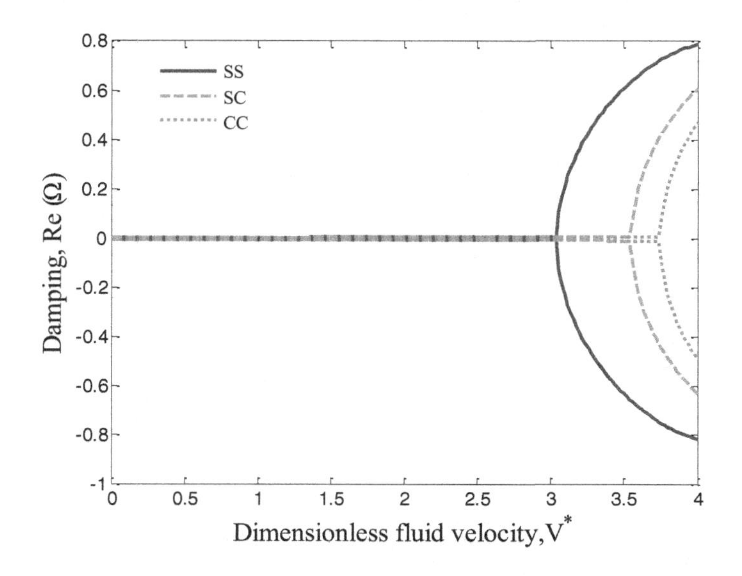


**FIGURE 13.9** Boundary condition effect on the real part of the frequency.

for the CC boundary condition are maximum since the rigidity and stiffness of the structure for this type of boundary condition are higher than those of two other cases.

Figures 13.10 and 13.11 indicate the dimensionless frequency and the damping of the structure versus the dimensionless flow velocity for different pipe lengths. With an increase in the pipe length, the dimensionless frequency and the damping of the structure are decreased. This is because with an increase in the pipe length, the stiffness of the system decreases.

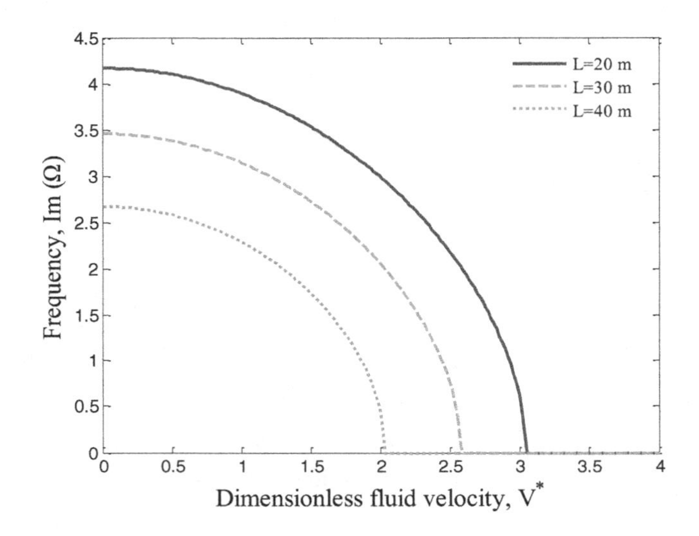


**FIGURE 13.10** Effect of pipe length on the imaginary part of the frequency.

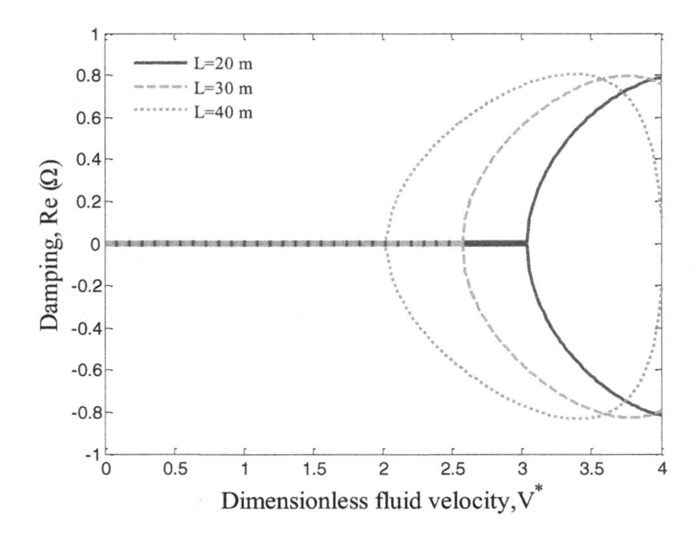


**FIGURE 13.11** Effect of pipe length on the real part of the frequency.

**Acknowledgments:** This chapter is a slightly modified version of Ref. [28] and has been reproduced here with the permission of the copyright holder.

## REFERENCES

[1] Paidoussis MP, Li GX. Pipes conveying fluid: A model dynamical problem. *J Fluids Struct* 1993;7:137–204.
[2] Toorani MH, Lakis AA. Dynamic analysis of anisotropic cylindrical shells containing flowing fluid. *J Press Ves Technol* 2001;123:454–460.
[3] Zhang YL, Reese JM, Gorman DG. Initially tensioned orthotropic cylindrical shells conveying fluid: A vibration analysis. *J Fluids Struct* 2002;16:53–70.
[4] Kadoli R, Ganesan N. Free vibration and buckling analysis of composite cylindrical shells conveying hot fluid. *Compos Struct* 2003;60:19–32.
[5] Zhang T, Ouyang H, Zhang YO, Lv BL. Nonlinear dynamics of straight fluid-conveying pipes with general boundary conditions and additional springs and masses. *Appl Math Model* 2016;40:7880–7900.
[6] Meng D, Guo H, Xu S. Non-linear dynamic model of a fluid-conveying pipe undergoing overall motions. *Appl Math Model* 2011;35:781–796.
[7] Dai HL, Wang L, Qian Q, Gan J. Vibration analysis of three-dimensional pipes conveying fluid with consideration of steady combined force by transfer matrix method. *Appl Math Comput* 2012;219:2453–2464.
[8] Gu J, Tianqi M, Menglan D. Effect of aspect ratio on the dynamic response of a fluid-conveying pipe using the Timoshenko beam model. *Ocean Eng* 2016;114:185–191.
[9] Zemri A, Houari MSA, Bousahla AA, Tounsi A. A mechanical response of functionally graded nanoscale beam: An assessment of a refined nonlocal shear deformation theory beam theory. *Struct Eng Mech* 2015;54:693–710.
[10] Larbi Chaht F, Kaci A, Houari MSA, Hassan S. Bending and buckling analyses of functionally graded material (FGM) size-dependent nanoscale beams including the thickness stretching effect. *Steel Compos Struct* 2015;18:425–442.
[11] Belkorissat I, Houari MSA, Tounsi A, Hassan S. On vibration properties of functionally graded nano-plate using a new nonlocal refined four variable model. *Steel Compos Struct* 2015;18:1063–1081.
[12] Ahouel M, Houari MSA, Adda Bedia EA, Tounsi A. Size-dependent mechanical behavior of functionally graded trigonometric shear deformable nanobeams including neutral surface position concept. *Steel Compos Struct* 2016;20:963–981.
[13] Bounouara F, Benrahou KH, Belkorissat I, Tounsi A. A nonlocal zeroth-order shear deformation theory for free vibration of functionally graded nanoscale plates resting on elastic foundation. *Steel Compos Struct* 2016;20:227–249.
[14] Bouafia KH, Kaci A, Houari MSA, Tounsi A. A nonlocal quasi-3D theory for bending and free flexural vibration behaviors of functionally graded nanobeams. *Smart Struct Syst* 2017;19:115–126.
[15] Besseghier A, Houari MSA, Tounsi A, Hassan S. Free vibration analysis of embedded nanosize FG plates using a new nonlocal trigonometric shear deformation theory. *Smart Struct Syst* 2017;19:601–614.
[16] Bellifa H, Benrahou KH, Bousahla AA, Tounsi A, Mahmoud SR. A nonlocal zeroth-order shear deformation theory for nonlinear postbuckling of nanobeams. *Struct Eng Mech* 2017;62:695–702.
[17] Mouffoki A, Adda Bedia EA, Houari MSA, Hassan S. Vibration analysis of nonlocal advanced nanobeams in hygro-thermal environment using a new two-unknown trigonometric shear deformation beam theory. *Smart Struct Syst* 2017;20:369–383.

[18] Khetir H, Bouiadjra MB, Houari MSA, Tounsi A, Mahmoud SR. A new nonlocal trigonometric shear deformation theory for thermal buckling analysis of embedded nanosize FG plates. *Struct Eng Mech* 2017;64:391–402.
[19] Vodenitcharova T, Zhang L. Bending and local buckling of a nanocomposite beam reinforced by a single-walled carbon nanotube. *Int J Solids Struct* 2006;43:3006–3024.
[20] Shen H, Xiang Y. Nonlinear vibration of nanotube-reinforced composite cylindrical shells in thermal environments. *Comput Methods Appl Mech Eng* 2012;213:196–211.
[21] Rafiee R, Moghadam RM. Simulation of impact and post-impact behavior of carbon nanotube reinforced polymer using multi-scale finite element modelling. *Comput Mater Sci* 2012;63:261–266.
[22] Kolahchi R, Rabani Bidgoli M, Beygipoor GH, Fakhar MH. A nonlocal nonlinear analysis for buckling in embedded FG-SWCNT-reinforced microplates subjected to magnetic field. *J Mech Sci Technol* 2015;29:3669–3677.
[23] Thomas B, Roy T. Vibration analysis of functionally graded carbon nanotube-reinforced composite shell structures. *Acta Mech* 2016;227:581–599.
[24] Mehri M, Asadi H, Wang Q. Buckling and vibration analysis of a pressurized CNT reinforced functionally graded truncated conical shell under an axial compression using HDQ method. *Comput Meth Appl Mech Eng* 2016;303:75–100.
[25] Bayat MR, Rahmani O, Mosavi Mashhadi M. Nonlinear low-velocity impact analysis of functionally graded nanotube-reinforced composite cylindrical shells in thermal environments. *Polymer Compos* 2018;39:730–745.
[26] Safari Bilouei B, Kolahchi R, Rabani Bidgoli M. Buckling of beams retrofitted with Nano-Fiber Reinforced Polymer (NFRP). *Comput Concr* 2016;18:1053–1063.
[27] Kolahchi R, Safari M, Esmailpour M. Dynamic stability analysis of temperature-dependent functionally graded CNT-reinforced visco-plates resting on orthotropic elastomeric medium. *Compos Struct* 2016;150:255–265.
[28] Golabchi H, Kolahchi R, Bidgoli MR. Vibration and instability analysis of pipes reinforced by $SiO_2$ nanoparticles considering agglomeration effects. *Comput Concrete* 2018;(21)4:431–440.
[29] Bessaim A, Houari MSA, Tounsi A. A new higher-order shear and normal deformation theory for the static and free vibration analysis of sandwich plates with functionally graded isotropic face sheets. *J Sandw Struct Mater* 2013;15:671–703.
[30] Bouderba B, Houari MSA, Tounsi A. Thermomechanical bending response of FGM thick plates resting on Winkler–Pasternak elastic foundations. *Steel Compos Struct* 2013;14:85–104.
[31] Belabed Z, Houari MSA, Tounsi A, Mahmoud SR, Bég OA. An efficient and simple higher order shear and normal deformation theory for functionally graded material (FGM) plates. *Compos: Part B* 2014;60:274–283.
[32] Attia A, Tounsi A, Adda Bedia EA, Mahmoud SR. Free vibration analysis of functionally graded plates with temperature-dependent properties using various four variable refined plate theories. *Steel Compos Struct* 2015;18:187–212.
[33] Zidi M, Tounsi A, Bég OA. Bending analysis of FGM plates under hygro-thermo-mechanical loading using a four variable refined plate theory. *Aerosp Sci Tech* 2014;34:24–34.
[34] Hamidi A, Houari MSA, Mahmoud SR, Tounsi A. A sinusoidal plate theory with 5–unknowns and stretching effect for thermomechanical bending of functionally graded sandwich plates. *Steel Compos Struct* 2015;18:235–253.
[35] Bourada M, Kaci A, Houari MSA, Tounsi A. A new simple shear and normal deformations theory for functionally graded beams. *Steel Compos Struct* 2015;18:409–423.

[36] Bousahla AA, Benyoucef S, Tounsi A, Mahmoud SR. On thermal stability of plates with functionally graded coefficient of thermal expansion. *Struct Eng Mech* 2016a;60:313–335.
[37] Chen W, Shih BJ, Chen YC, Hung JH, Hwang HH. Seismic response of natural gas and water pipelines in the Ji-Ji earthquake. *Soil Dyn Earthq Eng* 2002;22:1209–1214.
[38] Beldjelili Y, Tounsi A, Mahmoud SR. Hygro-thermo-mechanical bending of S-FGM plates resting on variable elastic foundations using a four-variable trigonometric plate theory. *Smart Struct Syst* 2016;18:755–786.
[39] Boukhari A, Atmane HA, Tounsi A, Adda Bedia EA, Mahmoud SR. An efficient shear deformation theory for wave propagation of functionally graded material plates. *Struct Eng Mech* 2016;57:837–859.
[40] Draiche K, Tounsi A, Mahmoud SR. A refined theory with stretching effect for the flexure analysis of laminated composite plates. *Geomech Eng* 2016;11:671–690.
[41] Bellifa H, Benrahou KH, Hadji L, Houari MSA, Tounsi A. Bending and free vibration analysis of functionally graded plates using a simple shear deformation theory and the concept the neutral surface position. *J Braz Soc Mech Sci Eng* 2016;38:265–275.
[42] Abdoun TH, Ha D, O'Rourke M, Symans M, O'Rourke T, Palmer M, Harry E. Factors influencing the behavior of buried pipelines subjected to earthquake faulting. *Soil Dyn Earthq Eng* 2009;29:415–427.
[43] Mahi A, Bedia EAA, Tounsi A. A new hyperbolic shear deformation theory for bending and free vibration analysis of isotropic, functionally graded, sandwich and laminated composite plates. *Appl Math Model* 2015;39:2489–2508.
[44] Shamsuddoha M, Islam MM, Aravinthan T, Manalo A, Lau KT. Effectiveness of using fibre-reinforced polymer composites for underwater steel pipeline repairs. *Compos Struct* 2013;100:40–54.
[45] Bennoun M, Houari MSA, Tounsi A. A novel five variable refined plate theory for vibration analysis of functionally graded sandwich plates. *Mech Advan Mat Struct* 2016;23:423–431.
[46] El-Haina F, Bakora A, Bousahla AA, Hassan S. A simple analytical approach for thermal buckling of thick functionally graded sandwich plates. *Struct Eng Mech* 2017;63:585–595.
[47] Menasria A, Bouhadra A, Tounsi A, Hassan S. A new and simple HSDT for thermal stability analysis of FG sandwich plates. *Steel Compos Struct* 2017;25:157–175.
[48] Chikh A, Tounsi A, Hebali H, Mahmoud SR. Thermal buckling analysis of cross-ply laminated plates using a simplified HSDT. *Smart Struct Syst* 2017;19:289–297.

# 14 Dynamic Response of Nanocomposite Shells Covered with a Piezoelectric Layer

## 14.1 INTRODUCTION

Cylindrical shells are used extensively in many engineering fields such as mechanical, chemical, aerospace, civil, nuclear, and so forth. For example, in the automotive industry, the body of an automobile and, in the oil and gas industry, pressure vessels are the obvious cases of applications for cylindrical shells.

The linear vibrations of cylindrical shells have been investigated by many researchers [1]. The first study of nonlinear vibrations of cylindrical shells was performed by Reissner [2]. He used the nonlinear Donnell's theory and concluded that considering the nonlinear terms can increase the accuracy of the calculations. Afterward, Grigolyuk and Prikladnaya [3] studied the nonlinear free vibration of cylindrical shells with simply supported boundary conditions. They determined the frequency of the system using nonlinear vibrational modes. Chu [4] continued with Reissner's work and probed nonlinear vibrations of closed cylindrical shells. Then, Nowinski [5] confirmed Chu's results. Cylindrical shells conveying fluid also are widely used in engineering applications. For instance, the thermal shielding of nuclear reactors and aircraft internal combustion engines, heat exchangers, oil and gas pipes, veins, the pulmonary system, and so on all are models with cylindrical shells conveying fluid. Most investigations on cylindrical shells conveying fluid were done by Amabili [6]. The effect of flowing fluid in cylindrical shells was studied by many researchers such as Païdoussis and Denise [7], Weaver and Unny [8], Païdoussis et al. [9], and Amabili and Garziera [10]. A static and free vibration analysis of three-layer composite shells was performed by Maturi et al. [11] using a radial basis function collocation, according to a new layer-wise theory that considers independent layer rotations, accounting for through-the-thickness deformation by considering a linear evolution of all displacements with each layer thickness coordinate.

A static and free vibration analysis of doubly curved laminated shells was performed by Ferreira et al. [12] using radial basis functions collocation.

In none of the mentioned investigations is the structure smart. Ghorbanpour et al. [13] studied the effect of material inhomogeneity on the behavior of the smart

DOI: 10.1201/9781003349525-14

piezoelectric cylinder by applying analytical method. Dynamic stability response of an embedded piezoelectric nanoplate made of polyvinylidene fluoride (PVDF) was presented by Kolahchi et al. [14] based on the differential cubature method (DCM) in conjunction with Bolotin's method. The dynamic elasticity solution for a clamped laminated cylindrical shell with two orthotropic layers bounded with a piezoelectric layer and subjected to an exponential dynamic load distributed on the inner surface was presented by Saviz [15]. The active vibration control of carbon nanotube (CNT)–reinforced functionally graded composite cylindrical shell was studied by Song et al. [16] using piezoelectric materials.

All the papers mentioned employed nanocomposite structures. Considering these structures, especially smart polymeric nanocomposites have many applications in major industries. Messina and Soldatos [17] studied the free vibration of a composite cylindrical shell. Tan and Tong [18] proposed micro-electro-mechanics models to obtain the effective properties of the composite. The nonlinear dynamic stability of embedded temperature-dependent viscoelastic plates reinforced by single-walled CNTs (SWCNTs) was investigated by Kolahchi et al. [19]. An accurate buckling analysis for piezoelectric fiber–reinforced composite (PFRC) cylindrical shells subjected to combined loads comprising compression, external voltage, and thermal load was presented by Sun et al. [20]. Zamani et al. [21] presented seismic response of pipes is investigated through the application of nanotechnology and piezoelectric materials. In this study, a pipe reinforced with carbon nanotubes (CNTs) and coated with a piezoelectric layer is considered. The structure is exposed to dynamic loads induced by earthquakes, and the governing equations of the system are derived using a mathematical model employing cylindrical shell elements and Mindlin theory.

Hitherto, no researcher has examined the dynamic behavior of the smart nanocomposite pipes conveying fluid under earthquake load. This problem is very significant in the fields of mechanical and civil engineering. In this chapter, the seismic response of the nanocomposite pipe covered with a piezoelectric layer and conveying fluid under earthquake load is investigated. The Mori–Tanaka approach is applied to estimate the effective material properties of the nanocomposite and to consider the effect of the agglomeration. The governing equations of the structure are derived using the energy method and according to the Mindlin theory. Given that the extracted equations are nonlinear, the discrete singular convolution method is employed to obtain the dynamic displacement of the structure caused by an earthquake. In the present chapter, the effect of various parameters like the volume percent and agglomeration of CNTs, the external voltage applied to the piezoelectric layer, the geometrical parameters of the pipe, and boundary conditions on the dynamic displacement of the structure are studied.

## 14.2 GEOMETRY OF THE PROBLEM

Figure 14.1 shows a nanocomposite pipe with average radius $R_s$, thickness $h_s$, length $L$, and density $\rho_s$ in a cylindrical coordinate system $(x,\theta,R)$.

The pipe is covered by a layer of piezoelectric with a thickness of $h_p$ and a density of $\rho_p$ and conveying a flow with a density $\rho_f$ and a viscosity $\mu_0$. In all the following equations, the subscripts $s$ and $p$ stand for pipe and piezoelectric layer, respectively. According to the Mindlin theory, the displacement field is given as presented in Section 1.5.2.

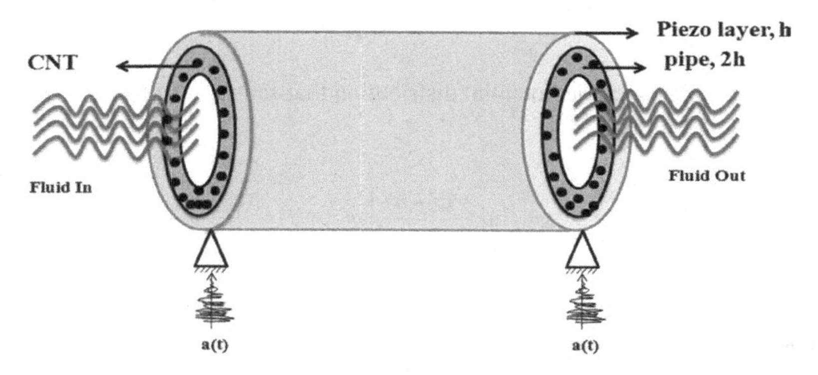


**FIGURE 14.1** Schematic of a nanocomposite pipe covered by a piezoelectric layer conveying fluid under seismic load.

## 14.3 CONSTITUTIVE EQUATIONS

### 14.3.1 Piezoelectric Layer

The piezoelectric layer is made of PVDF and is reinforced by CNTs. This material displays piezoelectric properties and is known as a smart material. In a piezoelectric material, stresses $\sigma$ and strains $\varepsilon$ tensors on the mechanical viewpoint, as well as flux density $D$ and field strength $E$ on the electrostatic viewpoint, can be arbitrarily combined as the following forms [15]:

$$\begin{bmatrix} \sigma_{xx} \\ \sigma_{\theta\theta} \\ \sigma_{zz} \\ \tau_{\theta z} \\ \tau_{xz} \\ \tau_{x\theta} \end{bmatrix}_p = \begin{bmatrix} Q_{11} & Q_{12} & Q_{13} & 0 & 0 & 0 \\ Q_{12} & Q_{22} & Q_{23} & 0 & 0 & 0 \\ Q_{13} & Q_{23} & Q_{33} & 0 & 0 & 0 \\ 0 & 0 & 0 & Q_{44} & 0 & 0 \\ 0 & 0 & 0 & 0 & Q_{55} & 0 \\ 0 & 0 & 0 & 0 & 0 & Q_{66} \end{bmatrix} \begin{Bmatrix} \varepsilon_{xx} \\ \varepsilon_{\theta\theta} \\ \varepsilon_{zz} \\ \gamma_{\theta z} \\ \gamma_{xz} \\ \gamma_{x\theta} \end{Bmatrix} - \begin{bmatrix} 0 & 0 & e_{31} \\ 0 & 0 & e_{32} \\ 0 & 0 & e_{33} \\ 0 & e_{24} & 0 \\ e_{15} & 0 & 0 \\ 0 & 0 & 0 \end{bmatrix} \begin{Bmatrix} E_x \\ E_\theta \\ E_z \end{Bmatrix}, \tag{14–1}$$

$$\begin{bmatrix} D_x \\ D_\theta \\ D_z \end{bmatrix} = \begin{bmatrix} 0 & 0 & 0 & 0 & e_{15} & 0 \\ 0 & 0 & 0 & e_{24} & 0 & 0 \\ e_{31} & e_{31} & e_{33} & 0 & 0 & 0 \end{bmatrix} \begin{Bmatrix} \varepsilon_{xx} \\ \varepsilon_{\theta\theta} \\ \varepsilon_{zz} \\ \gamma_{\theta z} \\ \gamma_{xz} \\ \gamma_{x\theta} \end{Bmatrix} + \begin{bmatrix} \in_{11} & 0 & 0 \\ 0 & \in_{22} & 0 \\ 0 & 0 & \in_{33} \end{bmatrix} \begin{Bmatrix} E_x \\ E_\theta \\ E_z \end{Bmatrix}, \tag{14–2}$$

where $Q_{ij}, e_{ijk}$, and $\in_{mk}$ are the elastic, piezoelectric, and dielectric constants, respectively. Also, the electric field $E_k$ in terms of electric potential $\varphi$ is given as follows [22]:

$$E_k = -\nabla \phi. \tag{14–3}$$

In this chapter, the electric potential distribution that satisfies the Maxwell equation is considered as follows [23]:

$$\phi(x,y,z,t) = \sin\left(\frac{\pi(z-h)}{h_p}\right)\varphi(x,y,t) + \frac{(z-h)V_0}{h_p}, \tag{14–4}$$

where $V_0$ is the external electric voltage applied to the structure. So the electric field components may be obtained by

$$E_x = -\frac{\partial \phi}{\partial x} = -\sin\left(\frac{\pi(z-h)}{h_p}\right)\frac{\partial \varphi}{\partial x}, \tag{14–5}$$

$$E_\theta = -\frac{\partial \phi}{R\partial \theta} = -\sin\left(\frac{\pi(z-h)}{h_p}\right)\frac{\partial \varphi}{R\partial \theta}, \tag{14–6}$$

$$E_z = -\frac{\partial \Phi}{\partial z} = -\frac{\pi}{h_p}\cos\left(\frac{\pi(z-h)}{h_p}\right)\varphi - \frac{V_0}{h_p}. \tag{14–7}$$

Finally, by applying the classical theory, the coupled electro-mechanical relations of the piezoelectric layer can be rewritten as follows:

$$\begin{bmatrix} \sigma_{xx} \\ \sigma_{\theta\theta} \\ \tau_{\theta z} \\ \tau_{xz} \\ \tau_{x\theta} \end{bmatrix}_p = \begin{bmatrix} Q_{11} & Q_{12} & 0 & 0 & 0 \\ Q_{12} & Q_{22} & 0 & 0 & 0 \\ 0 & 0 & Q_{44} & 0 & 0 \\ 0 & 0 & 0 & Q_{55} & 0 \\ 0 & 0 & 0 & 0 & Q_{66} \end{bmatrix} \begin{Bmatrix} \varepsilon_{xx} \\ \varepsilon_{\theta\theta} \\ \gamma_{\theta z} \\ \gamma_{xz} \\ \gamma_{x\theta} \end{Bmatrix} - \begin{bmatrix} 0 & 0 & e_{31} \\ 0 & 0 & e_{32} \\ 0 & e_{24} & 0 \\ e_{15} & 0 & 0 \\ 0 & 0 & 0 \end{bmatrix} \begin{Bmatrix} E_x \\ E_\theta \\ E_z \end{Bmatrix}, \tag{14–8}$$

$$\begin{bmatrix} D_x \\ D_\theta \\ D_z \end{bmatrix} = \begin{bmatrix} 0 & 0 & 0 & e_{15} & 0 \\ 0 & 0 & e_{24} & 0 & 0 \\ e_{31} & e_{32} & 0 & 0 & 0 \end{bmatrix} \begin{Bmatrix} \varepsilon_{xx} \\ \varepsilon_{\theta\theta} \\ \gamma_{\theta z} \\ \gamma_{xz} \\ \gamma_{x\theta} \end{Bmatrix} \begin{bmatrix} \in_{11} & 0 & 0 \\ 0 & \in_{22} & 0 \\ 0 & 0 & \in_{33} \end{bmatrix} \begin{Bmatrix} E_x \\ E_\theta \\ E_z \end{Bmatrix}. \tag{14–9}$$

### 14.3.2 Nanocomposite Pipe

According to Hook's law, the constitutive equation of nanocomposite pipe is expressed as follows:

$$\begin{bmatrix} \sigma_{xx} \\ \sigma_{\theta\theta} \\ \tau_{\theta z} \\ \tau_{xz} \\ \tau_{x\theta} \end{bmatrix}_s = \begin{bmatrix} C_{11} & C_{12} & 0 & 0 & 0 \\ C_{12} & C_{22} & 0 & 0 & 0 \\ 0 & 0 & C_{44} & 0 & 0 \\ 0 & 0 & 0 & C_{55} & 0 \\ 0 & 0 & 0 & 0 & C_{66} \end{bmatrix} \begin{Bmatrix} \varepsilon_{xx} \\ \varepsilon_{\theta\theta} \\ \gamma_{\theta z} \\ \gamma_{xz} \\ \gamma_{x\theta} \end{Bmatrix}. \quad (14\text{–}10)$$

It should be noted that the effective material properties of the nanocomposite pipe $C_{ij}$ are calculated based on the Mori–Tanaka approach and by considering the agglomeration effect, which is addressed in Section 4.3.

## 14.4 ENERGY METHOD

One of the general and comprehensive ways to obtain the governing equations of the structure is to equate the work done by external forces and the energy stored in the structure under load.

The strain energy of the structure is equal to the sum of the strain energy that stored in the pipe and the piezoelectric layer and is given as follows:

$$U = \int_V \begin{bmatrix} \left(\sigma_{xx}\varepsilon_{xx} + \sigma_{\theta\theta}\varepsilon_{\theta\theta} + \tau_{x\theta}\gamma_{x\theta} + \tau_{xz}\gamma_{xz} + \tau_{z\theta}\gamma_{z\theta}\right)_s \\ + \left(\sigma_{xx}\varepsilon_{xx} + \sigma_{\theta\theta}\varepsilon_{\theta\theta} + \tau_{x\theta}\gamma_{x\theta} + \tau_{xz}\gamma_{xz} + \tau_{z\theta}\gamma_{z\theta} \right. \\ \left. - D_x E_x - D_\theta E_\theta - D_z E_z\right)_p \end{bmatrix} \quad (14\text{–}11)$$

Substituting Equation 1–41 into Equation 13–11, the strain energy stored in the cylindrical shell can be expressed as follows:

$$\begin{aligned} U = \frac{1}{2}\int_z \int_0^{2\pi}\int_0^L \Bigg\{ & \left(\sigma_{xxs} + \sigma_{xxp}\right)\left[\frac{\partial u}{\partial x} + z\frac{\partial \psi_x}{\partial x} + \frac{1}{2}\left(\frac{\partial w}{\partial x}\right)^2\right] \\ & + \left(\sigma_{\theta\theta s} + \sigma_{\theta\theta p}\right)\left[\frac{1}{R}\left(w + \frac{\partial v}{\partial \theta}\right) + \frac{z}{R}\frac{\partial \psi_\theta}{\partial \theta} + \frac{1}{2}\left(\frac{\partial w}{R\partial \theta}\right)^2\right] \\ & + \left(\tau_{x\theta s} + \tau_{x\theta p}\right)\left(\frac{\partial v}{\partial x} + \frac{1}{R}\frac{\partial u}{\partial \theta} + z\frac{\partial \psi_\theta}{\partial x} + \frac{z}{R}\frac{\partial \psi_x}{\partial \theta} + \frac{\partial w}{R\partial \theta}\frac{\partial w}{\partial x}\right) \\ & + \left(\tau_{z\theta s} + \tau_{z\theta p}\right)\left[\frac{1}{R}\left(\frac{\partial w}{\partial \theta} - v\right) + \psi_\theta\right] + \left(\tau_{xzs} + \tau_{xzp}\right)\left(\psi_x + \frac{\partial w}{\partial x}\right) \\ & + D_x\left[\sin\left(\frac{\pi(z-h)}{h_p}\right)\frac{\partial \varphi}{\partial x}\right] + D_\theta\left[\sin\left(\frac{\pi(z-h)}{h_p}\right)\frac{\partial \varphi}{R\partial \theta}\right] \\ & + D_z\left[\frac{\pi}{h_p}\cos\left(\frac{\pi(z-h)}{h_p}\right)\varphi + \frac{V_0}{h_p}\right]\Bigg\} R\,dx\,d\theta\,dz, \end{aligned} \quad (14\text{–}12)$$

Introducing the stress resultants that follow,

$$\begin{Bmatrix} N_{xx} \\ N_{\theta\theta} \\ N_{x\theta} \end{Bmatrix} = \int_{-h}^{h} \begin{Bmatrix} \sigma_{xx} \\ \sigma_{\theta\theta} \\ \tau_{x\theta} \end{Bmatrix}_s dz + \int_{h}^{h+h_p} \begin{Bmatrix} \sigma_{xx} \\ \sigma_{\theta\theta} \\ \tau_{x\theta} \end{Bmatrix}_p dz, \tag{14–13}$$

$$\begin{Bmatrix} Q_x \\ Q_\theta \end{Bmatrix} = k' \left( \int_{-h}^{h} \begin{Bmatrix} \sigma_{xz} \\ \tau_{\theta z} \end{Bmatrix}_s dz + \int_{h}^{h+h_p} \begin{Bmatrix} \sigma_{xz} \\ \tau_{\theta z} \end{Bmatrix}_p dz \right), \tag{14–14}$$

$$\begin{Bmatrix} M_{xx} \\ M_{\theta\theta} \\ M_{x\theta} \end{Bmatrix} = \int_{-h}^{h} \begin{Bmatrix} \sigma_{xx} \\ \sigma_{\theta\theta} \\ \tau_{x\theta} \end{Bmatrix}_s zdz + \int_{h}^{h+h_p} \begin{Bmatrix} \sigma_{xx} \\ \sigma_{\theta\theta} \\ \tau_{x\theta} \end{Bmatrix}_p zdz, \tag{14–15}$$

we have

$$\begin{aligned} U = 0.5\int & \left( N_{xx}\left( \frac{\partial u}{\partial x} + \frac{1}{2}\left(\frac{\partial w}{\partial x}\right)^2 \right) + N_{\theta\theta}\left( \frac{\partial v}{R\partial\theta} + \frac{w}{R} + \frac{1}{2}\left(\frac{\partial w}{R\partial\theta}\right)^2 \right) \right. \\ & + Q_\theta\left( \frac{\partial w}{R\partial\theta} - \frac{v}{R} + \psi_\theta \right) + Q_x\left( \frac{\partial w}{\partial x} + \psi_x \right) + N_{x\theta}\left( \frac{\partial v}{\partial x} + \frac{\partial u}{R\partial\theta} + \frac{\partial w}{\partial x}\frac{\partial w}{R\partial\theta} \right) \\ & \left. + M_{xx}\frac{\partial \psi_x}{\partial x} + M_{\theta\theta}\frac{\partial \psi_\theta}{R\partial\theta} + M_{x\theta}\left( \frac{\partial \psi_x}{R\partial\theta} + \frac{\partial \psi_\theta}{\partial x} \right) \right) dA \\ & + 0.5\int \left( D_x\left( \sin\left( \frac{\pi(z-h)}{h_p} \right)\frac{\partial \varphi}{\partial x} \right) + D_\theta\left( \sin\left( \frac{\pi(z-h)}{h_p} \right)\frac{\partial \varphi}{R\partial\theta} \right) \right. \\ & \left. + D_z\left( \frac{\pi}{h_p}\cos\left( \frac{\pi(z-h)}{h_p} \right)\varphi + \frac{V_0}{h_p} \right) \right) dzdA, \end{aligned} \tag{14–16}$$

where $k'$ is the shear correction factor. The kinetic energy of the structure can be described as follows:

$$K = \frac{(\rho_s + \rho_p)}{2}\int \left( \left( \frac{\partial u}{\partial t} + z\frac{\partial \psi_x}{\partial t} \right)^2 + \left( \frac{\partial v}{\partial t} + z\frac{\partial \psi_\theta}{\partial t} \right)^2 + \left( \frac{\partial w}{\partial t} \right)^2 \right) dV, \tag{14–17}$$

where $\rho_s$ and $\rho_p$ are the equivalent density of the nanocomposite pipe and the piezoelectric layer, respectively. By defining the following terms,

$$\begin{Bmatrix} I_0 \\ I_1 \\ I_2 \end{Bmatrix} = \int_{-h}^{h} \begin{bmatrix} \rho \\ \rho z \\ \rho z^2 \end{bmatrix} dz + \int_{h}^{h+h_p} \begin{bmatrix} \rho \\ \rho z \\ \rho z^2 \end{bmatrix} dz, \tag{14–18}$$

Equation 14–17 can be rewritten as follows:

$$K = 0.5\int \left[ I_0 \left( \left(\frac{\partial u}{\partial t}\right)^2 + \left(\frac{\partial v}{\partial t}\right)^2 + \left(\frac{\partial w}{\partial t}\right)^2 \right) + 2I_1 \left( \frac{\partial u}{\partial t}\frac{\partial \psi_x}{\partial t} + \frac{\partial v}{\partial t}\frac{\partial \psi_\theta}{\partial t} \right) \right.$$
$$\left. + I_2 \left( \left(\frac{\partial \psi_x}{\partial t}\right)^2 + \left(\frac{\partial \psi_\theta}{\partial t}\right)^2 \right) \right] dA. \tag{14–19}$$

By assuming the Newtonian fluid, the governing equation of the fluid can be described by the well-known Navier–Stokes equation [14–26]:

$$\rho_f \frac{dV}{dt} = -\nabla P + \mu \nabla^2 V + F_{body}, \tag{14–20}$$

where $V \equiv (v_z, v_\theta, v_x)$ is the flow velocity vector in the cylindrical coordinate system with components in the longitudinal $x$, circumferential $\theta$ and radial $z$ directions. Also, P, $\mu$, and $\rho_f$ are the pressure, the viscosity, and the density of the fluid, respectively, and $F_{body}$ denotes the body forces. In the Navier–Stokes equation, the total derivative operator with respect to $t$ is

$$\frac{d}{dt} = \frac{\partial}{\partial t} + v_x \frac{\partial}{\partial x} + v_\theta \frac{\partial}{R\partial \theta} + v_z \frac{\partial}{\partial z}. \tag{14–21}$$

At the point of contact between the fluid and the core, the relative velocity and acceleration in the radial direction are equal. So [26]

$$v_z = \frac{dw}{dt}. \tag{14–22}$$

By employing Equations 14–21 and 14–22 and substituting them into Equation 14–20, the pressure inside the pipe can be computed as

$$\frac{\partial p_z}{\partial z} = -\rho_f \left( \frac{\partial^2 w}{\partial t^2} + 2v_x \frac{\partial^2 w}{\partial x \partial t} + v_x^2 \frac{\partial^2 w}{\partial x^2} \right)$$
$$+ \mu \left( \frac{\partial^3 w}{\partial x^2 \partial t} + \frac{\partial^3 w}{R^2 \partial \theta^2 \partial t} + v_x \left( \frac{\partial^3 w}{\partial x^3} + \frac{\partial^3 w}{R^2 \partial \theta^2 \partial x} \right) \right). \tag{14–23}$$

By multiplying two sides of Equation 14–23 in the inside area of the pipe ($A$), the radial force in the pipe is calculated as follows:

$$F_{fluid} = A\frac{\partial p_z}{\partial z} = -\rho_f \left( \frac{\partial^2 w}{\partial t^2} + 2v_x \frac{\partial^2 w}{\partial x \partial t} + v_x^2 \frac{\partial^2 w}{\partial x^2} \right)$$
$$+ \mu \left( \frac{\partial^3 w}{\partial x^2 \partial t} + \frac{\partial^3 w}{R^2 \partial \theta^2 \partial t} + v_x \left( \frac{\partial^3 w}{\partial x^3} + \frac{\partial^3 w}{R^2 \partial \theta^2 \partial x} \right) \right). \tag{14–24}$$

Finally, the external work due to the pressure of the fluid may be obtained as follows:

$$W_f = \int (F_{fluid}) w dA = \int \left( -\rho_f \left( \frac{\partial^2 w}{\partial t^2} + 2 v_x \frac{\partial^2 w}{\partial x \partial t} + v_x^{\ 2} \frac{\partial^2 w}{\partial x^2} \right) + \mu \left( \frac{\partial^3 w}{\partial x^2 \partial t} + \frac{\partial^3 w}{R^2 \partial \theta^2 \partial t} + v_x \left( \frac{\partial^3 w}{\partial x^3} + \frac{\partial^3 w}{R^2 \partial \theta^2 \partial x} \right) \right) \right) w dA. \tag{14–25}$$

The external work due to the earthquake loads can be computed:

$$W_s = \int \underbrace{(ma(t))}_{F_{Seismic}} w dA, \tag{14–26}$$

where $m$ and $a(t)$ are the mass and the acceleration of the earth.

## 14.5 HAMILTON'S PRINCIPLE

The governing equations of the structure are derived using Hamilton's principle, which is considered as follows:

$$\int_0^t (\delta U - \delta K - \delta W) dt = 0. \tag{14–27}$$

Now, by applying Hamilton's principle and after integration by part and some algebraic manipulation, six electro-mechanical equations of motion can be derived as follows:

$$\delta u: \quad \frac{\partial N_{xx}}{\partial x} + \frac{\partial N_{x\theta}}{R \partial \theta} = I_0 \frac{\partial^2 u}{\partial t^2} + I_1 \frac{\partial^2 \psi_x}{\partial t^2}, \tag{14–28}$$

$$\delta v: \quad \frac{\partial N_{x\theta}}{\partial x} + \frac{\partial N_{\theta\theta}}{R \partial \theta} + \frac{Q_\theta}{R} = I_0 \frac{\partial^2 v}{\partial t^2} + I_1 \frac{\partial^2 \psi_\theta}{\partial t^2}, \tag{14–29}$$

$$\delta w: \quad \frac{\partial Q_x}{\partial x} + \frac{\partial Q_\theta}{R \partial \theta} + \frac{\partial}{\partial x} \left( N_x^f \frac{\partial w}{\partial x} \right) + \frac{\partial}{R \partial \theta} \left( N_\theta^f \frac{\partial w}{R \partial \theta} \right) + F_{fluid} + F_{Seismic} = I_0 \frac{\partial^2 w}{\partial t^2}, \tag{14–30}$$

$$\delta \psi_x: \quad \frac{\partial M_{xx}}{\partial x} + \frac{\partial M_{x\theta}}{R \partial \theta} - Q_x = I_1 \frac{\partial^2 u}{\partial t^2} + I_2 \frac{\partial^2 \psi_x}{\partial t^2}, \tag{14–31}$$

$$\delta \psi_\theta: \quad \frac{\partial M_{x\theta}}{\partial x} + \frac{\partial M_{\theta\theta}}{R \partial \theta} - Q_\theta = I_1 \frac{\partial^2 v}{\partial t^2} + I_2 \frac{\partial^2 \psi_\theta}{\partial t^2}, \tag{14–32}$$

$$\delta\varphi : \quad \int_h^{h+h_p} \left( \left( \sin\left(\frac{\pi(z-h)}{h_p}\right) \frac{\partial D_x}{\partial x} \right) + \left( \sin\left(\frac{\pi(z-h)}{h_p}\right) \frac{\partial D_\theta}{R\partial\theta} \right) + D_z \left( \frac{\pi}{h_p} \cos\left(\frac{\pi(z-h)}{h_p}\right) \right) \right) dz. \tag{14–33}$$

By integrating the stress–strain relations of the structure from Equations 14–13 through 14–15, we have

$$N_{xx} = A_{11}\left(\frac{\partial u}{\partial x} + \frac{1}{2}\left(\frac{\partial w}{\partial x}\right)^2\right) + B_{11}\left(\frac{\partial \psi_x}{\partial x}\right) + A_{12}\left(\frac{\partial v}{R\partial\theta} + \frac{w}{R} + \frac{1}{2}\left(\frac{\partial w}{R\partial\theta}\right)^2\right) + B_{12}\left(\frac{\partial \psi_\theta}{R\partial\theta}\right) + E_{31}\varphi, \tag{14–34}$$

$$N_{\theta\theta} = A_{12}\left(\frac{\partial u}{\partial x} + \frac{1}{2}\left(\frac{\partial w}{\partial x}\right)^2\right) + B_{12}\left(\frac{\partial \psi_x}{\partial x}\right) + A_{22}\left(\frac{\partial v}{R\partial\theta} + \frac{w}{R} + \frac{1}{2}\left(\frac{\partial w}{R\partial\theta}\right)^2\right) + B_{22}\left(\frac{\partial \psi_\theta}{R\partial\theta}\right) + E_{32}\varphi, \tag{14–35}$$

$$N_{x\theta} = A_{66}\left(\frac{\partial u}{R\partial\theta} + \frac{\partial v}{\partial x} + \frac{\partial w}{\partial x}\frac{\partial w}{R\partial\theta}\right) + B_{66}\left(\frac{\partial \psi_x}{R\partial\theta} + \frac{\partial \psi_\theta}{\partial x}\right), \tag{14–36}$$

$$Q_x = A_{55}\left(\frac{\partial w}{\partial x} + \psi_x\right) - E_{15}\frac{\partial\varphi}{\partial x}, \tag{14–37}$$

$$Q_\theta = A_{44}\left(\frac{\partial w}{R\partial\theta} - \frac{v}{R} + \psi_\theta\right) - E_{24}\frac{\partial\varphi}{R\partial\theta}, \tag{14–38}$$

$$M_{xx} = B_{11}\left(\frac{\partial u}{\partial x} + \frac{1}{2}\left(\frac{\partial w}{\partial x}\right)^2\right) + D_{11}\left(\frac{\partial \psi_x}{\partial x}\right) + B_{12}\left(\frac{\partial v}{R\partial\theta} + \frac{w}{R} + \frac{1}{2}\left(\frac{\partial w}{R\partial\theta}\right)^2\right) + D_{12}\left(\frac{\partial \psi_\theta}{R\partial\theta}\right) + F_{31}\varphi, \tag{14–39}$$

$$M_{\theta\theta} = B_{12}\left(\frac{\partial u}{\partial x} + \frac{1}{2}\left(\frac{\partial w}{\partial x}\right)^2\right) + D_{12}\left(\frac{\partial \psi_x}{\partial x}\right) + B_{22}\left(\frac{\partial v}{R\partial\theta} + \frac{w}{R} + \frac{1}{2}\left(\frac{\partial w}{R\partial\theta}\right)^2\right) + D_{22}\left(\frac{\partial \psi_\theta}{R\partial\theta}\right) + F_{32}\varphi, \tag{14–40}$$

$$M_{x\theta} = B_{66}\left(\frac{\partial u}{R\partial\theta} + \frac{\partial v}{\partial x} + \frac{\partial w}{\partial x}\frac{\partial w}{R\partial\theta}\right) + D_{66}\left(\frac{\partial\psi_x}{R\partial\theta} + \frac{\partial\psi_\theta}{\partial x}\right), \tag{14–41}$$

where the constants $A_{ij}, B_{ij}, D_{ij}, E_{ij}$, and $F_{ij}$ are equal to

$$\begin{aligned}&\left(A_{11}, A_{12}, A_{22}, A_{44}, A_{55}, A_{66}\right) = \int_{-h}^{h}\left(C_{11}, C_{12}, C_{22}, C_{44}, C_{55}, C_{66}\right)dz \\ &+\int_{h}^{h+h_p}\left(Q_{11}, Q_{12}, Q_{22}, Q_{44}, Q_{55}, Q_{66}\right)dz,\end{aligned} \tag{14–42}$$

$$\left(B_{11}, B_{12}, B_{22}, B_{66}\right) = \int_{-h}^{h}\left(C_{11}, C_{12}, C_{22}, C_{66}\right)zdz + \int_{-h}^{h+h_p}\left(Q_{11}, Q_{12}, Q_{22}, Q_{66}\right)zdz, \tag{14–43}$$

$$\begin{aligned}&\left(D_{11}, D_{12}, D_{22}, D_{66}\right) = \int_{-h}^{h}\left(C_{11}, C_{12}, C_{22}, C_{66}\right)z^2dz \\ &+\int_{-h}^{h+h_p}\left(Q_{11}, Q_{12}, Q_{22}, Q_{66}\right)z^2dz,\end{aligned} \tag{14–44}$$

$$\left(E_{31}, E_{32}\right) = \frac{\pi}{h_p}\int_{h}^{h+h_p}\left(e_{31}, e_{32}\right)\cos\left(\frac{\pi(z-h)}{h_p}\right)dz, \tag{14–45}$$

$$\left(E_{15}, E_{24}\right) = \int_{h}^{h+h_p}\left(e_{15}, e_{24}\right)\sin\left(\frac{\pi(z-h)}{h_p}\right)dz, \tag{14–46}$$

$$\left(F_{31}, F_{32}\right) = \frac{\pi}{h_p}\int_{h}^{h+h_p}\left(e_{31}, e_{32}\right)\cos\left(\frac{\pi(z-h)}{h_p}\right)zdz. \tag{14–47}$$

Now, by substituting Equations 14–34 through 14–41 into the equations of motion (Equations 14–28 through 14–33), we have

$$\begin{aligned}&A_{11}\left(\frac{\partial^2 u}{\partial x^2} + \frac{\partial w}{\partial x}\frac{\partial^2 w}{\partial x^2}\right) + B_{11}\left(\frac{\partial^2\psi_x}{\partial x^2}\right) \\ &+A_{12}\left(\frac{\partial^2 v}{R\partial x\partial\theta} + \frac{\partial w}{R\partial x} + \frac{\partial w}{R\partial\theta}\frac{\partial^2 w}{R\partial x\partial\theta}\right) \\ &+B_{12}\left(\frac{\partial^2\psi_\theta}{R^2\partial x\partial\theta}\right) + E_{31}\frac{\partial\varphi}{\partial x} + \frac{A_{66}}{R} \\ &\left(\frac{\partial^2 u}{R\partial\theta^2} + \frac{\partial^2 v}{\partial x\partial\theta} + \frac{\partial^2 w}{\partial x\partial\theta}\frac{\partial w}{R\partial\theta} + \frac{\partial w}{\partial x}\frac{\partial^2 w}{R\partial\theta^2}\right) \\ &+\frac{B_{66}}{R}\left(\frac{\partial^2\psi_x}{R\partial\theta^2} + \frac{\partial^2\psi_\theta}{\partial x\partial\theta}\right) = I_0\frac{\partial^2 u}{\partial t^2} + I_1\frac{\partial^2\psi_x}{\partial t^2},\end{aligned} \tag{14–48}$$

$$
\begin{aligned}
&A_{66}\left(\frac{\partial^2 u}{R\partial\theta\partial x}+\frac{\partial^2 v}{\partial x^2}+\frac{\partial^2 w}{\partial x^2}\frac{\partial w}{R\partial\theta}+\frac{\partial w}{\partial x}\frac{\partial^2 w}{R\partial\theta\partial x}\right)\\
&+B_{66}\left(\frac{\partial^2\psi_x}{R\partial\theta\partial x}+\frac{\partial^2\psi_\theta}{\partial x^2}\right)+\frac{A_{12}}{R}\left(\frac{\partial^2 u}{\partial x\partial\theta}+\frac{\partial w}{\partial x}\frac{\partial^2 w}{\partial x\partial\theta}\right)\\
&+\frac{B_{12}}{R}\left(\frac{\partial^2\psi_x}{\partial x\partial\theta}\right)+\frac{A_{22}}{R}\left(\frac{\partial^2 v}{R\partial\theta^2}+\frac{\partial w}{R\partial\theta}+\frac{\partial w}{R\partial\theta}\frac{\partial^2 w}{R\partial\theta^2}\right)\\
&+\frac{B_{22}}{R}\left(\frac{\partial^2\psi_\theta}{R^2\partial\theta^2}\right)+E_{32}\frac{\partial\varphi}{R\partial\theta}=I_0\frac{\partial^2 v}{\partial t^2}+I_1\frac{\partial^2\psi_\theta}{\partial t^2},
\end{aligned}
\tag{14–49}
$$

$$
\begin{aligned}
&A_{55}\left(\frac{\partial^2 w}{\partial x^2}+\frac{\partial\psi_x}{\partial x}\right)-E_{15}\frac{\partial^2\varphi}{\partial x^2}+\frac{A_{44}}{R}\left(\frac{\partial^2 w}{R\partial\theta^2}-\frac{\partial v}{R\partial\theta}+\frac{\partial\psi_\theta}{\partial\theta}\right)\\
&-E_{24}\frac{\partial^2\varphi}{R^2\partial\theta^2}-\rho_f K_p\left(\frac{\partial^2 w}{\partial t^2}+2v_x\frac{\partial^2 w}{\partial x\,\partial t}+{v_x}^2\frac{\partial^2 w}{\partial x^2}\right)+\frac{\partial}{R\partial\theta}\left(N_\theta^f\frac{\partial w}{R\partial\theta}\right)\\
&+\frac{\partial}{\partial x}\left(N_x^f\frac{\partial w}{\partial x}\right)+\mu\left(\frac{\partial^3 w}{\partial x^2\,\partial t}+\frac{\partial^3 w}{R^2\,\partial\theta^2\,\partial t}+v_x\left(\frac{\partial^3 w}{\partial x^3}+\frac{\partial^3 w}{R^2\,\partial\theta^2\,\partial x}\right)\right)\\
&+F_{Seismic}=I_0\frac{\partial^2 w}{\partial t^2},
\end{aligned}
\tag{14–50}
$$

$$
\begin{aligned}
&B_{11}\left(\frac{\partial^2 u}{\partial x^2}+\frac{\partial w}{\partial x}\frac{\partial^2 w}{\partial x^2}\right)+D_{11}\left(\frac{\partial^2\psi_x}{\partial x^2}\right)+B_{12}\left(\frac{\partial^2 v}{R\partial\theta\partial x}+\frac{\partial w}{R\partial x}+\frac{\partial^2 w}{R\partial\theta\partial x}\right)\\
&+D_{12}\left(\frac{\partial^2\psi_\theta}{R\partial\theta\partial x}\right)+F_{31}\frac{\partial\varphi}{\partial x}+\frac{B_{66}}{R}\left(\frac{\partial^3 u}{R\partial\theta^2}+\frac{\partial^2 v}{\partial x\partial\theta}+\frac{\partial^2 w}{\partial x\partial\theta}\frac{\partial w}{R\partial\theta}+\frac{\partial w}{\partial x}\frac{\partial^2 w}{R\partial\theta^2}\right)\\
&+\frac{D_{66}}{R}\left(\frac{\partial^2\psi_x}{R\partial\theta^2}+\frac{\partial^2\psi_\theta}{\partial x\partial\theta}\right)-A_{55}\left(\frac{\partial w}{\partial x}+\psi_x\right)+E_{15}\frac{\partial\varphi}{\partial x}=I_1\frac{\partial^2 u}{\partial t^2}+I_2\frac{\partial^2\psi_x}{\partial t^2},
\end{aligned}
\tag{14–51}
$$

$$
\begin{aligned}
&B_{66}\left(\frac{\partial^2 u}{R\partial\theta\partial x}+\frac{\partial^2 v}{\partial x^2}+\frac{\partial^2 w}{\partial x^2}\frac{\partial w}{R\partial\theta}+\frac{\partial w}{\partial x}\frac{\partial^2 w}{R\partial\theta\partial x}\right)\\
&+D_{66}\left(\frac{\partial^2\psi_x}{R\partial\theta\partial x}+\frac{\partial^2\psi_\theta}{\partial x^2}\right)+\frac{B_{12}}{R}\left(\frac{\partial^2 u}{\partial x\partial\theta}+\frac{\partial w}{\partial x}\frac{\partial^2 w}{\partial x\partial\theta}\right)\\
&+\frac{D_{12}}{R}\left(\frac{\partial^2\psi_x}{\partial x\partial\theta}\right)+\frac{B_{22}}{R}\left(\frac{\partial^2 v}{R\partial\theta^2}+\frac{\partial w}{R\partial\theta}+\frac{\partial w}{R\partial\theta}\frac{\partial^2 w}{R\partial\theta^2}\right)\\
&+\frac{D_{22}}{R}\left(\frac{\partial^2\psi_\theta}{R\partial\theta^2}\right)+F_{32}\frac{\partial\varphi}{R\partial\theta}-A_{44}\left(\frac{\partial w}{R\partial\theta}-\frac{v}{R}+\psi_\theta\right)\\
&+E_{24}\frac{\partial\varphi}{R\partial\theta}=I_1\frac{\partial^2 v}{\partial t^2}+I_2\frac{\partial^2\psi_\theta}{\partial t^2},
\end{aligned}
\tag{14–52}
$$

$$\begin{aligned}&\frac{2e_{15}h}{\pi}\frac{\partial w}{\partial x}+\frac{2e_{15}h}{\pi}\psi_x+\frac{\in_{11}}{2}\frac{h}{\partial x}\frac{\partial \phi}{\partial x}+\frac{2e_{24}h}{\pi}\frac{\partial w}{R\partial \theta}-\frac{2e_{24}h}{\pi}\frac{v}{R}+\frac{2e_{24}h}{\pi}\psi_\theta\\&+\frac{\in_{22}}{2}\frac{h}{R\partial\theta}\frac{\partial \phi}{}-\frac{2he_{32}}{\pi}\frac{\partial^2 w}{R^2\partial\theta^2}+\frac{2h}{\pi}e_{31}\frac{\partial \psi_x}{\partial x}-\frac{\pi^2\in_{33}}{2h}\phi=0.\end{aligned} \tag{14–53}$$

Also, the boundary conditions are considered as follows:

- Clamped–clamped supported

$$x=0,L\Rightarrow u=v=w=\psi_x=\psi_\theta=0, \tag{14–54}$$

- Simply–simply supported

$$x=0,L\Rightarrow u=v=w=\psi_\theta=M_{xx}=0, \tag{14–55}$$

- Clamped–simply supported

$$\begin{aligned}&x=0\Rightarrow u=v=w=\psi_x=\psi_\theta=0,\\&x=L\Rightarrow u=v=w=\psi_x=M_{xx}=0.\end{aligned} \tag{14–56}$$

Given that the governing equations are nonlinear, so HDQM, along with the multiple scale method, is applied to achieve results with higher accuracy (see Sections 2.2.3 and 2.3.3).

## 14.6 NUMERICAL RESULTS

In this section, the numerical results for the dynamic response of the pipeline that is reinforced by CNTs and covered with a piezoelectric layer under the earthquake loads are examined. For this purpose, a polyethylene pipe of length $L$ = 2m, $R$ =20in., and thickness $h$ = 1in. is considered. The pipe is covered with a piezoelectric layer of PVDF with the thickness $h_p = 5mm$ and conveying a fluid flow of velocity $v_x = 40\,ft/s$ and viscosity $\mu = 63.6\,Pa.s$. The elastic and piezoelectric properties of these materials are given in Table 14.1 [23].

It is worth mentioning that the acceleration of the earthquake is considered according to the Bam earthquake; the distribution of acceleration in 20 seconds is shown in Figure 14.2.

### 14.6.1 VERIFICATION

Since this research has been defined for the first time in the world, there is no reference to validate the obtained results. Therefore, it has been tried to examine the results without considering the nonlinear terms of the governing equations and by comparing the linear dynamic response of the structure obtained by two various

**TABLE 14.1**
**Material Property of PE, PVDF, and CNT**

| PVDF | CNT | PE |
|---|---|---|
| $C_{11} = 238.24(\text{GPa})$ | $E = 1(\text{TPa})$ | $E = 125(\text{GPa})$ |
| $C_{22} = 23.6(\text{GPa})$ | $\upsilon = 0.34$ | $\upsilon = 0.30$ |
| $C_{12} = 3.98(\text{GPa})$ | $\rho = 1.4(gr / \text{cm}^3)$ | $\rho = 1.45(\text{kg}/\text{m}^3)$ |
| $C_{66} = 6.43(\text{GPa})$ | | |
| $e_{11} = -0.135(\text{C}/\text{m}^2)$ | | |
| $e_{12} = -0.145(\text{C}/\text{m}^2)$ | | |
| $\in = 1.1068 \times 10^{-8}(\text{F}/\text{m})$ | | |
| $\alpha_x = 7.1 \times 10^{-5}(1/\text{K})$ | | |
| $\alpha_\theta = 7.1 \times 10^{-5}(1/\text{K})$ | | |

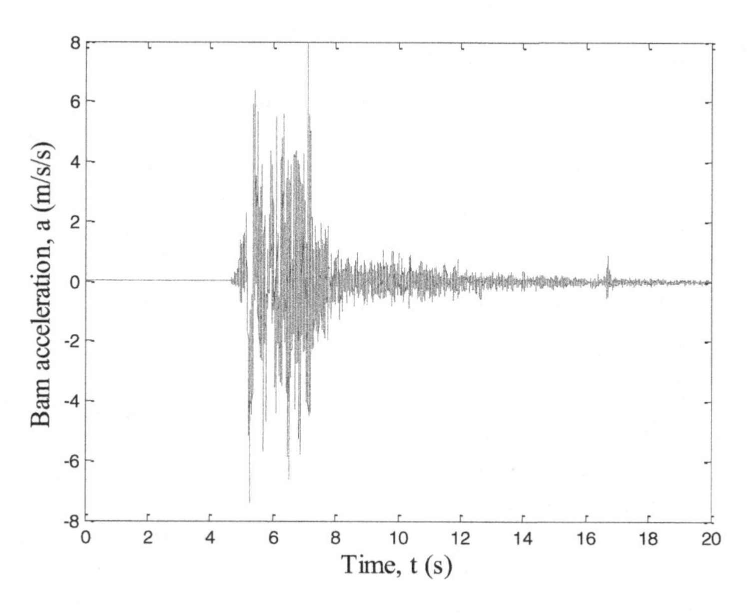


**FIGURE 14.2** Acceleration history of the Bam earthquake.

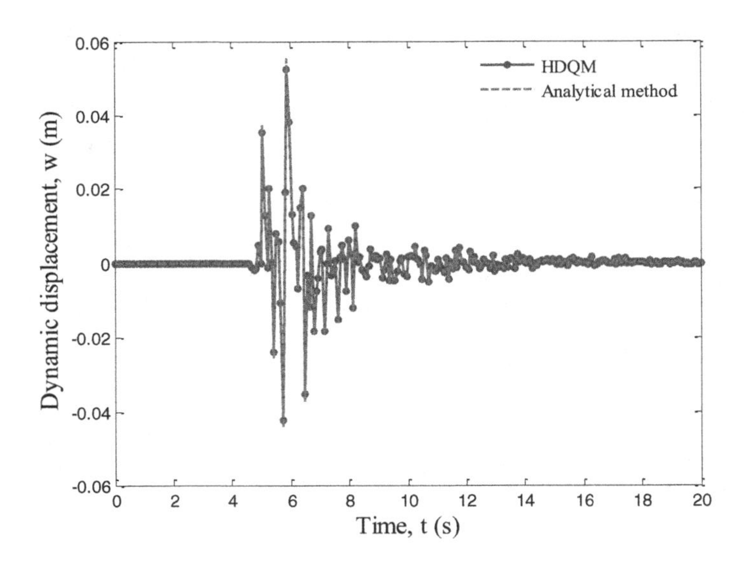


**FIGURE 14.3** Comparison of the analytical and numerical results.

solution methods. The results of the analytical and numerical (HDQ) methods are depicted in Figure 14.3. As can be seen, the difference between the analytical method and the HDQ method is negligible, and so the obtained results are accurate and acceptable.

### 14.6.2 Convergence of the Numerical Method

The convergence of HDQM in evaluating the maximum deflection of the structure versus the number of grid points is illustrated in Figure 14.4. As can be seen, by increasing the number of grid points, the maximum deflection of the structure decreases so far that, at $N = 15$, the deflection converges. So the following results are based on 15 grid points for the DQM solution.

### 14.6.3 Effects of Various Parameters

Figure 14.5 shows the effect of CNTs volume percent on the dynamic deflection of the structure versus time. The changes of the deflection are shown for $c_r = 0$, $c_r = 0.05$, and $c_r = 0.10$. It is apparent that by increasing the CNT volume percent, the dynamic deflection of the system is reduced because the stiffness of the structure increases.

The agglomeration effect of CNTs on the dynamic deflection of the structure versus time is indicated by Figure 14.6. It can be found that considering the agglomeration decreases the stiffness of the structure, and as a result, the displacement of the structure increases. Given that during the process of nanocomposite manufacturing, the uniform distribution of CNTs in the polymer matrix is impossible, so the results of this figure can be very significant. So it can be concluded that about CNT-reinforced pipes, as the agglomeration in various regions decreases, the displacement of the structure decreases.

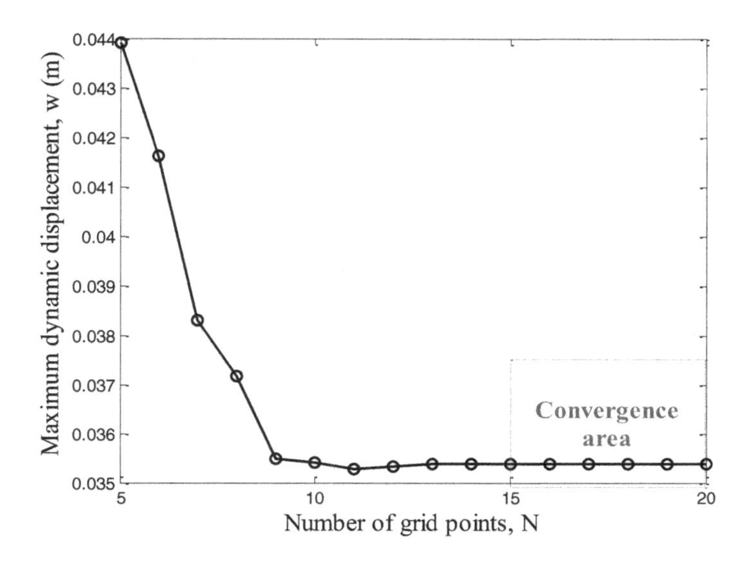


**FIGURE 14.4** Convergence and accuracy of HDQM.

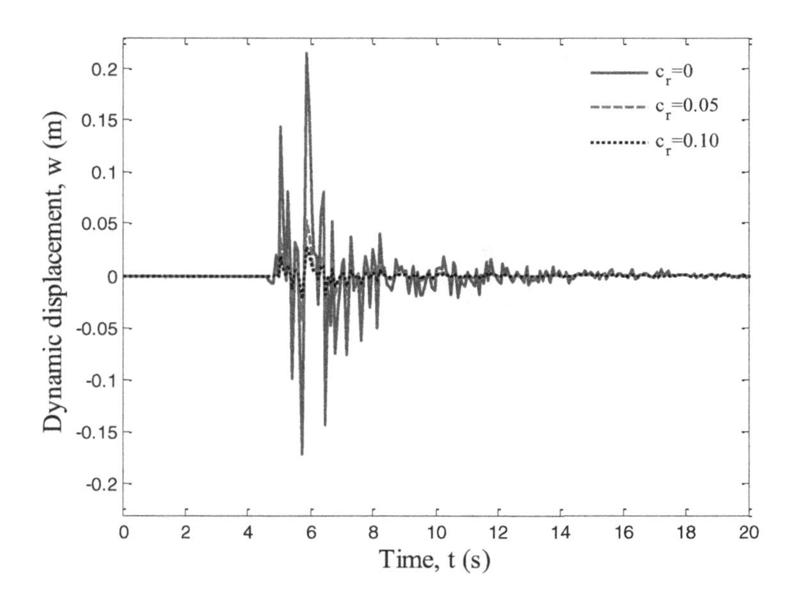


**FIGURE 14.5** Effect of CNT volume percent on the dynamic deflection of the structure.

Figure 14.7 illustrates the effect of various boundary conditions on the dynamic displacement of the structure versus time. It is found that the boundary conditions have a significant effect on the dynamic displacement of the system so that the pipe with clamped–clamped boundary condition has the lowest deflection. This is predictable because the constraint of the clamped boundary condition is stronger than the other ones and, consequently, the structure is stiffer.

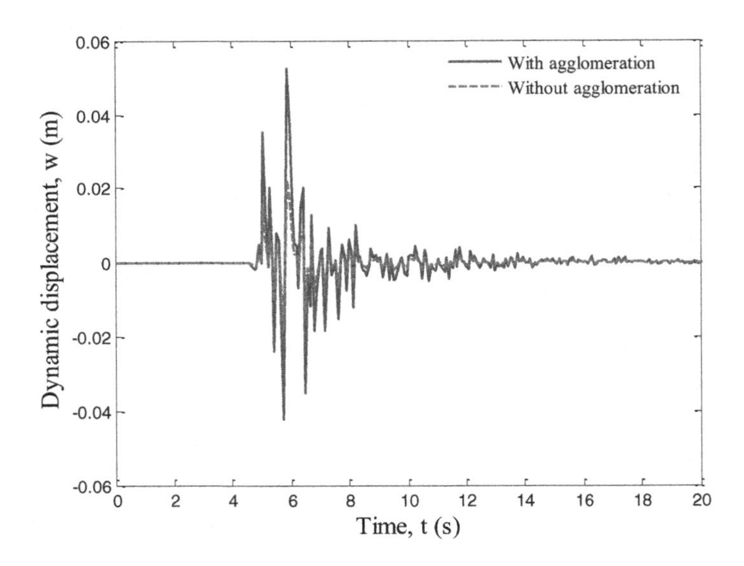


**FIGURE 14.6** Effect of CNT agglomeration on the dynamic deflection of the structure.

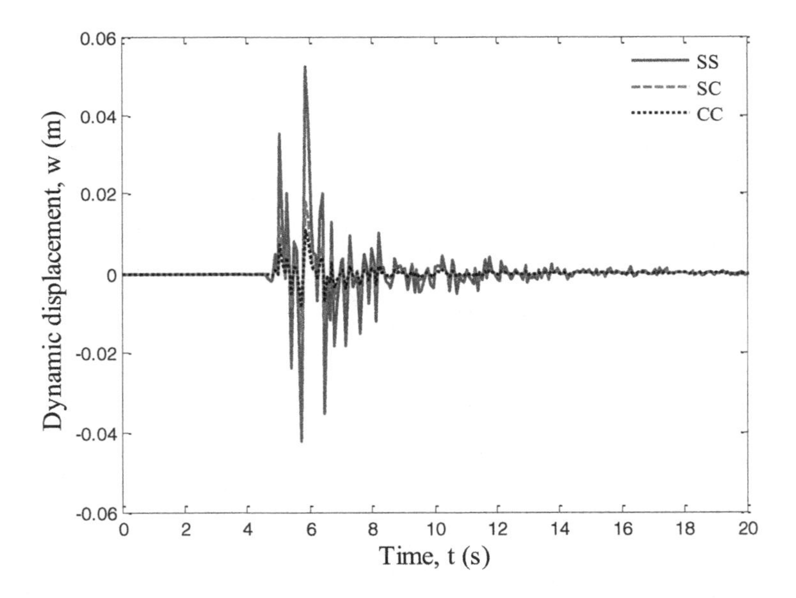


**FIGURE 14.7** Effect of boundary conditions on the dynamic deflection of the structure.

Figure 14.8. indicates the effect of the applied external voltage to the piezoelectric layer on the dynamic deflection versus time. It can be observed that by applying positive voltage to the structure, the dynamic deflection of the system increases, and this is because of the tensile force exerted on the structure, making the structure softer. Applying an external negative voltage has the reverse effect and leads

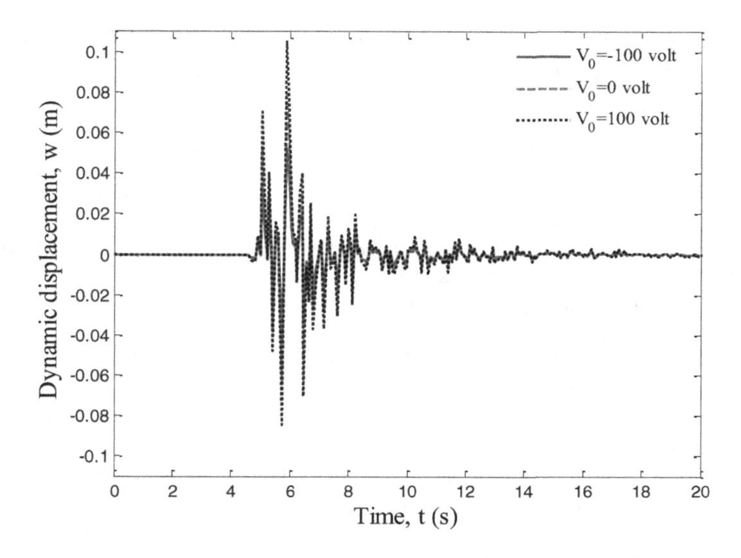


**FIGURE 14.8** Effect of applied external voltage on the dynamic deflection of the structure.

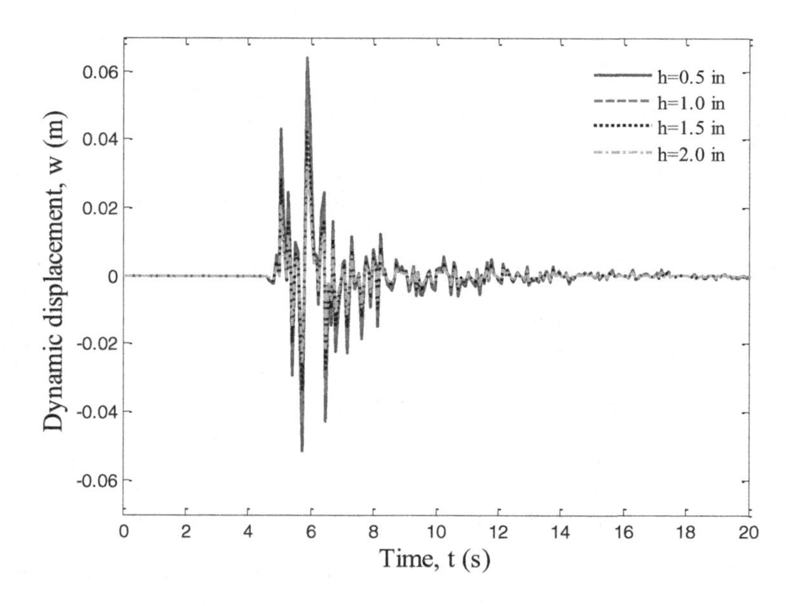


**FIGURE 14.9** Effect of pipe thickness on the dynamic deflection of the structure.

to a compressive force in the structure and decreases the dynamic deflection of the system.

The effect of pipe thickness on the deflection behavior versus time is shown in Figure 14.9. By increasing the thickness of the pipe, the stiffness of the structure increases, and so the dynamic deflection of the system decreases.

**Acknowledgments:** This chapter is a slightly modified version of Ref. [21] and has been reproduced here with the permission of the copyright holder.

## REFERENCES

[1] Love AEH. On the small free vibrations and deformations of a thin elastic shell. *Phil Trans Royal Soci A* 1888;179:491–549.
[2] Reissner E. The effect of transverse shear deformation on the bending of elastic plates. *ASME J Appl Mech* 1945;12:A68–A77.
[3] Grigolyuk EI, Prikladnaya A. Vibrations of circular cylindrical panels subjected to finite deflections. *Matematika* 1955;19:376–382.
[4] Chu HN. Influence of large-amplitudes on flexural vibrations of a thin circular cylindrical shell. *J Aerosp Sci* 1961;28:602–609.
[5] Nowinski J. Nonlinear transverse vibrations of orthotropic cylindrical shells. *AIAA J* 1963;1:617–620.
[6] Amabili M. A comparison of shell theories for large-amplitude vibrations of circular cylindrical shells: Lagrangian approach. *J Sound Vib* 2003;264:1091–1125.
[7] Päidoussis MP, Denise JP. Flutter of thin cylindrical shells conveying fluid. *J Sound Vib* 1972;20:9–26.
[8] Weaver DS, Unny TE. On the dynamic stability of fluid-conveying pipes. *J Appl Mech* 1973;40:48–52.
[9] Päidoussis MP. *Fluid-structure interactions: Slender structures and axial flow*. Elsevier Academic Press, London, 2003.
[10] Amabili M, Garziera R. Vibrations of circular cylindrical shells with nonuniform constraints, elastic bed and added mass. Part II: Shells containing or immersed in axial flow. *J Fluids Struct* 2002;16:31–51.
[11] Maturi DA, Ferreira AJM, Zenkour AM, Mashat DS. analysis of three-layer composite shells by a new layerwise theory and radial basis functions collocation, accounting for through-the-thickness deformations. *Mech Adv Mat Struct* 2015;22:722–730.
[12] Ferreira AJM, Carrera E, Cinefra M, Zenkour AM. Radial basis functions solution for the analysis of laminated doubly-curved shells by a Reissner-Mixed Variational Theorem. *Mech Adv Mat Struct* 2016;23:1068–1079.
[13] Ghorbanpour Arani A, Kolahchi R, Mosallaie Barzoki AA. Effect of material inhomogeneity on electro-thermo-mechanical behaviors of functionally graded piezoelectric rotating cylinder. *Appl Math Model* 2011;35:2771–2789.
[14] Kolahchi R, Hosseini H, Esmailpour M. Differential cubature and quadrature-Bolotin methods for dynamic stability of embedded piezoelectric nanoplates based on visco-nonlocal-piezoelasticity theories. *Compos Struct* 2016;157:174–186.
[15] Saviz MR. Dynamic analysis of a laminated cylindrical shell with piezoelectric layer and clamped boundary condition. *Finit Elem Anal Des* 2015;104:1–15.
[16] Song ZG, Zhang LW, Liew KM. Active vibration control of CNT-reinforced composite cylindrical shells via piezoelectric patches. *Compos Struct* 2016;158:92–100.
[17] Messina A, Soldatos KP. Vibration of completely free composite plates and cylindrical shell panels by a higher-order theory. *Int J Mech Sci* 1999;41:891–918.
[18] Tan P, Tong L. Micro-electromechanics models for piezoelectric-fiber-reinforced composite materials. *Compos Sci Tech* 2001;61:759–769.
[19] Kolahchi R, Safari M, Esmailpour M. Dynamic stability analysis of temperature-dependent functionally graded CNT-reinforced visco-plates resting on orthotropic elastomeric medium. *Compos Struct* 2016;150:255–265.

[20] Sun J, Xu X, Lim CW, Zhou ZH, Xiao SH. Accurate thermo-electro-mechanical buckling of shear deformable piezoelectric fiber-reinforced composite cylindrical shells. *Compos Struct* 2016;141:221–231.
[21] Zamani A, Kolahchi R, Bidgoli MR. Seismic response of smart nanocomposite cylindrical shell conveying fluid flow using HDQ-Newmark methods. *Comput Concrete* 2017;(20)6:671–682.
[22] Brush O, Almorth B. *Buckling of bars, plates and shells*. McGraw-Hill, New York, 1975.
[23] Kolahch R, Hosseini H, Esmailpour M. Differential cubature and quadrature-Bolotin methods for dynamic stability of embedded piezoelectric nanoplates based on visco-nonlocal-piezoelasticity theories. *Compos Struct* 2016;157:174–186.
[24] Mori T, Tanaka K. Average stress in Matrix and average elastic energy of materials with misfitting inclusions. *Acta Metall Mater* 1973;21:571–574.
[25] Shi DL, Feng X. The Effect of nanotube waviness and agglomeration on the elastic property of carbon nanotube-reinforced composites. *J Eng Mat Tech* 2004;126:250–270.
[26] Kolahchi R, Moniribidgoli AM. Size-dependent sinusoidal beam model for dynamic instability of single-walled carbon nanotubes. *Appl Math Mech* 2016;37:265–274.

# Appendix A
## *The MATLAB Code for Chapter 4*

```
clc
clear all

L=3;
m=1;
h=30e-2;
Ri=56e-3;
Ro=205e-3;
rhoc=2400;
rho=rhoc;
rhor=4506;
Ec=20e9;
nuc=.2;
Ef=25e9;
Er=1e12;
Cr=0.1;
zeta=0.1;
eta=0.1;Pi=pi;b=15e-2;
%Mori-Tanaka approach-------------------------
Em=Ef;
num=0.3;
Cm=1-Cr;
% Heal cofficent------------------
Kr=30e9;
Lr=10e9;
Mr=1e9;
Nr=450e9;
Pr=1e9;
Km=Em/(3*(1-2*num));
Gm=Em/(2*(1+num));
alphar=(3*(Km+Gm)+Kr-Lr)/(3*(Gm+Kr));
betar=.2*((4*Gm+2*Kr+Lr)/(3*(Gm+Kr))+4*Gm/(Gm+Pr)+(2*(Gm*(3*Km
+Gm)+Gm*(3*Km+7*Gm)))/(Gm*(3*Km+Gm)+Mr*(3*Km+7*Gm)));
deltar=1/3*(Nr+2*Lr+(2*Kr+Lr)*(3*Km+2*Gm-Lr)/(Gm+Kr));
etar=1/5*(2/3*(Nr-Lr)+(8*Gm*Pr)/(Gm+Pr)+(8*Mr*Gm*(3*Km+4*Gm))/
(3*Km*(Mr+Gm)+Gm*(7*Mr+Gm)));
Kin=Km+(deltar-3*Km*alphar)/(3*(zeta-Cr*eta+Cr*eta*alphar));
Kout=Km+(Cr*(deltar-3*Km*alphar)*(1-eta))/(3*(1-zeta-Cr*(1-eta)
+Cr*alphar*(1-eta)));
Gin=Gm+((etar-3*Gm*betar)*Cr*eta)/(2*(zeta-Cr*eta+Cr*eta*betar));
Gout=Gm+(Cr*(etar-3*Gm*betar)*(1-eta))/(2*(1-zeta-Cr*(1-eta)+
Cr*betar*(1-eta)));
nuout=(3*Kout-2*Gout)/(6*Kout+2*Gout);
```

```
alpha=(1+nuout)/(3*(1-nuout));
beta=(2*(4-5*nuout))/(15*(1-nuout));
K=Kout*(1+(zeta*(Kin/Kout-1))/(1+alpha*(1-zeta)*(Kin/
Kout-1)));
G=Gout*(1+(zeta*(Gin/Gout-1))/(1+beta*(1-zeta)*(Gin/Gout-1)));
Eff=(9*K*G)/(3*K+G);
nuff=(3*K-2*G)/(6*K+2*G);

Q11=Eff/(1-nuff^2);
Q55=Eff/(2*(1+nuff));
I0=rho*b*h;
I1=0;
I2=rho*h^3/12*b;
I=h^3/12*b;

%%%Stiffness and Mass matrices
K11=-I*Q11*m^4*Pi^4/L^4;
K12=(24*I)*Q11*m^3/L^3;
K13=0;

K21=(24*I)*Q11*m^3/L^3;
K22=-(6*I)*Q11*m^2/L^2-(1/2)*Q55*h;
K23=0;

K31=0;
K32=0;
K33=0;

M11=-24*I2/pi^3*(m*pi/L);
M12=6*I2/pi^2;
M13=0;

M21=I0-I2*m^2*pi^2/L^2;
M22=-24*I2/pi^3*(m*pi/L);
M23=0;

M31=0;
M32=0;
M33=0;

KL=[K11 K12;
    K21 K22];

ML=[M11 M12;
    M21 M22];

Eq1=-inv(K)*ML;
[dd Eq2]=eig(Eq1);
Eq2(Eq2==0)=[];
Eq2(Eq2<0)=[];
Ans=sqrt(min(Eq2));
```

```
Freq=Ans;
d=dd(:,((diag((Eq2))))==Ans^2);

Q=2300;
Nm=max(size(KL));
q=zeros(Nm,1);
qdot=zeros(Nm,1);
qddot=zeros(Nm,1);
time=30;
tt=linspace(0,time,Q);
kkk=2;
tt(Q+1)=time+time/Q;
for kk=1:Q;
    kk
t=tt(kk);
dt=time/Q*kk;
delta=0.5;
alpha=0.25;
a0=1/(alpha*dt^2);
a1=delta/(alpha*dt);
a2=1/(alpha*dt);
a3=1/(2*alpha)-1;
a4=delta/alpha-1;
a5=dt/2*(delta/alpha-2);
a6=dt*(1-delta);
a7=delta*dt;
kobe earthquake from peer ground motion database
Capp=kobe(:,1);
F=-ML*Capp(kk)*[1;0];
Keq=triu(KL+a0*ML);
Feq=F+ML*(a0*q(:,kk)+a2*qdot(:,kk)+a3*qddot(:,kk));
q(:,kkk)=Keq\Feq;
qddot(:,kkk)=a0*(q(:,kkk)-q(:,kk))-a2*qdot(:,kk)-
a3*qddot(:,kk);
qdot(:,kkk)=qdot(:,kk)+a6*qddot(:,kk)+a7*qddot(:,kkk);
kkk=kkk+1;
end
```

# Appendix B

## *The MATLAB Code for Chapter 6*

```
clc
clear all
syms P

Q=20;
QQ=linspace(.5,1,Q);
for i=1:Q
    i
    gg=QQ(i);
n=1;
m=1;
d=1;
h=d;
a=20;
b=10;
Pi=pi;
KK=.75;
NM=0;
rhom=7600;
KW=64000e3;
KG=0;
c1=4/(3*(2*d)^2);
c2=3*c1;

rho=2400;
Ec=20e9;
nuc=.2;
Er=70e9;
Cr=0.2;
zeta=-gg;
eta=0.1;
Pi=pi;
%Mori-Tanaka approach--------------------------
Em=Ec;
num=0.3;
Cm=1-Cr;
% Heal cofficent------------------
Kr=30e9;
Lr=10e9;
Mr=1e9;
Nr=450e9;
Pr=1e9;
```

```
Km=Em/(3*(1-2*num));
Gm=Em/(2*(1+num));
alphar=(3*(Km+Gm)+Kr-Lr)/(3*(Gm+Kr));
betar=.2*((4*Gm+2*Kr+Lr)/(3*(Gm+Kr))+4*Gm/(Gm+Pr)+(2*(Gm*(3*Km
+Gm)+Gm*(3*Km+7*Gm)))/(Gm*(3*Km+Gm)+Mr*(3*Km+7*Gm)));
deltar=1/3*(Nr+2*Lr+(2*Kr+Lr)*(3*Km+2*Gm-Lr)/(Gm+Kr));
etar=1/5*(2/3*(Nr-Lr)+(8*Gm*Pr)/(Gm+Pr)+(8*Mr*Gm*(3*Km+4*Gm))/
(3*Km*(Mr+Gm)+Gm*(7*Mr+Gm)));
Kin=Km+(deltar-3*Km*alphar)/(3*(zeta-Cr*eta+Cr*eta*alphar));
Kout=Km+(Cr*(deltar-3*Km*alphar)*(1-eta))/(3*(1-zeta-Cr*
(1-eta)+Cr*alphar*(1-eta)));
Gin=Gm+((etar-3*Gm*betar)*Cr*eta)/(2*(zeta-Cr*eta+Cr*eta*betar));
Gout=Gm+(Cr*(etar-3*Gm*betar)*(1-eta))/(2*(1-zeta-Cr*(1-eta)
+Cr*betar*(1-eta)));
nuout=(3*Kout-2*Gout)/(6*Kout+2*Gout);
alpha=(1+nuout)/(3*(1-nuout));
beta=(2*(4-5*nuout))/(15*(1-nuout));
K=Kout*(1+(zeta*(Kin/Kout-1))/(1+alpha*(1-zeta)*(Kin/Kout-1)));
G=Gout*(1+(zeta*(Gin/Gout-1))/(1+beta*(1-zeta)*(Gin/Gout-1)));
Eff=(9*K*G)/(3*K+G);
nuff=(3*K-2*G)/(6*K+2*G);

c11=Eff/(1-nuff^2);
c12=(nuff*Eff)/(1-nuff^2);
c21=c12;
c13=c12;
c23=c13;
c22=Eff/(1-nuff^2);
c33=c22;
c31=c13;
c32=c13;
c66=Eff/(2*(1+nuff));
c44=Eff/(2*(1+nuff));
c55=Eff/(2*(1+nuff));


Q11t=c11;
Q12t=c12;
Q13t=c13;
Q14t=0;
Q22t=c22;
Q23t=c13;
Q24t=0;
Q23t=c13;
Q34t=0;
Q44t=0;
Q55t=c55;
Q56t=0;
Q66t=c66;

Q11m=0;
Q12m=0;
```

```
Q13m=0;
Q14m=0;
Q22m=0;
Q23m=0;
Q24m=0;
Q23m=0;
Q34m=0;
Q44m=0;
Q55m=0;
Q56m=0;
Q66m=0;

Q11b=0;
Q12b=0;
Q13b=0;
Q14b=0;
Q22b=0;
Q23b=0;
Q24b=0;
Q23b=0;
Q34b=0;
Q44b=0;
Q55b=0;
Q56b=0;
Q66b=0;

I0=rho*d;
I1=0;
I2=rho*d^3/12;
I3=0;
I4=rho*d^4/80;
I6=rho*d^6/448;
J1=I1-(4/3/h^2)*I3;
J4=I4-(4/3/h^2)*I6;
K2=I2-(8/3/h^2)*I4+(4/3/h^2)^2*I6;

K11=-(2/3)*Q11m*m^2*Pi^2*d/a^2-(2/3)*Q11t*m^2*Pi^2*d/a^2-
(2/3)*Q66m*n^2*Pi^2*d/b^2-(2/3)*Q11b*...
    m^2*Pi^2*d/a^2-(2/3)*Q66t*n^2*Pi^2*d/b^2-
(2/3)*Q66b*n^2*Pi^2*d/b^2;
K12=-(2/3)*Q66m*m*Pi^2*n*d/(a*b)-(2/3)*Q66t*m*Pi^2*n*d/(a*b)-
(2/3)*Q12t*m*Pi^2*n*d/(a*b)-(2/3)*Q66b*m*...
    Pi^2*n*d/(a*b)-(2/3)*Q12m*m*Pi^2*n*d/(a*b)-
(2/3)*Q12b*m*Pi^2*n*d/(a*b);
K13=-(20/81*(Q12b*c1*m*Pi^3*n^2/(a*b^2)+Q11b*c1*m^3*Pi^3/a^3))
*d^4+(40/81)*Q66t*c1*m*Pi^3*n^2*d^4/(a*b^2)+...
 (20/81*(Q12t*c1*m*Pi^3*n^2/(a*b^2)+Q11t*c1*m^3*Pi^3/
a^3))*d^4-(40/81)*Q66b*c1*m*Pi^3*n^2*d^4/(a*b^2);
K14=(20/81)*Q11t*c1*m^2*Pi^2*d^4/a^2-(20/81)*Q66b*c1*n^2*Pi^2*
d^4/b^2-(4/9)*Q66t*n^2*Pi^2*d^2/b^2+(20/81)*Q66t*...
   c1*n^2*Pi^2*d^4/b^2+(4/9)*Q66b*n^2*Pi^2*d^2/b^2-(4/9)*Q11t*
m^2*Pi^2*d^2/a^2-(20/81)*Q11b*c1*m^2*Pi^2*d^4/a^2+...
```

```
    (4/9)*Q11b*m^2*Pi^2*d^2/a^2;
K15=-(20/81)*Q12b*c1*m*Pi^2*n*d^4/(a*b)+(4/9)*Q12b*m*Pi^2*n
*d^2/(a*b)+(20/81)*Q12t*c1*m*Pi^2*n*d^4/(a*b)-(20/81)*...
    Q66b*c1*m*Pi^2*n*d^4/(a*b)-(4/9)*Q66t*m*Pi^2*n*d^2/(a*b)+
(20/81)*Q66t*c1*m*Pi^2*n*d^4/(a*b)+(4/9)*Q66b*m*...
    Pi^2*n*d^2/(a*b)-(4/9)*Q12t*m*Pi^2*n*d^2/(a*b);

K21=-(2/3)*Q66m*m*Pi^2*n*d/(a*b)-(2/3)*Q66t*m*Pi^2*n*d/(a*b)-
(2/3)*Q12t*m*Pi^2*n*d/(a*b)-(2/3)*Q66b*m*Pi^2*n*...
    d/(a*b)-(2/3)*Q12m*m*Pi^2*n*d/(a*b)-(2/3)*Q12b*m*Pi^2*n*d/
(a*b);
K22=-(2/3)*Q22m*n^2*Pi^2*d/b^2-(2/3)*Q66m*m^2*Pi^2*d/a^2-
(2/3)*Q22b*n^2*Pi^2*d/b^2-(2/3)*Q66t*m^2*Pi^2*...
    d/a^2-(2/3)*Q66b*m^2*Pi^2*d/a^2-(2/3)*Q22t*n^2*Pi^2*d/b^2;
K23=(20/81*(Q22t*c1*n^3*Pi^3/b^3+Q12t*c1*m^2*Pi^3*n/
(a^2*b)))*d^4-(20/81*(Q22b*c1*n^3*Pi^3/b^3+Q12b*c1*...
    m^2*Pi^3*n/(a^2*b)))*d^4-(40/81)*Q66b*c1*m^2*Pi^3*n*d^4/
(a^2*b)+(40/81)*Q66t*c1*m^2*Pi^3*n*d^4/(a^2*b);
K24=-(20/81)*Q12b*c1*m*Pi^2*n*d^4/
(a*b)+(4/9)*Q12b*m*Pi^2*n*d^2/(a*b)+(20/81)*Q12t*c1*m*Pi^2*
n*d^4/(a*b)-(20/81)*...
    Q66b*c1*m*Pi^2*n*d^4/(a*b)-(4/9)*Q66t*m*Pi^2*n*d^2/(a*b)+
(20/81)*Q66t*c1*m*Pi^2*n*d^4/(a*b)+(4/9)*Q66b*...
    m*Pi^2*n*d^2/(a*b)-(4/9)*Q12t*m*Pi^2*n*d^2/(a*b);
K25=-(4/9)*Q22t*n^2*Pi^2*d^2/b^2+(4/9)*Q22b*n^2*Pi^2*d^2/
b^2-(20/81)*Q66b*c1*m^2*Pi^2*d^4/a^2+(20/81)*Q22t*...
    c1*n^2*Pi^2*d^4/b^2+(4/9)*Q66b*m^2*Pi^2*d^2/a^2-
(20/81)*Q22b*c1*n^2*Pi^2*d^4/b^2-(4/9)*Q66t*m^2*Pi^2*d^2/a^...
    2+(20/81)*Q66t*c1*m^2*Pi^2*d^4/a^2;

K31=c1*((20/81)*Q11t*m^3*Pi^3*d^4/a^3+(40/81)*Q66t*m*Pi^3*
n^2*d^4/(a*b^2)-(40/81)*Q66b*m*Pi^3*n^2*d^4/(a*b^2)-...
    (20/81)*Q12b*m*Pi^3*n^2*d^4/(a*b^2)-(20/81)*Q11b*m^3
*Pi^3*d^4/a^3+(20/81)*Q12t*m*Pi^3*n^2*d^4/(a*b^2));
K32=c1*(-(20/81)*Q22b*n^3*Pi^3*d^4/b^3+(40/81)*Q66t*m^2*
Pi^3*n*d^4/(a^2*b)-(40/81)*Q66b*m^2*Pi^3*n*d^4/(a^2*b)+...
(20/81)*Q12t*m^2*Pi^3*n*d^4/(a^2*b)+(20/81)*Q22t*n^3*Pi^3*d^4/
b^3-(20/81)*Q12b*m^2*Pi^3*n*d^4/(a^2*b));
K33=(26/27)*Q55b*c1*m^2*Pi^2*d^3/a^2+(2/27)*Q55m*c1*m^2*Pi^2*
d^3/a^2+(26/27)*Q55t*c1*m^2*Pi^2*d^3/a^2+(26/27)*Q44b*c1*n^2*P
i^2*d^3/b^2-(2/3)*Q55b*m^2*Pi^2*d/a^2+(2/27)*Q44m*...
    c1*n^2*Pi^2*d^3/b^2-(2/3)*Q44m*n^2*Pi^2*d/b^2+(26/27)*Q44t
*c1*n^2*Pi^2*d^3/b^2-(2/3)*Q55m*m^2*Pi^2*d/a^2+c1*...
    ((2186/15309*(-Q22t*c1*n^4*Pi^4/b^4-Q12t*c1*m^2*Pi^4*n^2/
(a^2*b^2)))*d^7+(2186/15309*(-Q22b*c1*n^4*Pi^4/b^4-...
    Q12b*c1*m^2*Pi^4*n^2/(a^2*b^2)))*d^7-(8744/15309)*Q66b*c1*
m^2*Pi^4*n^2*d^7/(a^2*b^2)+(2186/15309*(-Q12b*c1*...
    m^2*Pi^4*n^2/(a^2*b^2)-Q11b*c1*m^4*Pi^4/a^4))*d^7-(8744/
15309)*Q66t*c1*m^2*Pi^4*n^2*d^7/(a^2*b^2)+(2186/15309*...
    (-Q12t*c1*m^2*Pi^4*n^2/(a^2*b^2)-Q11t*c1*m^4*Pi^4/
a^4))*d^7+(2/15309*(-Q12m*c1*m^2*Pi^4*n^2/(a^2*b^2)-Q11m*...
```

```
    c1*m^4*Pi^4/a^4))*d^7+(2/15309*(-Q22m*c1*n^4*Pi^4/b^4-
Q12m*c1*m^2*Pi^4*n^2/(a^2*b^2)))*d^7-(8/15309)*Q66m*c1*...
    m^2*Pi^4*n^2*d^7/(a^2*b^2))-(2/3)*Q55t*m^2*Pi^2*d/a^2-
(2/3)*Q44b*n^2*Pi^2*d/b^2-c2*((242/405)*Q55b*c1*m^2*Pi^2*...
    d^5/a^2-(26/81)*Q55b*m^2*Pi^2*d^3/a^2+(2/405)*Q55m*c1*m^2*
Pi^2*d^5/a^2-(2/81)*Q55m*m^2*Pi^2*d^3/a^2+(242/405)*...
    Q55t*c1*m^2*Pi^2*d^5/a^2-(26/81)*Q55t*m^2*Pi^2*d^3/a^2)-
(2/3)*Q44t*n^2*Pi^2*d/b^2-c2*((242/405)*Q44b*c1*n^2*...
    Pi^2*d^5/b^2-(26/81)*Q44b*n^2*Pi^2*d^3/b^2+(2/405)*Q44m*c1
*n^2*Pi^2*d^5/b^2-(2/81)*Q44m*n^2*Pi^2*d^3/b^2+...
    (242/405)*Q44t*c1*n^2*Pi^2*d^5/b^2-
(26/81)*Q44t*n^2*Pi^2*d^3/b^2);
K34=-(2/3)*Q55b*m*Pi*d/a+(26/27)*Q55b*c1*m*Pi*d^3/a-
(2/3)*Q55t*m*Pi*d/a-(2/3)*Q55m*m*Pi*d/a+(26/27)*Q55t*c1*m*...
    Pi*d^3/a+c1*(-(2186/15309)*Q11b*c1*m^3*Pi^3*d^7/a^3-
(2186/15309)*Q12b*c1*m*Pi^3*n^2*d^7/
(a*b^2)+(242/1215)*Q11b*...
    m^3*Pi^3*d^5/a^3+(4/1215)*Q66m*m*Pi^3*n^2*d^5/(a*b^2)-(4372/
15309)*Q66t*c1*m*Pi^3*n^2*d^7/(a*b^2)+(242/1215)*Q11t*...
    m^3*Pi^3*d^5/a^3-(2186/15309)*Q11t*c1*m^3*Pi^3*d^7/a^3+(24
2/1215)*Q12b*m*Pi^3*n^2*d^5/(a*b^2)-(2/15309)*Q11m*c1*...
    m^3*Pi^3*d^7/a^3+(484/1215)*Q66b*m*Pi^3*n^2*d^5/(a*b^2)+(4
84/1215)*Q66t*m*Pi^3*n^2*d^5/(a*b^2)+(2/1215)*Q12m*...
    m*Pi^3*n^2*d^5/(a*b^2)+(2/1215)*Q11m*m^3*Pi^3*d^5/a^3-
(4372/15309)*Q66b*c1*m*Pi^3*n^2*d^7/(a*b^2)-(2186/15309)*...
    Q12t*c1*m*Pi^3*n^2*d^7/(a*b^2)-(2/15309)*Q12m*c1*m*Pi^3*n^
2*d^7/(a*b^2)+(242/1215)*Q12t*m*Pi^3*n^2*d^5/(a*...
    b^2)-(4/15309)*Q66m*c1*m*Pi^3*n^2*d^7/
(a*b^2))+(2/27)*Q55m*c1*m*Pi*d^3/a-c2*((242/405)*Q55b*c1*m*Pi*
d^5/a-(26/81)*...
    Q55b*m*Pi*d^3/a+(2/405)*Q55m*c1*m*Pi*d^5/a-(2/81)*Q55m*m*P
i*d^3/a+(242/405)*Q55t*c1*m*Pi*d^5/a-(26/81)*Q55t*...
    m*Pi*d^3/a)-KW-KG*((m*pi/a)^2+(n*pi/b)^2);
K35=(2/27)*Q44m*c1*n*Pi*d^3/b-(2/3)*Q44b*n*Pi*d/
b+(26/27)*Q44b*c1*n*Pi*d^3/b-(2/3)*Q44t*n*Pi*d/
b+c1*(-(2186/15309)*...
    Q12b*c1*m^2*Pi^3*n*d^7/(a^2*b)+(484/1215)*Q66t*m^2*Pi^3*
n*d^5/(a^2*b)-(2186/15309)*Q22b*c1*n^3*Pi^3*d^7/b^3+...
    (2/1215)*Q12m*m^2*Pi^3*n*d^5/(a^2*b)+(242/1215)*Q12b*m^2*P
i^3*n*d^5/(a^2*b)-(2186/15309)*Q22t*c1*n^3*Pi^3*...
    d^7/b^3+(4/1215)*Q66m*m^2*Pi^3*n*d^5/(a^2*b)+(242/1215)*Q2
2t*n^3*Pi^3*d^5/b^3-(4372/15309)*Q66t*c1*m^2*Pi^3*...
    n*d^7/(a^2*b)+(242/1215)*Q12t*m^2*Pi^3*n*d^5/(a^2*b)-(2186/
15309)*Q12t*c1*m^2*Pi^3*n*d^7/(a^2*b)+(242/1215)*Q22b*...
    n^3*Pi^3*d^5/b^3-(2/15309)*Q12m*c1*m^2*Pi^3*n*d^7/(a^2*b)+
(484/1215)*Q66b*m^2*Pi^3*n*d^5/(a^2*b)+(2/1215)*Q22m*...
    n^3*Pi^3*d^5/b^3-(4372/15309)*Q66b*c1*m^2*Pi^3*n*d^7/
(a^2*b)-(2/15309)*Q22m*c1*n^3*Pi^3*d^7/b^3-(4/15309)*Q66m*...
    c1*m^2*Pi^3*n*d^7/(a^2*b))-(2/3)*Q44m*n*Pi*d/b+(26/27)
*Q44t*
c1*n*Pi*d^3/b-c2*((242/405)*Q44b*c1*n*Pi*d^5/b-(26/81)*...
```

```
    Q44b*n*Pi*d^3/b+(2/405)*Q44m*c1*n*Pi*d^5/b-
(2/81)*Q44m*n*Pi*d^3/b+(242/405)*Q44t*c1*n*Pi*d^5/b-(26/81)*
Q44t*...
    n*Pi*d^3/b);
K41=(4/9)*Q66b*n^2*Pi^2*d^2/b^2-c2*((20/81)*Q66b*n^2*Pi^2*d^4/
b^2-(20/81)*Q66t*n^2*Pi^2*d^4/b^2)-c1*((20/81)*Q11b*...
    m^2*Pi^2*d^4/a^2-(20/81)*Q11t*m^2*Pi^2*d^4/a^2)+(4/9)*
Q11b*m^2*Pi^2*d^2/a^2-(4/9)*Q66t*n^2*Pi^2*d^2/b^2-(4/9)*...
    Q11t*m^2*Pi^2*d^2/a^2;
K42=(4/9)*Q12b*m*Pi^2*n*d^2/(a*b)-(4/9)*Q12t*m*Pi^2*n*d^2/
(a*b)-(4/9)*Q66t*m*Pi^2*n*d^2/(a*b)-c2*((20/81)*Q66b*m*...
    Pi^2*n*d^4/(a*b)-(20/81)*Q66t*m*Pi^2*n*d^4/(a*b))+(4/9)*
Q66b*m*Pi^2*n*d^2/(a*b)-c1*((20/81)*Q12b*m*Pi^2*n*d^4/(a*...
    b)-(20/81)*Q12t*m*Pi^2*n*d^4/(a*b));
K43=-c2*((4372/15309)*Q66b*c1*m*Pi^3*n^2*d^7/(a*b^2)+(4/15309)
*Q66m*c1*m*Pi^3*n^2*d^7/(a*b^2)+(4372/15309)*Q66t*...
    c1*m*Pi^3*n^2*d^7/(a*b^2))-(2/3)*Q55m*m*Pi*d/a-(2/3)*
Q55b*m*Pi*d/a+(4/1215)*Q66m*c1*m*Pi^3*n^2*d^5/(a*b^2)+...
    (26/27)*Q55b*c1*m*Pi*d^3/a+(484/1215)*Q66b*c1*m*Pi^3*n^
2*d^5/(a*b^2)-c1*((2186/15309*(Q12b*c1*m*Pi^3*n^2/(a*...
    b^2)+Q11b*c1*m^3*Pi^3/a^3))*d^7+(2/15309*(Q12m*c1*m*Pi^
3*n^2/(a*b^2)+Q11m*c1*m^3*Pi^3/a^3))*d^7+(2186/15309*...
    (Q12t*c1*m*Pi^3*n^2/(a*b^2)+Q11t*c1*m^3*Pi^3/a^3))*d^7)
+c2*(-(242/405)*Q55b*c1*m*Pi*d^5/a+(26/81)*Q55b*m*Pi*...
    d^3/a-(2/405)*Q55m*c1*m*Pi*d^5/a+(2/81)*Q55m*m*Pi*d^3/a-
(242/405)*Q55t*c1*m*Pi*d^5/a+(26/81)*Q55t*m*Pi*d^3/a)+...
    (2/27)*Q55m*c1*m*Pi*d^3/a-(2/3)*Q55t*m*Pi*d/a+(484/1215)*Q
66t*c1*m*Pi^3*n^2*d^5/(a*b^2)+(2/1215*(Q12m*c1*m*...
    Pi^3*n^2/(a*b^2)+Q11m*c1*m^3*Pi^3/a^3))*d^5+(26/27)*Q55t*c
1*m*Pi*d^3/a+(242/1215*(Q12b*c1*m*Pi^3*n^2/(a*b^2)+...
    Q11b*c1*m^3*Pi^3/a^3))*d^5+(242/1215*(Q12t*c1*m*Pi^3*n^2/
(a*b^2)+Q11t*c1*m^3*Pi^3/a^3))*d^5;
K44=-(26/81)*Q66b*n^2*Pi^2*d^3/b^2-c2*((2186/15309)*Q66b*c1*n^
2*Pi^2*d^7/b^2-(242/1215)*Q66t*n^2*Pi^2*d^5/b^2-...
(242/1215)*Q66b*n^2*Pi^2*d^5/b^2+(2186/15309)*Q66t*c1*n^2*Pi^
2*d^7/b^2-(2/1215)*Q66m*n^2*Pi^2*d^5/b^2+(2/15309)*...
Q66m*c1*n^2*Pi^2*d^7/b^2)-(2/81)*Q11m*m^2*Pi^2*d^3/a^2+(242/12
15)*Q11b*c1*m^2*Pi^2*d^5/a^2+(2/1215)*Q66m*...
    c1*n^2*Pi^2*d^5/b^2+(26/27)*Q55b*c1*d^3+(2/1215)*Q11m*c1*m
^2*Pi^2*d^5/a^2+(242/1215)*Q66t*c1*n^2*Pi^2*d^5/b^2+...
    (242/1215)*Q66b*c1*n^2*Pi^2*d^5/b^2-(2/3)*Q55b*d-(26/81)*
Q11b*m^2*Pi^2*d^3/a^2-(2/3)*Q55m*d-(2/81)*Q66m*n^2*...
    Pi^2*d^3/b^2+(2/27)*Q55m*c1*d^3-c1*((2186/15309)*Q11b*c1*m
^2*Pi^2*d^7/a^2-(242/1215)*Q11t*m^2*Pi^2*d^5/a^2-...
(242/1215)*Q11b*m^2*Pi^2*d^5/a^2+(2186/15309)*Q11t*c1*m^2*Pi^
2*d^7/a^2-(2/1215)*Q11m*m^2*Pi^2*d^5/a^2+(2/15309)*...
    Q11m*c1*m^2*Pi^2*d^7/a^2)-(26/81)*Q66t*n^2*Pi^2*d^3/b^2+(2
42/1215)*Q11t*c1*m^2*Pi^2*d^5/a^2+(26/27)*Q55t*c1*...
    d^3-(2/3)*Q55t*d-(26/81)*Q11t*m^2*Pi^2*d^3/a^2+c2*(-
(242/405)*Q55b*c1*d^5+(26/81)*Q55b*d^3-(2/405)*Q55m*c1*...
    d^5+(2/81)*Q55m*d^3-(242/405)*Q55t*c1*d^5+(26/81)*Q55
t*d^3);
```

```
K45=-(26/81)*Q66b*m*Pi^2*n*d^3/(a*b)-(2/81)*Q12m*m*Pi^2*n*d^3/
(a*b)-(26/81)*Q66t*m*Pi^2*n*d^3/(a*b)+(242/1215)*...
    Q66t*c1*m*Pi^2*n*d^5/(a*b)-c2*((2186/15309)*Q66b*c1*m*
Pi^2*n*d^7/(a*b)-(242/1215)*Q66t*m*Pi^2*n*d^5/(a*b)-(242/
1215)*...
    Q66b*m*Pi^2*n*d^5/(a*b)+(2186/15309)*Q66t*c1*m*Pi^2*n*d^7/
(a*b)-(2/1215)*Q66m*m*Pi^2*n*d^5/(a*b)+(2/15309)*Q66m*...
    c1*m*Pi^2*n*d^7/(a*b))+(242/1215)*Q66b*c1*m*Pi^2*n*d^5/
(a*b)+(242/1215)*Q12b*c1*m*Pi^2*n*d^5/(a*b)-(26/81)*Q12b*m*...
    Pi^2*n*d^3/(a*b)+(2/1215)*Q66m*c1*m*Pi^2*n*d^5/(a*b)+(2/12
15)*Q12m*c1*m*Pi^2*n*d^5/(a*b)-c1*((2186/15309)*Q12b*...
    c1*m*Pi^2*n*d^7/(a*b)-(242/1215)*Q12t*m*Pi^2*n*d^5/(a*b)-
(242/1215)*Q12b*m*Pi^2*n*d^5/(a*b)+(2186/15309)*Q12t*c1*...
    m*Pi^2*n*d^7/(a*b)-(2/1215)*Q12m*m*Pi^2*n*d^5/(a*b)+(2/153
09)*Q12m*c1*m*Pi^2*n*d^7/(a*b))-(2/81)*Q66m*m*Pi^2*n*...
    d^3/(a*b)+(242/1215)*Q12t*c1*m*Pi^2*n*d^5/(a*b)-
(26/81)*Q12t*m*Pi^2*n*d^3/(a*b);
K51=-c2*((20/81)*Q12b*m*Pi^2*n*d^4/(a*b)-
(20/81)*Q12t*m*Pi^2*n*d^4/(a*b))-(4/9)*Q12t*m*Pi^2*n*d^2/
(a*b)-(4/9)*Q66t*...
    m*Pi^2*n*d^2/(a*b)+(4/9)*Q12b*m*Pi^2*n*d^2/
(a*b)+(4/9)*Q66b*m*Pi^2*n*d^2/(a*b)-
c1*((20/81)*Q66b*m*Pi^2*n*d^4/(a*...
    b)-(20/81)*Q66t*m*Pi^2*n*d^4/(a*b));
K52=-c2*((20/81)*Q22b*n^2*Pi^2*d^4/b^2-
(20/81)*Q22t*n^2*Pi^2*d^4/b^2)+(4/9)*Q66b*m^2*Pi^2*d^2/
a^2-c1*((20/81)*Q66b*...
    m^2*Pi^2*d^4/a^2-(20/81)*Q66t*m^2*Pi^2*d^4/
a^2)+(4/9)*Q22b*n^2*Pi^2*d^2/b^2-(4/9)*Q22t*n^2*Pi^2*d^2/
b^2-(4/9)*...
    Q66t*m^2*Pi^2*d^2/a^2;
K53=(2/1215*(Q22m*c1*n^3*Pi^3/b^3+Q12m*c1*m^2*Pi^3*n/
(a^2*b)))*d^5+c2*(-(242/405)*Q44b*c1*n*Pi*d^5/b+(26/81)*...
    Q44b*n*Pi*d^3/b-(2/405)*Q44m*c1*n*Pi*d^5/
b+(2/81)*Q44m*n*Pi*d^3/b-(242/405)*Q44t*c1*n*Pi*d^5/
b+(26/81)*Q44t*...
    n*Pi*d^3/b)+(242/1215*(Q22t*c1*n^3*Pi^3/
b^3+Q12t*c1*m^2*Pi^3*n/(a^2*b)))*d^5+(26/27)*Q44b*c1*n*P
i*d^3/b-c2*...
    ((2186/15309*(Q22b*c1*n^3*Pi^3/b^3+Q12b*c1*m^2*Pi^3*n/(a^2
*b)))*d^7+(2/15309*(Q22m*c1*n^3*Pi^3/b^3+Q12m*...
    c1*m^2*Pi^3*n/(a^2*b)))*d^7+(2186/15309*(Q22t*c1*n^3*Pi^3/
b^3+Q12t*c1*m^2*Pi^3*n/(a^2*b)))*d^7)+(242/1215*(Q22b*...
    c1*n^3*Pi^3/b^3+Q12b*c1*m^2*Pi^3*n/(a^2*b)))*d^5+(4/1215)*
Q66m*c1*m^2*Pi^3*n*d^5/(a^2*b)-c1*((4372/15309)*Q66b*...
    c1*m^2*Pi^3*n*d^7/(a^2*b)+(4/15309)*Q66m*c1*m^2*Pi^3*
n*d^7/(a^2*b)+(4372/15309)*Q66t*c1*m^2*Pi^3*n*d^7/(a^2*b))
+...
    (484/1215)*Q66t*c1*m^2*Pi^3*n*d^5/(a^2*b)+(484/1215)*Q66b*
c1*m^2*Pi^3*n*d^5/(a^2*b)-(2/3)*Q44t*n*Pi*d/b-(2/3)*Q44b*...
    n*Pi*d/b+(26/27)*Q44t*c1*n*Pi*d^3/b-(2/3)*Q44m*n*Pi*d/
b+(2/27)*Q44m*c1*n*Pi*d^3/b;
```

```
K54=-c1*((2186/15309)*Q66b*c1*m*Pi^2*n*d^7/(a*b)-(242/1215)*
Q66t*m*Pi^2*n*d^5/(a*b)-(242/1215)*Q66b*m*Pi^2*n*d^5/(a*b)+...
    (2186/15309)*Q66t*c1*m*Pi^2*n*d^7/(a*b)-
(2/1215)*Q66m*m*Pi^2*n*d^5/(a*b)+(2/15309)*Q66m*c1*m*Pi^2*
n*d^7/(a*b))+...
    (242/1215)*Q66b*c1*m*Pi^2*n*d^5/(a*b)+(242/1215)*Q12t*c1*m
*Pi^2*n*d^5/(a*b)+(242/1215)*Q12b*c1*m*Pi^2*n*d^5/(a*b)+...
    (2/1215)*Q12m*c1*m*Pi^2*n*d^5/(a*b)-
(26/81)*Q12t*m*Pi^2*n*d^3/(a*b)-(26/81)*Q12b*m*Pi^2*n*d^3/
(a*b)+(242/1215)*...
    Q66t*c1*m*Pi^2*n*d^5/(a*b)-(2/81)*Q12m*m*Pi^2*n*d^3/(a*b)-
(2/81)*Q66m*m*Pi^2*n*d^3/(a*b)-(26/81)*Q66t*m*Pi^2*...
    n*d^3/(a*b)+(2/1215)*Q66m*c1*m*Pi^2*n*d^5/(a*b)-
(26/81)*Q66b*m*Pi^2*n*d^3/(a*b)-c2*((2186/15309)*Q12b*c1*m
*Pi^2*...
    n*d^7/(a*b)-(242/1215)*Q12t*m*Pi^2*n*d^5/(a*b)-(242/1215)*
Q12b*m*Pi^2*n*d^5/(a*b)+(2186/15309)*Q12t*c1*m*Pi^2*n*...
    d^7/(a*b)-(2/1215)*Q12m*m*Pi^2*n*d^5/(a*b)+(2/15309)*Q12m*
c1*m*Pi^2*n*d^7/(a*b));
K55=c2*(-(242/405)*Q44b*c1*d^5+(26/81)*Q44b*d^3-(2/405)*Q44m*c
1*d^5+(2/81)*Q44m*d^3-(242/405)*Q44t*c1*d^5+...
    (26/81)*Q44t*d^3)-c1*((2186/15309)*Q66b*c1*m^2*Pi^2*d^7/a^2-
(242/1215)*Q66t*m^2*Pi^2*d^5/a^2-(242/1215)*Q66b*...
    m^2*Pi^2*d^5/a^2+(2186/15309)*Q66t*c1*m^2*Pi^2*d^7/a^2-
(2/1215)*Q66m*m^2*Pi^2*d^5/a^2+(2/15309)*Q66m*c1*m^2*...
    Pi^2*d^7/a^2)+(242/1215)*Q22b*c1*n^2*Pi^2*d^5/b^2+(2/1215)
*Q22m*c1*n^2*Pi^2*d^5/b^2-(2/3)*Q44t*d+(242/1215)*...
    Q66b*c1*m^2*Pi^2*d^5/a^2-(26/81)*Q22b*n^2*Pi^2*d^3/b^2-
(2/81)*Q22m*n^2*Pi^2*d^3/b^2+(2/27)*Q44m*c1*d^3+...
    (242/1215)*Q22t*c1*n^2*Pi^2*d^5/b^2-
(26/81)*Q66t*m^2*Pi^2*d^3/
a^2+(26/27)*Q44t*c1*d^3-(26/81)*Q22t*n^2*Pi^2*...
d^3/b^2-(2/3)*Q44b*d+(242/1215)*Q66t*c1*m^2*Pi^2*d^5/a^2-
(2/81)*Q66m*m^2*Pi^2*d^3/a^2-(2/3)*Q44m*d+(2/1215)*...
    Q66m*c1*m^2*Pi^2*d^5/a^2-c2*((2186/15309)*Q22b*c1*n^2*Pi^
2*d^7/b^2-(242/1215)*Q22t*n^2*Pi^2*d^5/b^2-(242/1215)*...
    Q22b*n^2*Pi^2*d^5/b^2+(2186/15309)*Q22t*c1*n^2*Pi^2*d^7/
b^2-(2/1215)*Q22m*n^2*Pi^2*d^5/b^2+(2/15309)*Q22m*...
    c1*n^2*Pi^2*d^7/b^2)+(26/27)*Q44b*c1*d^3-
(26/81)*Q66b*m^2*Pi^2*d^3/a^2;

KK=[K11 K12 K13 K14 K15;
        K21 K22 K23 K24 K25;
        K31 K32 K33 K34 K35;
        K41 K42 K43 K44 K45;
        K51 K52 K53 K54 K55];

MM=[I0 0 4*I3/h^3*(m^2*Pi^2/a^2) J1 0;
         0 I0 4*I3/h^3*(n^2*Pi^2/b^2) 0 J1;
```

```
          4/3/h^2*I3*(m*Pi/a) 4/3/h^2*I3*(n*Pi/b) I0+(16/9/
h^4*I6)*(m^2*Pi^2/a^2+n^2*Pi^2/b^2) 4/3/h^2*J4*(m*Pi/a) 4/3/
h^2*J4*(n*Pi/b);
          J1 0 4*J4/h^3*(m^2*Pi^2/a^2) K2 0;
          0 J1 4*J4/h^3*(m^2*Pi^2/a^2) 0 K2];
EQ1=-inv(MM)*KK;
[EQ2 Eq2]=eig(EQ1);
Eq2(Eq2==0)=[];
Eq2(Eq2<0)=[];
Freq(i)=min(Eq2).^.5*b*(rho/Em).^.5;
end
```

# Appendix C
## *The MATLAB Code for Chapter 7*

```
clc
clear all

Q=20;
QQ=linspace(0,200e-2,Q);
for i=1:Q;
    i
    gg=QQ(i);
n=1;
m=1;
h=1;
hf=gg;
b=10;
a=20;
Pi=pi;
Kw=24000e3;

rhom=1900;
Er=1e12;
Cr=5;
zeta=.5;
eta=.5;
Pi=pi;
%Mori-Tanaka approach-------------------------
Em=20e9;
num=0.3;
Cm=1-Cr;
% Heal cofficent------------------
Kr=30e9;
Lr=10e9;
Mr=1e9;
Nr=450e9;
Pr=1e9;

Km=Em/(3*(1-2*num));
Gm=Em/(2*(1+num));
alphar=(3*(Km+Gm)+Kr-Lr)/(3*(Gm+Kr));
betar=.2*((4*Gm+2*Kr+Lr)/(3*(Gm+Kr))+4*Gm/(Gm+Pr)+(2*(Gm*(3*Km
+Gm)+Gm*(3*Km+7*Gm)))/(Gm*(3*Km+Gm)+Mr*(3*Km+7*Gm)));
deltar=1/3*(Nr+2*Lr+(2*Kr+Lr)*(3*Km+2*Gm-Lr)/(Gm+Kr));
etar=1/5*(2/3*(Nr-Lr)+(8*Gm*Pr)/(Gm+Pr)+(8*Mr*Gm*(3*Km+4*Gm))/
(3*Km*(Mr+Gm)+Gm*(7*Mr+Gm)));
```

```
Kin=Km+(deltar-3*Km*alphar)/(3*(zeta-Cr*eta+Cr*eta*alphar));
Kout=Km+(Cr*(deltar-3*Km*alphar)*(1-eta))/
(3*(1-zeta-Cr*(1-eta)+Cr*alphar*(1-eta)));
Gin=Gm+((etar-3*Gm*betar)*Cr*eta)/
(2*(zeta-Cr*eta+Cr*eta*betar));
Gout=Gm+(Cr*(etar-3*Gm*betar)*(1-eta))/
(2*(1-zeta-Cr*(1-eta)+Cr*betar*(1-eta)));
nuout=(3*Kout-2*Gout)/(6*Kout+2*Gout);
alpha=(1+nuout)/(3*(1-nuout));
beta=(2*(4-5*nuout))/(15*(1-nuout));
K=Kout*(1+(zeta*(Kin/Kout-1))/(1+alpha*(1-zeta)*(Kin/
Kout-1)));
G=Gout*(1+(zeta*(Gin/Gout-1))/(1+beta*(1-zeta)*(Gin/Gout-1)));
Eff=(9*K*G)/(3*K+G);
nuff=(3*K-2*G)/(6*K+2*G);

C11=Eff/(1-nuff^2);
C12=(nuff*Eff)/(1-nuff^2);
C21=C12;
C13=C12;
C23=C13;
C22=Eff/(1-nuff^2);
C33=C22;
C31=C13;
C32=C13;
C66=Eff/(2*(1+nuff));
C44=Eff/(2*(1+nuff));
C55=Eff/(2*(1+nuff));

Ec=20e9;
nuc=.2;
rhoc=2400;
Q11=Ec/(1-nuc^2);
Q12=Ec*nuc/(1-nuc^2);
Q21=Ec*nuc/(1-nuc^2);
Q22=Ec/(1-nuc^2);
Q44=Ec/(2*(1+nuc));
Q55=Ec/(2*(1+nuc));
Q66=Ec/(2*(1+nuc));

A11=C11*hf+Q11*h;
A12=C12*hf+Q12*h;
A21=C12*hf+Q12*h;
A22=C22*hf+Q22*h;
A44=C44*hf+Q44*h;
A55=C55*hf+Q55*h;
A66=C66*hf+Q66*h;

A11z=0;
A21z=0;
A12z=0;
```

```
A22z=0;
A44z=0;
A11f=0;
A12f=0;
A21f=0;
A22f=0;
A44f=0;

A55g=2*hf/Pi*C44+Q44*h;
A66g=2*hf/Pi*C66+Q66*h;

B11=C11*hf^3/12+Q11*h^3/12;
B12=C12*hf^3/12+Q12*h^3/12;
B21=C12*hf^3/12+Q12*h^3/12;
B22=C22*hf^3/12+Q22*h^3/12;
B44=C44*hf^3/12+Q44*h^3/12;

A11zf=C11*(1/12)*hf^3*(-24+Pi^3)/Pi^3+Q11*(1/12)*h^3*(-
24+Pi^3)/Pi^3;
A12zf=C12*(1/12)*hf^3*(-24+Pi^3)/Pi^3+Q12*(1/12)*h^3*(-
24+Pi^3)/Pi^3;
A21zf=C12*(1/12)*hf^3*(-24+Pi^3)/Pi^3+Q12*(1/12)*h^3*(-
24+Pi^3)/Pi^3;
A22zf=C22*(1/12)*hf^3*(-24+Pi^3)/Pi^3+Q22*(1/12)*h^3*(-
24+Pi^3)/Pi^3;
A44zf=C44*(1/12)*hf^3*(-24+Pi^3)/Pi^3+Q44*(1/12)*h^3*(-
24+Pi^3)/Pi^3;

E11=C11*(1/12)*hf^3*(-48+6*Pi+Pi^3)/Pi^3+Q11*(1/12)*h^3*(-
48+6*Pi+Pi^3)/Pi^3;
E12=C12*(1/12)*hf^3*(-48+6*Pi+Pi^3)/Pi^3+Q12*(1/12)*h^3*(-
48+6*Pi+Pi^3)/Pi^3;
E21=C12*(1/12)*hf^3*(-48+6*Pi+Pi^3)/Pi^3+Q12*(1/12)*h^3*(-
48+6*Pi+Pi^3)/Pi^3;
E22=C22*(1/12)*hf^3*(-48+6*Pi+Pi^3)/Pi^3+Q22*(1/12)*h^3*(-
48+6*Pi+Pi^3)/Pi^3;

I0=rhom*hf+rhoc*h;
I1=0;
I2=0;
J1=(1/12)*hf^3*(-24+Pi^3)/Pi^3*rhom+(1/12)*h^3*(-24+Pi^3)/
Pi^3*rhoc;
J2=rhom*hf^3/12+rhoc*h^3/12;
K2=(1/12)*hf^3*(-48+6*Pi+Pi^3)/Pi^3*rhom+(1/12)*h^3*(-
48+6*Pi+Pi^3)/Pi^3*rhoc;

K11=-A11*m^2*Pi^2/a^2-A44*n^2*Pi^2/b^2;
K12=-A12*m*Pi^2*n/(a*b);
K13=A12z*m*Pi^3*n^2/(a*b^2)+A11z*m^3*Pi^3/
a^3+2*A44z*m*Pi^3*n^2/(a*b^2);
```

```
K14=2*A44f*m*Pi^3*n^2/(a*b^2)+A12f*m*Pi^3*n^2/
(a*b^2)+A11f*m^3*Pi^3/a^3;

M11=+I0;
M12=0;
M13=-I1*m*Pi/a;
M14=-J1*m*Pi/a;

K21=-A44*m*Pi^2*n/(a*b)-A21*m*Pi^2*n/(a*b);
K22=-A22*n^2*Pi^2/b^2;
K23=A22z*n^3*Pi^3/b^3+A21z*m^2*Pi^3*n/
(a^2*b)+2*A44z*m^2*Pi^3*n/(a^2*b);
K24=A22f*n^3*Pi^3/b^3+A21f*m^2*Pi^3*n/
(a^2*b)+2*A44f*m^2*Pi^3*n/(a^2*b);

M21=0;
M22=I0;
M23=-I1*n*Pi/b;
M24=-J1*n*Pi/b;

K31=A21z*m*Pi^3*n^2/(a*b^2)+4*A44z*m*Pi^3*n^2/
(a*b^2)+A11z*m^3*Pi^3/a^3;
K32=4*A44z*m^2*Pi^3*n/(a^2*b)+A12z*m^2*Pi^3*n/
(a^2*b)+A22z*n^3*Pi^3/b^3;
K33=-B12*m^2*Pi^4*n^2/(a^2*b^2)-B22*n^4*Pi^4/b^4-Kw-
4*B44*m^2*Pi^4*n^2/(a^2*b^2)-B21*m^2*Pi^4*n^2/(a^2*b^2)-
B11*m^4*Pi^4/a^4;
K34=-A22zf*n^4*Pi^4/b^4-A11zf*m^4*Pi^4/a^4-Kw-
A21zf*m^2*Pi^4*n^2/(a^2*b^2)-A12zf*m^2*Pi^4*n^2/(a^2*b^2)-
4*A44zf*m^2*Pi^4*n^2/(a^2*b^2);

M31=-I1*m*Pi/a;
M32=-I1*n*Pi/b;
M33=I2*(m^2*Pi^2/a^2+n^2*Pi^2/b^2)+I0;
M34=I0+J2*(m^2*Pi^2/a^2+n^2*Pi^2/b^2);

K41=A11f*m^3*Pi^3/a^3+4*A44f*m*Pi^3*n^2/
(a*b^2)+A21f*m*Pi^3*n^2/(a*b^2);
K42=A12f*m^2*Pi^3*n/(a^2*b)+A22f*n^3*Pi^3/
b^3+4*A44f*m^2*Pi^3*n/(a^2*b);
K43=-A21zf*m^2*Pi^4*n^2/(a^2*b^2)-A12zf*m^2*Pi^4*n^2/
(a^2*b^2)-A11zf*m^4*Pi^4/a^4-Kw-...
 A22zf*n^4*Pi^4/b^4-4*A44zf*m^2*Pi^4*n^2/(a^2*b^2);
K44=-A66g*n^2*Pi^2/b^2-E12*m^2*Pi^4*n^2/(a^2*b^2)-
E11*m^4*Pi^4/a^4-A55g*m^2*Pi^2/a^2-E21*m^2*Pi^4*n^2/(a^2*b^2)-
4*A44zf*m^2*Pi^4*n^2/(a^2*b^2)-E22*n^4*Pi^4/b^4-Kw;

M41=-J1*m*Pi/a;
M42=-J1*n*Pi/b;
M43=J2*(m^2*Pi^2/a^2+n^2*Pi^2/b^2)+K2*(m^2*Pi^2/a^2+n^2*Pi^2/
b^2)+I0;
M44=I0;
```

```
KK=[K11 K12 K13 K14;
    K21 K22 K23 K24;
    K31 K32 K33 K34;
    K41 K42 K43 K44;];

MM=[M11 M12 M13 M14;
    M21 M22 M23 M24;
    M31 M32 M33 M34;
    M41 M42 M43 M44];

[eigve eigv]=eig(-inv(MM)*KK);
eigv(eigv==0)=[];
eigv(eigv<0)=[];
eigv=sort(eigv);
Freq(i)=eigv(1).^.5*b*sqrt(rhom/Em);
end
```

# Appendix D

## *The MATLAB Code for Chapter 8*

```
clc
clear all

n=1;
m=1;
R=1;
h=1;
hp=0.1;
L=20;
Pi=pi;
KK=.75;
NM=0;
V0=-5000;
DeltaT=0;
nuCN=0.175;
rhom=7600;
rhoCN=1400;
eta=4*pi*10^-7;
Hx=0;
Kw=64000e3;
Kg=0;

rho=2400;
Ec=20e9;
nuc=.2;
Er=70e9;
Cr=0.1;
zeta=0.5;
eta=0.5;
Pi=pi;
%Mori-Tanaka approach-------------------------
Em=Ec;
num=0.3;
Cm=1-Cr;
% Heal cofficent------------------
Kr=30e9;
Lr=10e9;
Mr=1e9;
Nr=450e9;
Pr=1e9;

Km=Em/(3*(1-2*num));
Gm=Em/(2*(1+num));
alphar=(3*(Km+Gm)+Kr-Lr)/(3*(Gm+Kr));
```

```
betar=.2*((4*Gm+2*Kr+Lr)/(3*(Gm+Kr))+4*Gm/(Gm+Pr)+(2*(Gm*(3*Km
+Gm)+Gm*(3*Km+7*Gm)))/(Gm*(3*Km+Gm)+Mr*(3*Km+7*Gm)));
deltar=1/3*(Nr+2*Lr+(2*Kr+Lr)*(3*Km+2*Gm-Lr)/(Gm+Kr));
etar=1/5*(2/3*(Nr-Lr)+(8*Gm*Pr)/(Gm+Pr)+(8*Mr*Gm*(3*Km+4*Gm))/
(3*Km*(Mr+Gm)+Gm*(7*Mr+Gm)));
Kin=Km+(deltar-3*Km*alphar)/(3*(zeta-Cr*eta+Cr*eta*alphar));
Kout=Km+(Cr*(deltar-3*Km*alphar)*(1-eta))/
(3*(1-zeta-Cr*(1-eta)+Cr*alphar*(1-eta)));
Gin=Gm+((etar-3*Gm*betar)*Cr*eta)/
(2*(zeta-Cr*eta+Cr*eta*betar));
Gout=Gm+(Cr*(etar-3*Gm*betar)*(1-eta))/
(2*(1-zeta-Cr*(1-eta)+Cr*betar*(1-eta)));
nuout=(3*Kout-2*Gout)/(6*Kout+2*Gout);
alpha=(1+nuout)/(3*(1-nuout));
beta=(2*(4-5*nuout))/(15*(1-nuout));
K=Kout*(1+(zeta*(Kin/Kout-1))/(1+alpha*(1-zeta)*(Kin/
Kout-1)));
G=Gout*(1+(zeta*(Gin/Gout-1))/(1+beta*(1-zeta)*(Gin/Gout-1)));
Eff=(9*K*G)/(3*K+G);
nuff=(3*K-2*G)/(6*K+2*G);

Q11=Eff/(1-nuff^2);
Q12=(nuff*Eff)/(1-nuff^2);
Q21=Q12;
Q22=Eff/(1-nuff^2);
Q66=Eff/(2*(1+nuff));
Q44=Eff/(2*(1+nuff));
Q55=Eff/(2*(1+nuff));

%%%%Actuator%%%%
Ca11=207e9;
Ca12=117.7e9;
Ca13=106.1e9;
Ca31=Ca13;
Ca22=Ca11;
Ca23=Ca13;
Ca32=Ca13;
Ca33=209.5e9;
Ca44=44.8e9;
Ca55=44.6e9;
Ca66=(Ca11-Ca12)/2;

e31=-0.51;
e32=-0.51;
e33=1.22;
e15=-0.45;
e24=-0.45;
KK11=7.77e-8;
KK22=7.77e-8;
KK33=8.91e-8;
rhop=5610;
```

```
syms z
A11=eval(int(Q11,z,-h/2,h/2))+eval(int(Ca11,z,h/2,h/2+hp));
B11=eval(int(Q11*z,-h/2,h/2))+eval(int(Ca11*z,z,h/2,h/2+hp));
D11=eval(int(Q11*z^2,-h/2,h/2))+eval(int(Ca11*z^2,z,h/2,h/2
+hp));

A12=eval(int(Q12,z,-h/2,h/2))+eval(int(Ca12,z,h/2,h/2+hp));
B12=eval(int(Q12*z,-h/2,h/2))+eval(int(Ca12*z,z,h/2,h/2+hp));
D12=eval(int(Q12*z^2,-h/2,h/2))+eval(int(Ca12*z^2,z,h/2,h/2
+hp));

A22=eval(int(Q22,z,-h/2,h/2))+eval(int(Ca22,z,h/2,h/2+hp));
B22=eval(int(Q22*z,-h/2,h/2))+eval(int(Ca22*z,z,h/2,h/2+hp));
D22=eval(int(Q22*z^2,-h/2,h/2))+eval(int(Ca22*z^2,z,h/2,h/2
+hp));

A44=eval(int(Q44,z,-h/2,h/2))+eval(int(Ca44,z,h/2,h/2+hp));
B44=eval(int(Q44*z,z,-h/2,h/2))+eval(int(Ca44*z,z,h/2,h/2
+hp));

A55=eval(int(Q55,z,-h/2,h/2))+eval(int(Ca55,z,h/2,h/2+hp));
B55=eval(int(Q55*z,z,-h/2,h/2))+eval(int(Ca55*z,z,h/2,h/2
+hp));

A66=eval(int(Q66,z,-h/2,h/2))+eval(int(Ca66,z,h/2,h/2+hp));
B66=eval(int(Q66*z,-h/2,h/2))+eval(int(Ca66*z,z,h/2,h/2+hp));
D66=eval(int(Q66*z^2,-h/2,h/2))+eval(int(Ca66*z^2,z,h/2,h/2
+hp));

NT1=0;
NT2=0;
NT=NT1+NT2;

II0=eval(int(rho,z,-h/2,h/2))+eval(int(rhop,z,-h/2,h/2+hp));
II2=eval(int(rho*z,z,-h/2,h/2))+eval(int(rhop*z,z,-
h/2,h/2+hp));
II3=eval(int(rho*z^2,z,-h/2,h/2))+eval(int(rhop*z^2,z,-
h/2,h/2+hp));

Q=20;
QQ=linspace(200000000,-200000000,Q);
for i=1:Q;
    i
V0=QQ(i);

K11=-A11*n^2*Pi^2/L^2-A66*m^2/R^2;
K12=-A12*n*Pi*m/(R*L)-A66*n*Pi*m/(R*L);
K13=A12*n*Pi/(L*R);
K14=-B66*m^2/R^2-B11*n^2*Pi^2/L^2;
K15=-B66*n*Pi*m/(R*L)-B12*n*Pi*m/(R*L);
K16=0;
```

```
M11=+II0;
M12=0;
M13=0;
M14=+II2;
M15=0;
M16=0;

K21=-A12*n*Pi*m/(R*L)-A66*n*Pi*m/(R*L);
K22=-KK*h*A44/R^2-A66*n^2*Pi^2/L^2-A22*m^2/R^2;
K23=KK*h*A44*m/R^2+A22*m/R^2;
K24=-B66*n*Pi*m/(R*L)-B12*n*Pi*m/(R*L);
K25=-B22*m^2/R^2+KK*h*A44/R-B66*n^2*Pi^2/L^2;
K26=-2*e15*m*h/(R^2*Pi);

M21=0;
M22=+II0;
M23=0;
M24=0;
M25=+II2;
M26=0;

K31=A12*n*Pi/(L*R);
K32=KK*h*A44*m/R^2+A22*m/R^2;
K33=-KK*A55*n^2*Pi^2/L^2-KK*h*A44*m^2/R^2-Kw+Kg*(-n^2*Pi^2/
L^2-m^2/R^2)-A22/R^2+(NM+2*V0*e32+2*V0*e31+NT*DeltaT)*n^2
*Pi^2/L^2+eta*Hx^2*h*(-n^2*pi^2/L^2);
K34=-KK*A55*n*Pi/L+B12*n*Pi/(L*R);
K35=-KK*h*A44*m/R+B22*m/R^2;
K36=2*e24*n^2*Pi*h/L^2+2*e15*m^2*h/(R^2*Pi);

M31=0;
M32=0;
M33=+II0;
M34=0;
M35=0;
M36=0;

K41=-B66*m^2/R^2-B11*n^2*Pi^2/L^2;
K42=-B66*n*Pi*m/(R*L)-B12*n*Pi*m/(R*L);
K43=-KK*A55*n*Pi/L+B12*n*Pi/(L*R);
K44=-D11*n^2*Pi^2/L^2-D66*m^2/R^2-KK*A55;
K45=-D12*n*Pi*m/(R*L)-D66*n*Pi*m/(R*L);
K46=2*e31*h*n/L+2*e24*n*h/L;

M41=+II2;
M42=0;
M43=0;
M44=+II3;
M45=0;
M46=0;
```

```
K51=-B66*n*Pi*m/(R*L)-B12*n*Pi*m/(R*L);
K52=-B22*m^2/R^2+KK*h*A44/R-B66*n^2*Pi^2/L^2;
K53=-KK*h*A44*m/R+B22*m/R^2;
K54=-D12*n*Pi*m/(R*L)-D66*n*Pi*m/(R*L);
K55=-D22*m^2/R^2-D66*n^2*Pi^2/L^2-KK*h*A44;
K56=2*e32*h*m/(R*Pi)+2*e15*m*h/(R*Pi);

M51=0;
M52=+II2;
M53=0;
M54=0;
M55=+II3;
M56=0;

K61=0;
K62=2*h*m*e24/(Pi*R^2);
K63=-(1/2)*(4*h^2*m^2*e24*L^2+4*Pi^2*h^2*n^2*e15*R^2)/
(L^2*Pi*R^2*h);
K64=-2*h*e15*n/L;
K65=-2*h*e24*m/(R*Pi);
K66=-(1/2)*(Pi^3*KK33*L^2*R^2+Pi^3*h^2*n^2*KK11*R^2+h^2*m^2*KK
22*L^2*Pi*R)/(L^2*Pi*R^2*h);

M61=0;
M62=0;
M63=0;
M64=0;
M65=0;
M66=0;

K=[K11 K12 K13 K14 K15 K16;
       K21 K22 K23 K24 K25 K26;
       K31 K32 K33 K34 K35 K36;
       K41 K42 K43 K44 K45 K46;
       K51 K52 K53 K54 K55 K56;
       K61 K62 K63 K64 K65 K66];

M=[M11 M12 M13 M14 M15 M16;
       M21 M22 M23 M24 M25 M26;
       M31 M32 M33 M34 M35 M36;
       M41 M42 M43 M44 M45 M46;
       M51 M52 M53 M54 M55 M56;
       M61 M62 M63 M64 M65 M66];

Nmm=5;
KLmm=K(1:Nmm,1:Nmm);
KLme=K(1:Nmm,Nmm+1:end);
KLem=K(Nmm+1:end,1:Nmm);
KLee=K(Nmm+1:end,Nmm+1:end);
MLmm=M(1:Nmm,1:Nmm);
MLme=M(1:Nmm,Nmm+1:end);
```

```
KLnew=KLmm-KLme*inv(KLee)*KLem;
MLnew=MLmm-MLme*inv(KLee)*KLem;

Eq1=-inv((MLnew))*(KLnew);
[eigve eigva]=eig(Eq1);
eigva(eigva==0)=[];
eigva(eigva<0)=[];
eigvaa=sort(eigva.^.5);
freq(i)=eigvaa(1)*L*(rho/Em).^.5;
end
```

# Appendix E
## *The MATLAB Code for Chapter 11*

```
clc
clear all

syms r FF1 FF2 FF4 FF5
gamma=.0001;
betaa=2;
FF3=0;
FF6=0;

Cr=0.1;
ksi=0.1;
zeta=0.1;
%Mori-Tanaka approach------------------------
Em=17e9;
num=0.3;

Cm=1-Cr;
% Heal cofficent-----------------
Kr=30e9;
Lr=10e9;
Mr=1e9;
Nr=211e9;
Pr=1e9;


Km=Em/(2*(1-num));
Gm=Em/(2*(1+num));
alphar=(3*(Km+Gm)+Kr-Lr)/(3*(Gm+Kr));
betar=.2*((4*Gm+2*Kr+Lr)/(3*(Gm+Kr))+4*Gm/(Gm+Pr)+(2*(Gm*(3*Km
+Gm)+Gm*(3*Km+7*Gm)))/(Gm*(3*Km+Gm)+Mr*(3*Km+7*Gm)));
deltar=1/3*(Nr+2*Lr+(2*Kr+Lr)*(3*Km+2*Gm-Lr)/(Gm+Kr));
etar=1/5*(2/3*(Nr-Lr)+(8*Gm*Pr)/(Gm+Pr)+(8*Mr*Gm*(3*Km+4*Gm))/
(3*Km*(Mr+Gm)+Gm*(7*Mr+Gm)));

Kin=Km+(((deltar-3*Km*alphar)*Cr*ksi)/
(3*(zeta-Cr*ksi+Cr*ksi*alphar)));
Kout=Km+(((deltar-3*Km*alphar)*Cr*(1-ksi))/
(3*(1-zeta-Cr*(1-ksi)+Cr*(1-ksi)*alphar)));
Gin=Gm+(((etar-3*Gm*betar)*Cr*ksi)/
(3*(zeta-Cr*ksi+Cr*ksi*betar)));
Gout=Gm+(((etar-3*Gm*betar)*Cr*(1-ksi))/
(2*(1-zeta-Cr*(1-ksi)+Cr*(1-ksi)*betar)));
nuout=(3*Kout-2*Gout)/(6*Kout+2*Gout);
alpha=(1+nuout)/(3*(1-nuout));
```

```
beta=2*(4-5*nuout)/(15*(1-nuout));
K=Kout*(1+(zeta*(Kin/Kout-1))/(1+alpha*(1-zeta)*(Kin/
Kout-1)));
G=Gout*(1+(zeta*(Gin/Gout-1))/(1+beta*(1-zeta)*(Gin/Gout-1)));

Em=9*K*G/(3*K+G);
num=(3*K-2*G)/(6*K+2*G);

Kc=Em*(Em*Cm+2*Kr*(1+num)*(1+Cr*(1-2*num)))/...
     (2*(1+num)*(Em*(1+Cr-2*num)+2*Cm*Kr*(1-num-2*num^2)));
Lc=Em*(num*Cm*(Em+2*Kr*(1+num))+2*Cr*Lr*(1-num)^2)/...
     ((1+num)*(Em*(1+Cr-2*num)+2*Cm*Kr*(1-num-2*num^2)));
Nc=(Em^2*Cm*(1+Cr-Cm*num)+2*Cm*Cr*(Kr*Nr-Lr^2)*(1+num)^2*...
(1-2*num))/((1+num)*(Em*(1+Cr-2*num)+2*Cm*Kr*...
     (1-num-2*num^2)))+(Em*(2*Cm^2*Kr*(1-num)+Cr*Nr...
     *(1+Cr-2*num)-4*Cm*Lr*num))/((Em*(1+Cr-2*num)+2*Cm*Kr*...
     (1-num-2*num^2)));
Mc=Em*(Em*Cm+2*Pr*(1+num)*(3+Cr-4*num))/
(2*(1+num)*(Em*(Cm+4*...
     Cr*(1-num)))+2*Cm*Mr*(3-num-4*num^2));
Pc=Em*(Em*Cm+2*Pr*(1+num)*(1+Cr))/(2*(1+num)*(Em*(1+Cr))...
     +2*Cm*Pr*(1+num));

C11=Kc+Mc;
C12=Lc;
C13=Kc-Mc;
C23=Lc;
C22=Nc;
C33=Kc+Mc;
C44=Pc;
C55=Mc;
C66=Pc;

e31=0;
e32=0;
e33=0;
epsilon11=0;
epsilon22=0;
epsilon33=0;
alphar=1e-6;
alphat=1e-6;

C1=C11/C22;
C2=C12/C22;
E1=0;
E2=0;

Q1=20;
Q2=1;
j=1;
```

```
    ecr=0;
    ect=0;
    decr=0;
    dect=0;

K1=1;
K2=(1+gamma);
K3=(gamma*(C2+E1*E2)-1-E2^2)/(C1+E1^2);

A1=(((gamma+1)*(C1+E1^2)-(C2+E1*E2))*alphar+((gamma+1)*(C2+E
1*E2)-(1+E2^2))*alphat)/(C1+E1^2);
A2=(((-2*gamma-1)*(C1+E1^2)+(C2+E1*E2))*alphar+((-2*gamma-
1)*(C2+E1*E2)+(1+E2^2))*alphat)/(C1+E1^2);
A3=E2/(C1+E1^2);
A4=((-gamma+1)*(C1+E1^2)+(C2+E1*E2))/(C1+E1^2);
A5=((-gamma-1)*(C2+E1*E2)+(1+E1^2))/(C1+E1^2);
A6=(-C2-E1*E2)/(C1+E1^2);

B1=(-A1)/(K2+K3);
B11=(-A4)/(K2+K3);
B12=(-A5)/(K2+K3);
B2=-A2/(gamma*(1+gamma)+(1+gamma)*(K2)+K3);
B3=-A3/(-gamma*(-1-gamma)-(gamma)*(K2)+K3);
B4=1/(2+2*K2+K3);
B41=A6/(2+2*K2+K3);


N1=((1-K2)+((K2-2)^2-4*K3)^.5)/2;
N2=((1-K2)-((K2-2)^2-4*K3)^.5)/2;


U=FF4*r^(N1)+FF5*r^(N2)+B1*FF1*r+B2*r^(1+gamma)+B3*FF
3*r^(-gamma)+B4*r^3+B11*ecr*r+...
B12*ect*r+B4*decr*r^2+B41*dect*r^2;

Sr=(C1+E1^2)*r^gamma*(diff(U,r)-ecr-alphar*r^gamma*(-FF1*r^(-
gamma)/gamma+FF2))+...
 (C2+E1*E2)*(U/r-ect-alphat*r^gamma*(-FF1*r^(-gamma)/
gamma+FF2))-FF3*E1*r^(-1);

St=(C2+E1*E2)*r^gamma*(diff(U,r)-ecr-alphar*r^gamma*(-FF1*r^(-
gamma)/gamma+FF2))+...
 (1+E2^2)*(U/r-ect-alphat*r^gamma*(-FF1*r^(-gamma)/
gamma+FF2))-FF3*E1*r^(-1);

Temp=-FF1*r^(-gamma)/gamma+FF2;

srin=-1;
srout=0;
Tin=7355525+273;
```

```
Tout=7355550+273;
%BC1: free-free
Eq1=subs(Sr,r,1)-srin;
Eq2=subs(Sr,r,betaa)-srout;
Eq5=subs(Temp,r,1)-Tin;
Eq6=subs(Temp,r,betaa)-Tout;
[FFF1 FFF2 FFF4 FFF5]=solve(Eq1,Eq2,Eq5,Eq6);
F1=eval(FFF1);
F2=eval(FFF2);
F3=0;
F4=eval(FFF4);
F5=eval(FFF5);
F6=0;
rr=linspace(1,betaa,Q1);
for i=1:Q1;
i
beta=rr(i);

diffu=F4*N1*beta^(N1-1)+F5*N2*beta^(N2-1)+(B1*F1+B2*(1+gamma)*
beta^(gamma)+B3*F3*(-gamma)*beta^(-gamma-
1)+3*B4*beta^2+B11*ecr+...
    B12*ect+2*B4*decr*beta+2*B4*dect*beta);

iint=B1*F1*beta+B2*beta^(1+gamma)/(1+gamma)+B3*F3*beta^(-
gamma)/(-gamma)+B4*beta^3/3+B1*ecr*beta+...
    B1*ect*beta+B4*decr*beta^2/2+B4*dect*beta^2/2;

KK4=((B1*F1*beta+B2*beta^(1+gamma)/(1+gamma)+B3*F3*beta^(-
gamma)/(-gamma)+B4*beta^3/3+B11*ecr*beta+...
B12*ect*beta+B4*decr*beta^2/2+B41*dect*beta^2/2));

displacement(i,j)=(F4*beta^(N1)+F5*beta^(N2)+B1*F1*beta+B2*bet
a^(1+gamma)+B3*F3*beta^(-gamma)+B4*beta^3+B11*ecr*beta+...
    B12*ect*beta+B4*decr*beta^2+B41*dect*beta^2);

sigmar(i,j)=(C1+E1^2)*beta^gamma*(diffu-ecr-
alphar*beta^gamma*(-F1*beta^(-gamma)/gamma+F2))+...
    (C2+E1*E2)*(displacement(i,j)/beta-ect-
alphat*beta^gamma*(-F1*beta^(-gamma)/
gamma+F2))-F3*E1*beta^(-1);

sigmat(i,j)=-((C2+E1*E2)*beta^gamma*(diffu-ecr-
alphar*beta^gamma*(-F1*beta^(-gamma)/gamma+F2))+...
    (1+E2^2)*(displacement(i,j)/beta-ect-alphat*beta^gamma*(-
F1*beta^(-gamma)/gamma+F2))-F3*E1*beta^(-1));

Tempe(i,j)=-F1*beta^(-gamma)/gamma+F2;
sigmae(i,j)=sqrt(3)/2*(sigmat(i,j)-sigmar(i,j));
end
```

# Appendix F
## *The MATLAB Code for Chapter 12*

```
clc
clear all
N=9;
T=300;
T0=300;
deltaT=T-T0;
DeltaT=deltaT;
R=500e-3;
L=4;
h=12.3e-3;
alphax=1e-6;
alphat=1e-6;
vm=0.34;
v12cnt=0.175;
pm=1000;
pcnt=2000;
T=300;
Vs=0.05;
type='U';
c1=0/(3*h^2);
c2=0/(h^2);
rhof=1000;
hf=h;
mu=1e-4;
Vx=12;
Cd=10;
V0=0;

Hs=1;
nus=0.48;
Es=(3.22-0.0034*T)*10^9;
nu0=nus/(1-nus);
E0=Es/(1-nus^2);
gamma1=Hs/L;
cc1=(gamma1+2)*exp(-gamma1);
kw=0*(E0/
(4*L*(1-nu0^2)*(2-cc1)^2))*(5-(2*gamma1^2+6*gamma1+5)*exp(-
2*gamma1));
G1=.5*kw/1e4;
G2=.5*kw/1e4;
theta=pi/4;

Em=(3.52-0.0034*T)*(1e9);
E11c=((-2.093e-6)*T^3+0.003882*T^2-2.659*T+6151)*(1e9);
```

```
E22c=((-2.626e-6)*T^3+0.00487*T^2-3.336*T+7713)*(1e9);
G12c=(2100*exp((-6.803e-5)*T)-256.7*exp(-0.002739*T))*(1e9);
E11=(794.626312538604 1*Vs+1.815750463248898)*(1e9);
E22=(1.033*atan(60.21*Vs-9.068)+4.009)*(1e9);
eta1=(E11-Em*(1-Vs))/E11c/Vs;
eta2=E22*(Vs/E22c+(1-Vs)/Em);
eta3=0.7*eta2;
% type: U V A O or X
syms z
if strcmp(type,'U')==1
    Vcnt=Vs;
elseif strcmp(type,'V')==1
    Vcnt=(1-2*z/h)*Vs;
elseif strcmp(type,'A')==1
    Vcnt=(1+2*z/h)*Vs;
elseif strcmp(type,'O')==1
Vcnt=2*(1-2*abs(z)/h)*Vs;
elseif strcmp(type,'X')==1
    Vcnt=2*abs(2*z)/h*Vs;
end
Gm=Em/2/(1+vm);
E11=eta1*Vcnt*E11c+(1-Vcnt)*Em;
E22=eta2*Em*E22c/(Vcnt*(Em-E22c)+E22c);
G12=eta3*Gm*G12c/(Vcnt*(Gm-G12c)+G12c);
G13=G12;
G23=1.2*G12;
v12=Vcnt*v12cnt+(1-Vcnt)*vm;
v21=E22/E11*v12;
p=Vcnt*pcnt+(1-Vcnt)*pm;
%=======
Q11=E11/(1-v12*v21);
Q22=E22/(1-v12*v21);
Q12=v21*Q11;
Q44=G23;
Q55=G13;
Q66=G12;
%=======
zz=linspace(-h/2,h/2,50);
if strcmp(type,'U')==1
    A11=Q11*h;
    A12=Q12*h;
    A22=Q22*h;
    A44=Q44*h;
    A55=Q55*h;
    A66=Q66*h;
    Nxm=(Q11*alphax+Q12*alphat)*h+2*V0;
    Nym=(Q12*alphax+Q22*alphat)*h+2*V0;
else
    A11=double(trapz(zz,subs(Q11,z,zz)));
    A12=double(trapz(zz,subs(Q12,z,zz)));
    A22=double(trapz(zz,subs(Q22,z,zz)));
```

```
    A44=double(trapz(zz,subs(Q44,z,zz)));
    A55=double(trapz(zz,subs(Q55,z,zz)));
    A66=double(trapz(zz,subs(Q66,z,zz)));

Nxm=double(trapz(zz,subs((Q11*alphax+Q12*alphat),z,zz)));
Nym=double(trapz(zz,subs((Q12*alphax+Q22*alphat),z,zz)));
end
B11=double(trapz(zz,zz.*subs(Q11,z,zz)));
B12=double(trapz(zz,zz.*subs(Q12,z,zz)));
B22=double(trapz(zz,zz.*subs(Q22,z,zz)));
B44=double(trapz(zz,zz.*subs(Q44,z,zz)));
B55=double(trapz(zz,zz.*subs(Q55,z,zz)));
B66=double(trapz(zz,zz.*subs(Q66,z,zz)));
D11=double(trapz(zz,(zz.^2).*subs(Q11,z,zz)));
D12=double(trapz(zz,(zz.^2).*subs(Q12,z,zz)));
D22=double(trapz(zz,(zz.^2).*subs(Q22,z,zz)));
D44=double(trapz(zz,(zz.^2).*subs(Q44,z,zz)));
D55=double(trapz(zz,(zz.^2).*subs(Q55,z,zz)));
D66=double(trapz(zz,(zz.^2).*subs(Q66,z,zz)));
E11=double(trapz(zz,(zz.^3).*subs(Q11,z,zz)));
E12=double(trapz(zz,(zz.^3).*subs(Q12,z,zz)));
E22=double(trapz(zz,(zz.^3).*subs(Q22,z,zz)));
E44=double(trapz(zz,(zz.^3).*subs(Q44,z,zz)));
E55=double(trapz(zz,(zz.^3).*subs(Q55,z,zz)));
E66=double(trapz(zz,(zz.^3).*subs(Q66,z,zz)));
F11=double(trapz(zz,(zz.^4).*subs(Q11,z,zz)));
F12=double(trapz(zz,(zz.^4).*subs(Q12,z,zz)));
F22=double(trapz(zz,(zz.^4).*subs(Q22,z,zz)));
F44=double(trapz(zz,(zz.^4).*subs(Q44,z,zz)));
F55=double(trapz(zz,(zz.^4).*subs(Q55,z,zz)));
F66=double(trapz(zz,(zz.^4).*subs(Q66,z,zz)));
H11=double(trapz(zz,(zz.^6).*subs(Q11,z,zz)));
H12=double(trapz(zz,(zz.^6).*subs(Q12,z,zz)));
H22=double(trapz(zz,(zz.^6).*subs(Q22,z,zz)));
H44=double(trapz(zz,(zz.^6).*subs(Q44,z,zz)));
H55=double(trapz(zz,(zz.^6).*subs(Q55,z,zz)));
H66=double(trapz(zz,(zz.^6).*subs(Q66,z,zz)));
%========
if strcmp(type,'U')==1
    II0=p*h;
else
    II0=double(trapz(zz,subs(p,z,zz)));
end
II1=double(trapz(zz,zz.*subs(p,z,zz)));
II2=double(trapz(zz,(zz.^2).*subs(p,z,zz)));
II3=double(trapz(zz,(zz.^3).*subs(p,z,zz)));
II4=double(trapz(zz,(zz.^4).*subs(p,z,zz)));
II5=double(trapz(zz,(zz.^5).*subs(p,z,zz)));
II6=double(trapz(zz,(zz.^6).*subs(p,z,zz)));
J1=II1-4/3/(h^2)*II3;
J4=II4-4/3/(h^2)*II6;
```

```
K2=II2-8/3/(h^2)*II4+((4/3/(h^2))^2)*II6;

for i=1:N
    for j=1:N
    x(i)=(1/2)*(1-cos(pi*((i-1)/(N-1))));
    y(j)=(2*pi)*(1-cos(pi*((j-1)/(N-1))));
    end
end
for i=1:N
    M(i)=1;
    P(i)=1;
    for j=1:N
        if i~=j
            M(i)=M(i)*(x(i)-x(j));
            P(i)=P(i)*(y(i)-y(j));
        end
    end
end
for i=1:N
    C1(i,i)=0;
    A1(i,i)=0;
    for j=1:N
        if i~=j
        C1(i,j)=M(i)/(M(j)*(x(i)-x(j)));
        C1(i,i)=C1(i,i)-C1(i,j);
        A1(i,j)=P(i)/(P(j)*(y(i)-y(j)));
        A1(i,i)=A1(i,i)-A1(i,j);
        end
    end
end
for i=1:N
    W1(i,i)=C1(i,i)+1/(x(i)-0.01);
    for j=1:N
        if i~=j
        W1(i,j)=((x(i)-0.01)/(x(j)-0.01))*C1(i,j);
        end
    end
end
W2=inv(W1);
AA=C1*W2;
BB=A1*W2;

II=eye(N^2);
ZZ=zeros(N^2);
A=kron(AA,eye(N));
B=kron(eye(N),BB);

K11=A11*A^2+B^2*A66/R^2;
K12=A*A12*B/R+B*A66*A/R;
K13=A*(-E12*c1*B^2/R^2-c1*E11*A^2+A12/R)-2*B^2*E66*c1*A/R^2;
```

```
K14=A*(-c1*E11*A+B11*A)+B*(B66*B/R-E66*c1*B/R)/R;
K15=A*(B12*B/R-E12*c1*B/R)+B*(B66*A-E66*c1*A)/R;

K21=A*A12*B/R+B*A66*A/R;
K22=A66*A^2+B^2*A22/R^2;
K23=-2*A^2*E66*c1*B/R+B*(-E22*c1*B^2/R^2-c1*E12*A^2+A22/
R)/R;
K24=A*(B66*B/R-E66*c1*B/R)+B*(-c1*E12*A+B12*A)/R;
K25=A*(B66*A-E66*c1*A)+B*(B22*B/R-E22*c1*B/R)/R;

K31=-A12*A/R+c1*(A^3*E11+2*A*B^2*E66/R^2+B^2*E12*A/R^2);
K32=-A22*B/R^2+c1*(A^2*E12*B/R+2*A^2*B*E66/R+B^3*E22/R^3);
K33=A*(A55*A-3*D55*c1*A-c2*(D55*A-
3*F55*c1*A))+Nxm*A^2+Nym*B^2/R^2+B*(A44*B/R-3*D44*c1*B/R-
c2*(D44*B/R-3*F44*c1*B/R))/R-(-E22*c1*B^2/
R^2-c1*E12*A^2+A22/R)/R+G2*(sin(theta)^2*A^2+2*cos(theta)*sin(
theta)*A*B/R+cos(theta)^2*B^2/R^2)+c1*(A^2*(-(H12*c1*B^2)/
(R^2)-c1*H11*A^2+(E12)/(R))-(4* A^2 *B^2* H66* c1)/(R^2)+(B^2
*(-(H22* c1* B^2)/(R^2)-H12* c1 *A^2+(E22)/(R)))/(R^2))-
kw*II+mu*hf*(Vx^2*(A^3+(A*B^2)/(R^2))) *G1*(cos(theta)^2*A^2+2
*cos(theta)*sin(theta)*A*B/R+sin(theta)^2*B^2/
R^2)-rhof*hf*(Vx^2*A^2);
K34=-(-c1*E12*A+B12*A)/R+A*(A55-3*D55*c1-c2*(D55-
3*F55*c1))+c1*(A^2*(-c1*H11*A+F11*A)+B^2*(-H12*c1*A+F12*A)/
R^2+2*A*B*(F66*B/R-H66*c1*B/R)/R);
K35=c1*(A^2*(F12*B/R-H12*c1*B/R)+B^2*(F22*B/R-H22*c1*B/R)/
R^2+2*A*B*(F66*A-H66*c1*A)/R)+B*(A44-3*D44*c1-c2*(D44-
3*F44*c1))/R-(B22*B/R-E22*c1*B/R)/R;

K41=B*(B66*B/R-c2*E66*B/R)/R+A*(-c1*E11*A+B11*A);
K42=B*(B66*A-c2*E66*A)/R+A*(B12*B/R-E12*c1*B/R);
K43=-A55*A+A*(B12/R-c1*F11*A^2-F12*c1*B^2/R^2-c1*(-H12*c1*B^2/
R^2-c1*H11*A^2+E12/R))+3*D55*c1*A+B*(-2*F66*c1*A*B/
R+2*c2*H66*c1*A*B/R)/R+c2*(D55*A-3*F55*c1*A);
K44=3*D55*c1+c2*(D55-3*F55*c1)-A55+A*(D11*A-c1*(-
c1*H11*A+F11*A)-c1*F11*A)+B*(-c2*(F66*B/R-H66*c1*B/R)+D66*B/R-
F66*c1*B/R)/R;
K45=A*(D12*B/R-F12*c1*B/R-c1*(F12*B/R-H12*c1*B/R))+B*(-
c2*(F66*A-H66*c1*A)+D66*A-F66*c1*A)/R;

K51=B*(B12*A-c2*E12*A)/R+A*(B66*B/R-E66*c1*B/R);
K52=B*(B22*B/R-c2*E22*B/R)/R+A*(B66*A-E66*c1*A);
K53=-A44*B/R+A*(-2*F66*c1*A*B/R+2*c1^2*H66*A*B/R)+3*D44*c1*B/
R+B*(B22/R-F12*c1*A^2-F22*c1*B^2/R^2-c2*(-H22*c1*B^2/R^2-
H12*c1*A^2+E22/R))/R+c2*(D44*B/R-3*F44*c1*B/R);
K54=A*(-c1*(F66*B/R-H66*c1*B/R)+D66*B/R-F66*c1*B/R)+B*(D12*A-
c2*(-H12*c1*A+F12*A)-F12*c1*A)/R;
K55=A*(-c1*(F66*A-H66*c1*A)+D66*A-F66*c1*A)+B*(D22*B/R-
F22*c1*B/R-c2*(F22*B/R-H22*c1*B/R))/
R+3*D44*c1+c2*(D44-3*F44*c1)-A44;
```

```
M11=-II0*II;
M12=ZZ;
M13=c1*II3*A;
M14=-J1*II;
M15=ZZ;

M21=ZZ;
M22=-II0*II;
M23=c1*II3*B/R;
M24=ZZ;
M25=-J1*II;

M31=-c1*II3*II;
M32=-c1*II3*B/R*II;
M33=-(II0-c1^2*II6*(A^2+B^2/R^2))*II;
M34=-c1*J4*A;
M35=-c1*J4*B/R*II;

M41=-J1*II;
M42=ZZ;
M43=c1*J4*A;
M44=K2*II;
M45=ZZ;

M51=ZZ;
M52=-J1*II;
M53=c1*J4*B/R;
M54=ZZ;
M55=-K2*II;

C11=ZZ;
C12=ZZ;
C13=ZZ;
C14=ZZ;
C15=ZZ;

C21=ZZ;
C22=ZZ;
C23=ZZ;
C24=ZZ;
C25=ZZ;

C31=ZZ;
C32=ZZ;
C33=mu*hf*(A^2+B^2/R^2)-rhof*hf*2*Vx*A-Cd;
C34=ZZ;
C35=ZZ;

C41=ZZ;
C42=ZZ;
C43=ZZ;
```

```
C44=ZZ;
C45=ZZ;

C51=ZZ;
C52=ZZ;
C53=ZZ;
C54=ZZ;
C55=ZZ;

%Boundary Conditions:
C_1_2=A(N+1:2*N,:);
C_1_N1=A(N^2-2*N+1:N^2-N,:);

Klw_2b=[C_1_2(:,1:N) C_1_2(:,N+1:2*N) C_1_2(:,N^2-2*N+1:N^2-N)
C_1_2(:,N^2-N+1:N^2)];
Klw_2d=C_1_2(:,2*N+1:N^2-2*N);
Klw_N1b=[C_1_N1(:,1:N) C_1_N1(:,N+1:2*N) C_1_N1(:,N^2-
2*N+1:N^2-N) C_1_N1(:,N^2-N+1:N^2)];
Klw_N1d=C_1_N1(:,2*N+1:N^2-2*N);

%-------------------------------------
bmKl1u=[K11(N+1:N^2-N,1:N),K11(N+1:N^2-N,N^2-N+1:N^2)];
bmKl1v=[K12(N+1:N^2-N,1:N),K12(N+1:N^2-N,N^2-N+1:N^2)];
bmKl1w=[K13(N+1:N^2-N,1:2*N),K13(N+1:N^2-N,N^2-2*N+1:N^2)];
bmKl1sx=[K14(N+1:N^2-N,1:N),K14(N+1:N^2-N,N^2-N+1:N^2)];
bmKl1st=[K15(N+1:N^2-N,1:N),K15(N+1:N^2-N,N^2-N+1:N^2)];

bmKl2u=[K21(N+1:N^2-N,1:N),K21(N+1:N^2-N,N^2-N+1:N^2)];
bmKl2v=[K22(N+1:N^2-N,1:N),K22(N+1:N^2-N,N^2-N+1:N^2)];
bmKl2w=[K23(N+1:N^2-N,1:2*N),K23(N+1:N^2-N,N^2-2*N+1:N^2)];
bmKl2sx=[K24(N+1:N^2-N,1:N),K24(N+1:N^2-N,N^2-N+1:N^2)];
bmKl2st=[K25(N+1:N^2-N,1:N),K25(N+1:N^2-N,N^2-N+1:N^2)];

bmKl3u=[K31(2*N+1:N^2-2*N,1:N),K31(2*N+1:N^2-2*N,N^2-
N+1:N^2)];
bmKl3v=[K32(2*N+1:N^2-2*N,1:N),K32(2*N+1:N^2-2*N,N^2-
N+1:N^2)];
bmKl3w=[K33(2*N+1:N^2-2*N,1:2*N),K33(2*N+1:N^2-2*N,N^2-
2*N+1:N^2)];
bmKl3sx=[K34(2*N+1:N^2-2*N,1:N),K34(2*N+1:N^2-2*N,N^2-
N+1:N^2)];
bmKl3st=[K35(2*N+1:N^2-2*N,1:N),K35(2*N+1:N^2-2*N,N^2-
N+1:N^2)];

bmKl4u=[K41(N+1:N^2-N,1:N),K41(N+1:N^2-N,N^2-N+1:N^2)];
bmKl4v=[K42(N+1:N^2-N,1:N),K42(N+1:N^2-N,N^2-N+1:N^2)];
bmKl4w=[K43(N+1:N^2-N,1:2*N),K43(N+1:N^2-N,N^2-2*N+1:N^2)];
bmKl4sx=[K44(N+1:N^2-N,1:N),K44(N+1:N^2-N,N^2-N+1:N^2)];
bmKl4st=[K45(N+1:N^2-N,1:N),K45(N+1:N^2-N,N^2-N+1:N^2)];
```

```
bmKl5u=[K51(N+1:N^2-N,1:N),K51(N+1:N^2-N,N^2-N+1:N^2)];
bmKl5v=[K52(N+1:N^2-N,1:N),K52(N+1:N^2-N,N^2-N+1:N^2)];
bmKl5w=[K53(N+1:N^2-N,1:2*N),K53(N+1:N^2-N,N^2-2*N+1:N^2)];
bmKl5sx=[K54(N+1:N^2-N,1:N),K54(N+1:N^2-N,N^2-N+1:N^2)];
bmKl5st=[K55(N+1:N^2-N,1:N),K55(N+1:N^2-N,N^2-N+1:N^2)];

bmMl1u=[M11(N+1:N^2-N,1:N),M11(N+1:N^2-N,N^2-N+1:N^2)];
bmMl1v=[M12(N+1:N^2-N,1:N),M12(N+1:N^2-N,N^2-N+1:N^2)];
bmMl1w=[M13(N+1:N^2-N,1:2*N),M13(N+1:N^2-N,N^2-2*N+1:N^2)];
bmMl1sx=[M14(N+1:N^2-N,1:N),M14(N+1:N^2-N,N^2-N+1:N^2)];
bmMl1st=[M15(N+1:N^2-N,1:N),M15(N+1:N^2-N,N^2-N+1:N^2)];

bmMl2u=[M21(N+1:N^2-N,1:N),M21(N+1:N^2-N,N^2-N+1:N^2)];
bmMl2v=[M22(N+1:N^2-N,1:N),M22(N+1:N^2-N,N^2-N+1:N^2)];
bmMl2w=[M23(N+1:N^2-N,1:2*N),M23(N+1:N^2-N,N^2-2*N+1:N^2)];
bmMl2sx=[M24(N+1:N^2-N,1:N),M24(N+1:N^2-N,N^2-N+1:N^2)];
bmMl2st=[M25(N+1:N^2-N,1:N),M25(N+1:N^2-N,N^2-N+1:N^2)];

bmMl3u=[M31(2*N+1:N^2-2*N,1:N),M31(2*N+1:N^2-2*N,N^2-
N+1:N^2)];
bmMl3v=[M32(2*N+1:N^2-2*N,1:N),M32(2*N+1:N^2-2*N,N^2-
N+1:N^2)];
bmMl3w=[M33(2*N+1:N^2-2*N,1:2*N),M33(2*N+1:N^2-2*N,N^2-
2*N+1:N^2)];
bmMl3sx=[M34(2*N+1:N^2-2*N,1:N),M34(2*N+1:N^2-2*N,N^2-
N+1:N^2)];
bmMl3st=[M35(2*N+1:N^2-2*N,1:N),M35(2*N+1:N^2-2*N,N^2-
N+1:N^2)];

bmMl4u=[M41(N+1:N^2-N,1:N),M41(N+1:N^2-N,N^2-N+1:N^2)];
bmMl4v=[M42(N+1:N^2-N,1:N),M42(N+1:N^2-N,N^2-N+1:N^2)];
bmMl4w=[M43(N+1:N^2-N,1:2*N),M43(N+1:N^2-N,N^2-2*N+1:N^2)];
bmMl4sx=[M44(N+1:N^2-N,1:N),M44(N+1:N^2-N,N^2-N+1:N^2)];
bmMl4st=[M45(N+1:N^2-N,1:N),M45(N+1:N^2-N,N^2-N+1:N^2)];

bmMl5u=[M51(N+1:N^2-N,1:N),M51(N+1:N^2-N,N^2-N+1:N^2)];
bmMl5v=[M52(N+1:N^2-N,1:N),M52(N+1:N^2-N,N^2-N+1:N^2)];
bmMl5w=[M53(N+1:N^2-N,1:2*N),M53(N+1:N^2-N,N^2-2*N+1:N^2)];
bmMl5sx=[M54(N+1:N^2-N,1:N),M54(N+1:N^2-N,N^2-N+1:N^2)];
bmMl5st=[M55(N+1:N^2-N,1:N),M55(N+1:N^2-N,N^2-N+1:N^2)];

bmCl1u=[C11(N+1:N^2-N,1:N),C11(N+1:N^2-N,N^2-N+1:N^2)];
bmCl1v=[C12(N+1:N^2-N,1:N),C12(N+1:N^2-N,N^2-N+1:N^2)];
bmCl1w=[C13(N+1:N^2-N,1:2*N),C13(N+1:N^2-N,N^2-2*N+1:N^2)];
bmCl1sx=[C14(N+1:N^2-N,1:N),C14(N+1:N^2-N,N^2-N+1:N^2)];
bmCl1st=[C15(N+1:N^2-N,1:N),C15(N+1:N^2-N,N^2-N+1:N^2)];

bmCl2u=[C21(N+1:N^2-N,1:N),C21(N+1:N^2-N,N^2-N+1:N^2)];
bmCl2v=[C22(N+1:N^2-N,1:N),C22(N+1:N^2-N,N^2-N+1:N^2)];
bmCl2w=[C23(N+1:N^2-N,1:2*N),C23(N+1:N^2-N,N^2-2*N+1:N^2)];
```

```
bmCl2sx=[C24(N+1:N^2-N,1:N),C24(N+1:N^2-N,N^2-N+1:N^2)];
bmCl2st=[C25(N+1:N^2-N,1:N),C25(N+1:N^2-N,N^2-N+1:N^2)];

bmCl3u=[C31(2*N+1:N^2-2*N,1:N),C31(2*N+1:N^2-2*N,N^2-
N+1:N^2)];
bmCl3v=[C32(2*N+1:N^2-2*N,1:N),C32(2*N+1:N^2-2*N,N^2-
N+1:N^2)];
bmCl3w=[C33(2*N+1:N^2-2*N,1:2*N),C33(2*N+1:N^2-2*N,N^2-
2*N+1:N^2)];
bmCl3sx=[C34(2*N+1:N^2-2*N,1:N),C34(2*N+1:N^2-2*N,N^2-
N+1:N^2)];
bmCl3st=[C35(2*N+1:N^2-2*N,1:N),C35(2*N+1:N^2-2*N,N^2-
N+1:N^2)];

bmCl4u=[C41(N+1:N^2-N,1:N),C41(N+1:N^2-N,N^2-N+1:N^2)];
bmCl4v=[C42(N+1:N^2-N,1:N),C42(N+1:N^2-N,N^2-N+1:N^2)];
bmCl4w=[C43(N+1:N^2-N,1:2*N),C43(N+1:N^2-N,N^2-2*N+1:N^2)];
bmCl4sx=[C44(N+1:N^2-N,1:N),C44(N+1:N^2-N,N^2-N+1:N^2)];
bmCl4st=[C45(N+1:N^2-N,1:N),C45(N+1:N^2-N,N^2-N+1:N^2)];

bmCl5u=[C51(N+1:N^2-N,1:N),C51(N+1:N^2-N,N^2-N+1:N^2)];
bmCl5v=[C52(N+1:N^2-N,1:N),C52(N+1:N^2-N,N^2-N+1:N^2)];
bmCl5w=[C53(N+1:N^2-N,1:2*N),C53(N+1:N^2-N,N^2-2*N+1:N^2)];
bmCl5sx=[C54(N+1:N^2-N,1:N),C54(N+1:N^2-N,N^2-N+1:N^2)];
bmCl5st=[C55(N+1:N^2-N,1:N),C55(N+1:N^2-N,N^2-N+1:N^2)];

mKl1u=K11(N+1:N^2-N,N+1:N^2-N);
mKl1v=K12(N+1:N^2-N,N+1:N^2-N);
mKl1w=K13(N+1:N^2-N,2*N+1:N^2-2*N);
mKl1sx=K14(N+1:N^2-N,N+1:N^2-N);
mKl1st=K15(N+1:N^2-N,N+1:N^2-N);

mKl2u=K21(N+1:N^2-N,N+1:N^2-N);
mKl2v=K22(N+1:N^2-N,N+1:N^2-N);
mKl2w=K23(N+1:N^2-N,2*N+1:N^2-2*N);
mKl2sx=K24(N+1:N^2-N,N+1:N^2-N);
mKl2st=K25(N+1:N^2-N,N+1:N^2-N);

mKl3u=K31(2*N+1:N^2-2*N,N+1:N^2-N);
mKl3v=K32(2*N+1:N^2-2*N,N+1:N^2-N);
mKl3w=K33(2*N+1:N^2-2*N,2*N+1:N^2-2*N);
mKl3sx=K34(2*N+1:N^2-2*N,N+1:N^2-N);
mKl3st=K35(2*N+1:N^2-2*N,N+1:N^2-N);

mKl4u=K41(N+1:N^2-N,N+1:N^2-N);
mKl4v=K42(N+1:N^2-N,N+1:N^2-N);
mKl4w=K43(N+1:N^2-N,2*N+1:N^2-2*N);
mKl4sx=K44(N+1:N^2-N,N+1:N^2-N);
mKl4st=K45(N+1:N^2-N,N+1:N^2-N);
```

```
mKl5u=K51(N+1:N^2-N,N+1:N^2-N);
mKl5v=K52(N+1:N^2-N,N+1:N^2-N);
mKl5w=K53(N+1:N^2-N,2*N+1:N^2-2*N);
mKl5sx=K54(N+1:N^2-N,N+1:N^2-N);
mKl5st=K55(N+1:N^2-N,N+1:N^2-N);

mMl1u=M11(N+1:N^2-N,N+1:N^2-N);
mMl1v=M12(N+1:N^2-N,N+1:N^2-N);
mMl1w=M13(N+1:N^2-N,2*N+1:N^2-2*N);
mMl1sx=M14(N+1:N^2-N,N+1:N^2-N);
mMl1st=M15(N+1:N^2-N,N+1:N^2-N);

mMl2u=M21(N+1:N^2-N,N+1:N^2-N);
mMl2v=M22(N+1:N^2-N,N+1:N^2-N);
mMl2w=M23(N+1:N^2-N,2*N+1:N^2-2*N);
mMl2sx=M24(N+1:N^2-N,N+1:N^2-N);
mMl2st=M25(N+1:N^2-N,N+1:N^2-N);

mMl3u=M31(2*N+1:N^2-2*N,N+1:N^2-N);
mMl3v=M32(2*N+1:N^2-2*N,N+1:N^2-N);
mMl3w=M33(2*N+1:N^2-2*N,2*N+1:N^2-2*N);
mMl3sx=M34(2*N+1:N^2-2*N,N+1:N^2-N);
mMl3st=M35(2*N+1:N^2-2*N,N+1:N^2-N);

mMl4u=M41(N+1:N^2-N,N+1:N^2-N);
mMl4v=M42(N+1:N^2-N,N+1:N^2-N);
mMl4w=M43(N+1:N^2-N,2*N+1:N^2-2*N);
mMl4sx=M44(N+1:N^2-N,N+1:N^2-N);
mMl4st=M45(N+1:N^2-N,N+1:N^2-N);

mMl5u=M51(N+1:N^2-N,N+1:N^2-N);
mMl5v=M52(N+1:N^2-N,N+1:N^2-N);
mMl5w=M53(N+1:N^2-N,2*N+1:N^2-2*N);
mMl5sx=M54(N+1:N^2-N,N+1:N^2-N);
mMl5st=M55(N+1:N^2-N,N+1:N^2-N);

mCl1u=C11(N+1:N^2-N,N+1:N^2-N);
mCl1v=C12(N+1:N^2-N,N+1:N^2-N);
mCl1w=C13(N+1:N^2-N,2*N+1:N^2-2*N);
mCl1sx=C14(N+1:N^2-N,N+1:N^2-N);
mCl1st=C15(N+1:N^2-N,N+1:N^2-N);

mCl2u=C21(N+1:N^2-N,N+1:N^2-N);
mCl2v=C22(N+1:N^2-N,N+1:N^2-N);
mCl2w=C23(N+1:N^2-N,2*N+1:N^2-2*N);
mCl2sx=C24(N+1:N^2-N,N+1:N^2-N);
mCl2st=C25(N+1:N^2-N,N+1:N^2-N);

mCl3u=C31(2*N+1:N^2-2*N,N+1:N^2-N);
mCl3v=C32(2*N+1:N^2-2*N,N+1:N^2-N);
```

```
mCl3w=C33(2*N+1:N^2-2*N,2*N+1:N^2-2*N);
mCl3sx=C34(2*N+1:N^2-2*N,N+1:N^2-N);
mCl3st=C35(2*N+1:N^2-2*N,N+1:N^2-N);

mCl4u=C41(N+1:N^2-N,N+1:N^2-N);
mCl4v=C42(N+1:N^2-N,N+1:N^2-N);
mCl4w=C43(N+1:N^2-N,2*N+1:N^2-2*N);
mCl4sx=C44(N+1:N^2-N,N+1:N^2-N);
mCl4st=C45(N+1:N^2-N,N+1:N^2-N);

mCl5u=C51(N+1:N^2-N,N+1:N^2-N);
mCl5v=C52(N+1:N^2-N,N+1:N^2-N);
mCl5w=C53(N+1:N^2-N,2*N+1:N^2-2*N);
mCl5sx=C54(N+1:N^2-N,N+1:N^2-N);
mCl5st=C55(N+1:N^2-N,N+1:N^2-N);

Kl=[eye(5*N,5*N^2)
    %-----------------------------------
    zeros(N,4*N),Klw_2b,zeros(N,4*N),zeros(N,2
*(N^2-2*N)),Klw_2d,zeros(N,((2*(N^2-2*N)))));...
    zeros(N,4*N),Klw_N1b,zeros(N,4*N),zeros(N,2*(N^2-2*N)),
Klw_N1d,zeros(N,((2*(N^2-2*N)))));...
    zeros(N,7*N) eye(N,5*N^2-7*N);...
    %-----------------------------------
    zeros(N,8*N) eye(N,5*N^2-8*N);...
    zeros(N,9*N) eye(N,5*N^2-9*N);...
    %-----------------------------------
    zeros(N,10*N) eye(N,5*N^2-10*N);...
    zeros(N,11*N) eye(N,5*N^2-11*N);...
    %-----------------------------------
bmKl1u,bmKl1v,bmKl1w,bmKl1sx,bmKl1st,mKl1u,mKl1v,mKl1w,mKl1sx,
mKl1st;...
bmKl2u,bmKl2v,bmKl2w,bmKl2sx,bmKl2st,mKl2u,mKl2v,mKl2w,mKl2sx,
mKl2st;...
bmKl3u,bmKl3v,bmKl3w,bmKl3sx,bmKl3st,mKl3u,mKl3v,mKl3w,mKl3sx,
mKl3st;...
bmKl4u,bmKl4v,bmKl4w,bmKl4sx,bmKl4st,mKl4u,mKl4v,mKl4w,mKl4sx,
mKl4st;...
bmKl5u,bmKl5v,bmKl5w,bmKl5sx,bmKl5st,mKl5u,mKl5v,mKl5w,mKl5sx,
mKl5st];

Ml=[
zeros(12*N,5*N^2)
;...
bmMl1u,bmMl1v,bmMl1w,bmMl1sx,bmMl1st,mMl1u,mMl1v,mMl1w,mMl1sx,
mMl1st;...
bmMl2u,bmMl2v,bmMl2w,bmMl2sx,bmMl2st,mMl2u,mMl2v,mMl2w,mMl2sx,
mMl2st;...
bmMl3u,bmMl3v,bmMl3w,bmMl3sx,bmMl3st,mMl3u,mMl3v,mMl3w,mMl3sx,
mMl3st;...
```

```
bmMl4u,bmMl4v,bmMl4w,bmMl4sx,bmMl4st,mMl4u,mMl4v,mMl4w,mMl4sx,
mMl4st;...
bmMl5u,bmMl5v,bmMl5w,bmMl5sx,bmMl5st,mMl5u,mMl5v,mMl5w,mMl5sx,
mMl5st];

Dl=[
zeros(12*N,5*N^2)
;...
bmCl1u,bmCl1v,bmCl1w,bmCl1sx,bmCl1st,mCl1u,mCl1v,mCl1w,mCl1sx,
mCl1st;...
bmCl2u,bmCl2v,bmCl2w,bmCl2sx,bmCl2st,mCl2u,mCl2v,mCl2w,mCl2sx,
mCl2st;...
bmCl3u,bmCl3v,bmCl3w,bmCl3sx,bmCl3st,mCl3u,mCl3v,mCl3w,mCl3sx,
mCl3st;...
bmCl4u,bmCl4v,bmCl4w,bmCl4sx,bmCl4st,mCl4u,mCl4v,mCl4w,mCl4sx,
mCl4st;...
bmCl5u,bmCl5v,bmCl5w,bmCl5sx,bmCl5st,mCl5u,mCl5v,mCl5w,mCl5sx,
mCl5st];

KLbb=Kl(1:12*N,1:12*N);
KLbd=Kl(1:12*N,12*N+1:end);
KLdb=Kl(12*N+1:end,1:12*N);
KLdd=Kl(12*N+1:end,12*N+1:end);
DLdb=Dl(12*N+1:end,1:12*N);
DLdd=Dl(12*N+1:end,12*N+1:end);
MLdb=Ml(12*N+1:end,1:12*N);
MLdd=Ml(12*N+1:end,12*N+1:end);

KL=KLdd-KLdb*inv(KLbb)*KLbd;
DL=DLdd-DLdb*inv(KLbb)*KLbd;
ML=MLdd-MLdb*inv(KLbb)*KLbd;
Q=266;
Nm=max(size(KL));
q=zeros(Nm,1);
qdot=zeros(Nm,1);
qddot=zeros(Nm,1);
velocity=4;
alphaa=-0;
time=20;
tt=linspace(0,time,Q);
kkk=2;
tt(Q+1)=time+time/Q;
for kk=1:Q;
 kk
t=tt(kk);
dt=time/Q*kk;
delta=0.5;
alpha=0.25;
a0=1/(alpha*dt^2);
a1=delta/(alpha*dt);
a2=1/(alpha*dt);
```

```
a3=1/(2*alpha)-1;
a4=delta/alpha-1;
a5=dt/2*(delta/alpha-2);
a6=dt*(1-delta);
a7=delta*dt;
acceleration from PEER ground motion database
F=-ML*acce(kk*7)*[zeros(2*(N^2-2*N),1);ones(1*(N^2-
4*N),1);zeros(2*(N^2-2*N),1)];
Keq=triu(KL+a0*ML+a1*DL);
Feq=F+ML*(a0*q(:,kk)+a2*qdot(:,kk)+a3*qddot(:,kk))+DL*(a1*q(:,
kk)+a4*qdot(:,kk)+a5*qddot(:,kk));
q(:,kkk)=Keq\Feq;
qddot(:,kkk)=a0*(q(:,kkk)-q(:,kk))-a2*qdot(:,kk)-
a3*qddot(:,kk);
qdot(:,kkk)=qdot(:,kk)+a6*qddot(:,kk)+a7*qddot(:,kkk);
kkk=kkk+1;
end
```

# Index

Printed in the United States
by Baker & Taylor Publisher Services